Efficient Processing of Deep Neural Networks

Synthesis Lectures on Computer Architecture

Editors
Natalie Enright Jerger, *University of Toronto*
Margaret Martonosi, *Princeton University*

Founding Editor Emeritus
Mark D. Hill, *University of Wisconsin, Madison*

Synthesis Lectures on Computer Architecture publishes 50- to 100-page publications on topics pertaining to the science and art of designing, analyzing, selecting and interconnecting hardware components to create computers that meet functional, performance and cost goals. The scope will largely follow the purview of premier computer architecture conferences, such as ISCA, HPCA, MICRO, and ASPLOS.

Efficient Processing of Deep Neural Networks
Vivienne Sze, Yu-Hsin Chen, Tien-Ju Yang, and Joel S. Emer
2020

Quantum Computer Systems: Research for Noisy Intermediate-Scale Quantum Computers
Yonshan Ding and Frederic T. Chong
2020

A Primer on Memory Consistency and Cache Coherence, Second Edition
Vijay Nagarajan, Daniel J. Sorin, Mark D. Hill, and David A. Wood
2020

Innovations in the Memory System
Rajeev Balasubramonian
2019

Cache Replacement Policies
Akanksha Jain and Calvin Lin
2019

The Datacenter as a Computer: Designing Warehouse-Scale Machines, Third Edition
Luiz André Barroso, Urs Hölzle, and Parthasarathy Ranganathan
2018

Principles of Secure Processor Architecture Design
Jakub Szefer
2018

General-Purpose Graphics Processor Architectures
Tor M. Aamodt, Wilson Wai Lun Fung, and Timothy G. Rogers
2018

Compiling Algorithms for Heterogenous Systems
Steven Bell, Jing Pu, James Hegarty, and Mark Horowitz
2018

Architectural and Operating System Support for Virtual Memory
Abhishek Bhattacharjee and Daniel Lustig
2017

Deep Learning for Computer Architects
Brandon Reagen, Robert Adolf, Paul Whatmough, Gu-Yeon Wei, and David Brooks
2017

On-Chip Networks, Second Edition
Natalie Enright Jerger, Tushar Krishna, and Li-Shiuan Peh
2017

Space-Time Computing with Temporal Neural Networks
James E. Smith
2017

Hardware and Software Support for Virtualization
Edouard Bugnion, Jason Nieh, and Dan Tsafrir
2017

Datacenter Design and Management: A Computer Architect's Perspective
Benjamin C. Lee
2016

A Primer on Compression in the Memory Hierarchy
Somayeh Sardashti, Angelos Arelakis, Per Stenström, and David A. Wood
2015

Research Infrastructures for Hardware Accelerators
Yakun Sophia Shao and David Brooks
2015

Efficient Processing of Deep Neural Networks

Vivienne Sze, Yu-Hsin Chen, Tien-Ju Yang, and Joel S. Emer

ISBN: 978-3-031-00638-8 paperback
ISBN: 978-3-031-01766-7 ebook
ISBN: 978-3-031-00063-8 hardcover

DOI 10.1007/978-3-031-01766-7

A Publication in the Springer series
SYNTHESIS LECTURES ON ADVANCES IN AUTOMOTIVE TECHNOLOGY

Lecture #50
Series Editors: Natalie Enright Jerger, *University of Toronto*
                Margaret Martonosi, *Princeton University*
Founding Editor Emeritus: Mark D. Hill, *University of Wisconsin, Madison*
Series ISSN
Print 1935-3235   Electronic 1935-3243

Efficient Processing of Deep Neural Networks

Vivienne Sze, Yu-Hsin Chen, and Tien-Ju Yang
Massachusetts Institute of Technology

Joel S. Emer
Massachusetts Institute of Technology and Nvidia Research

SYNTHESIS LECTURES ON COMPUTER ARCHITECTURE #50

ABSTRACT

This book provides a structured treatment of the key principles and techniques for enabling efficient processing of deep neural networks (DNNs). DNNs are currently widely used for many artificial intelligence (AI) applications, including computer vision, speech recognition, and robotics. While DNNs deliver state-of-the-art accuracy on many AI tasks, it comes at the cost of high computational complexity. Therefore, techniques that enable efficient processing of deep neural networks to improve key metrics—such as energy-efficiency, throughput, and latency—without sacrificing accuracy or increasing hardware costs are critical to enabling the wide deployment of DNNs in AI systems.

The book includes background on DNN processing; a description and taxonomy of hardware architectural approaches for designing DNN accelerators; key metrics for evaluating and comparing different designs; features of DNN processing that are amenable to hardware/algorithm co-design to improve energy efficiency and throughput; and opportunities for applying new technologies. Readers will find a structured introduction to the field as well as formalization and organization of key concepts from contemporary work that provide insights that may spark new ideas.

KEYWORDS

deep learning, neural network, deep neural networks (DNN), convolutional neural networks (CNN), artificial intelligence (AI), efficient processing, accelerator architecture, hardware/software co-design, hardware/algorithm co-design, domain-specific accelerators

Contents

PART II   Design of Hardware for Processing DNNs . . . 41

Preface

Deep neural networks (DNNs) have become extraordinarily popular; however, they come at the cost of high computational complexity. As a result, there has been tremendous interest in enabling efficient processing of DNNs. The challenge of DNN acceleration is threefold:

- to achieve high performance and efficiency,

- to provide sufficient flexibility to cater to a wide and rapidly changing range of workloads, and

- to integrate well into existing software frameworks.

In order to understand the current state of art in addressing this challenge, this book aims to provide an overview of DNNs, the various tools for understanding their behavior, and the techniques being explored to efficiently accelerate their computation. It aims to explain foundational concepts and highlight key design considerations when building hardware for processing DNNs rather than trying to cover all possible design configurations, as this is not feasible given the fast pace of the field (see Figure 1). It is targeted at researchers and practitioners who are familiar with computer architecture who are interested in how to efficiently process DNNs or how to design DNN models that can be efficiently processed. We hope that this book will provide a structured introduction to readers who are new to the field, while also formalizing and organizing key concepts to provide insights that may spark new ideas for those who are already in the field.

Organization

This book is organized into three modules that each consist of several chapters. The first module aims to provide an overall background to the field of DNN and insight on characteristics of the DNN workload.

- Chapter 1 provides background on the context of why DNNs are important, their history, and their applications.

- Chapter 2 gives an overview of the basic components of DNNs and popular DNN models currently in use. It also describes the various resources used for DNN research and development. This includes discussion of the various software frameworks and the public datasets that are used for training and evaluation.

The second module focuses on the design of hardware for processing DNNs. It discusses various architecture design decisions depending on the degree of customization (from general

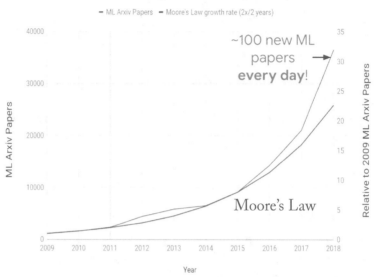

Machine Learning Arxiv Papers per Year

Figure 1: It's been observed that the number of ML publications are growing exponentially at a faster rate than Moore's law! (Figure from [1].)

purpose platforms to full custom hardware) and design considerations when mapping the DNN workloads onto these architectures. Both temporal and spatial architectures are considered.

- Chapter 3 describes the key metrics that should be considered when designing or comparing various DNN accelerators.

- Chapter 4 describes how DNN kernels can be processed, with a focus on temporal architectures such as CPUs and GPUs. To achieve greater efficiency, such architectures generally have a cache hierarchy and coarser-grained computational capabilities, e.g., vector instructions, making the resulting computation more efficient. Frequently for such architectures, DNN processing can be transformed into a matrix multiplication, which has many optimization opportunities. This chapter also discusses various software and hardware optimizations used to accelerate DNN computations on these platforms without impacting application accuracy.

- Chapter 5 describes the design of specialized hardware for DNN processing, with a focus on spatial architectures. It highlights the processing order and resulting data movement in the hardware used to process a DNN and the relationship to a loop nest representation of a DNN. The order of the loops in the loop nest is referred to as the *dataflow*, and it determines how often each piece of data needs to be moved. The limits of the loops in the loop nest describe how to break the DNN workload into smaller pieces, referred to as *tiling/blocking* to account for the limited storage capacity at different levels of the memory hierarchy.

- Chapter 6 presents the process of *mapping* a DNN workload on to a DNN accelerator. It describes the steps required to find an optimized mapping, including enumerating all legal mappings and searching those mappings by employing models that project throughput and energy efficiency.

The third module discusses how additional improvements in efficiency can be achieved either by moving up the stack through the co-design of the algorithms and hardware or down the stack by using mixed signal circuits and new memory or device technology. In the cases where the algorithm is modified, the impact on accuracy must be carefully evaluated.

- Chapter 7 describes how reducing the precision of data and computation can result in increased throughput and energy efficiency. It discusses how to reduce precision using quantization and the associated design considerations, including hardware cost and impact on accuracy.

- Chapter 8 describes how exploiting sparsity in DNNs can be used to reduce the footprint of the data, which provides an opportunity to reduce storage requirements, data movement, and arithmetic operations. It describes various sources of sparsity and techniques to increase sparsity. It then discusses how sparse DNN accelerators can translate sparsity into improvements in energy-efficiency and throughput. It also presents a new abstract data representation that can be used to express and obtain insight about the dataflows for a variety of sparse DNN accelerators.

- Chapter 9 describes how to optimize the structure of the DNN models (i.e., the 'network architecture' of the DNN) to improve both throughput and energy efficiency while trying to minimize impact on accuracy. It discusses both manual design approaches as well as automatic design approaches (i.e., neural architecture search).

- Chapter 10, on advanced technologies, discusses how mixed-signal circuits and new memory technologies can be used to bring the compute closer to the data (e.g., processing in memory) to address the expensive data movement that dominates throughput and energy consumption of DNNs. It also briefly discusses the promise of reducing energy consumption and increasing throughput by performing the computation and communication in the optical domain.

What's New?

This book is an extension of a tutorial paper written by the same authors entitled "Efficient Processing of Deep Neural Networks: A Tutorial and Survey" that appeared in the *Proceedings of the IEEE* in 2017 and slides from short courses given at ISCA and MICRO in 2016, 2017, and 2019 (slides available at http://eyeriss.mit.edu/tutorial.html). This book includes recent works since the publication of the tutorial paper along with a more in-depth treatment of topics such as dataflow, mapping, and processing in memory. We also provide updates on the fast-moving field of co-design of DNN models and hardware in the areas of reduced precision, sparsity, and efficient DNN model design. As part of this effort, we present a new way of thinking about sparse representations and give a detailed treatment of how to handle and exploit sparsity. Finally, we touch upon recurrent neural networks, auto encoders, and transformers, which we did not discuss in the tutorial paper.

Scope of book

The main goal of this book is to teach the reader how to tackle the computational challenge of efficiently processing DNNs rather than how to design DNNs for increased accuracy. As a result, this book does not cover training (only touching on it lightly), nor does it cover the theory of deep learning or how to design DNN models (though it discusses how to make them efficient) or use them for different applications. For these aspects, please refer to other references such as Goodfellow's book [2], Amazon's book [3], and Stanford cs231n course notes [4].

Vivienne Sze, Yu-Hsin Chen, Tien-Ju Yang, and Joel S. Emer
June 2020

Acknowledgments

The authors would like to thank Margaret Martonosi for her persistent encouragement to write this book. We would also like to thank Liane Bernstein, Davis Blalock, Natalie Enright Jerger, Jose Javier Gonzalez Ortiz, Fred Kjolstad, Yi-Lun Liao, Andreas Moshovos, Boris Murmann, James Noraky, Angshuman Parashar, Michael Pellauer, Clément Pit-Claudel, Sophia Shao, Mahmhut Ersin Sinangil, Po-An Tsai, Marian Verhelst, Tom Wenisch, Diana Wofk, Nellie Wu, and students in our "Hardware Architectures for Deep Learning" class at MIT, who have provided invaluable feedback and discussions on the topics described in this book. We would also like to express our deepest appreciation to Robin Emer for her suggestions, support, and tremendous patience during the writing of this book.

As mentioned earlier in the Preface, this book is an extension of an earlier tutorial paper, which was based on tutorials we gave at ISCA and MICRO. We would like to thank David Brooks for encouraging us to do the first tutorial at MICRO in 2016, which sparked the effort that led to this book.

This work was funded in part by DARPA YFA, the DARPA contract HR0011-18-3-0007, the MIT Center for Integrated Circuits and Systems (CICS), the MIT-IBM Watson AI Lab, the MIT Quest for Intelligence, the NSF E2CDA 1639921, and gifts/faculty awards from Nvidia, Facebook, Google, Intel, and Qualcomm.

Vivienne Sze, Yu-Hsin Chen, Tien-Ju Yang, and Joel S. Emer
June 2020

PART I

Understanding Deep Neural Networks

CHAPTER 1

Introduction

Deep neural networks (DNNs) are currently the foundation for many modern artificial intelligence (AI) applications [5]. Since the breakthrough application of DNNs to speech recognition [6] and image recognition[1] [7], the number of applications that use DNNs has exploded. These DNNs are employed in a myriad of applications from self-driving cars [8], to detecting cancer [9], to playing complex games [10]. In many of these domains, DNNs are now able to exceed human accuracy. The superior accuracy of DNNs comes from their ability to extract high-level features from raw sensory data by using statistical learning on a large amount of data to obtain an effective representation of an input space. This is different from earlier approaches that use hand-crafted features or rules designed by experts.

The superior accuracy of DNNs, however, comes at the cost of high computational complexity. To date, general-purpose compute engines, especially graphics processing units (GPUs), have been the mainstay for much DNN processing. Increasingly, however, in these waning days of Moore's law, there is a recognition that more specialized hardware is needed to keep improving compute performance and energy efficiency [11]. This is especially true in the domain of DNN computations. This book aims to provide an overview of DNNs, the various tools for understanding their behavior, and the techniques being explored to efficiently accelerate their computation.

1.1 BACKGROUND ON DEEP NEURAL NETWORKS

In this section, we describe the position of DNNs in the context of artificial intelligence (AI) in general and some of the concepts that motivated the development of DNNs. We will also present a brief chronology of the major milestones in the history of DNNs, and some current domains to which it is being applied.

1.1.1 ARTIFICIAL INTELLIGENCE AND DEEP NEURAL NETWORKS

DNNs, also referred to as deep learning, are a part of the broad field of AI. AI is the science and engineering of creating intelligent machines that have the ability to achieve goals like humans do, according to John McCarthy, the computer scientist who coined the term in the 1950s. The relationship of deep learning to the whole of AI is illustrated in Figure 1.1.

---

[1]Image recognition is also commonly referred to as image classification.

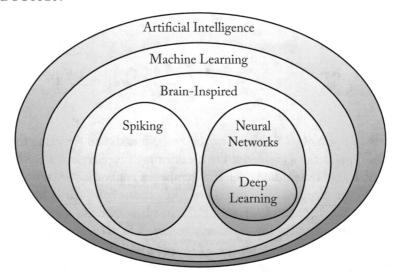

Figure 1.1: Deep learning in the context of artificial intelligence.

Within AI is a large sub-field called machine learning, which was defined in 1959 by Arthur Samuel [12] as "the field of study that gives computers the ability to learn without being explicitly programmed." That means a single program, once created, will be able to learn how to do some intelligent activities outside the notion of programming. This is in contrast to purpose-built programs whose behavior is defined by hand-crafted heuristics that explicitly and statically define their behavior.

The advantage of an effective machine learning algorithm is clear. Instead of the laborious and hit-or-miss approach of creating a distinct, custom program to solve each individual problem in a domain, a single machine learning algorithm simply needs to learn, via a process called *training*, to handle each new problem.

Within the machine learning field, there is an area that is often referred to as brain-inspired computation. Since the brain is currently the best "machine" we know of for learning and solving problems, it is a natural place to look for inspiration. Therefore, a brain-inspired computation is a program or algorithm that takes some aspects of its basic form or functionality from the way the brain works. This is in contrast to attempts to create a brain, but rather the program aims to emulate some aspects of how we understand the brain to operate.

Although scientists are still exploring the details of how the brain works, it is generally believed that the main computational element of the brain is the *neuron*. There are approximately 86 billion neurons in the average human brain. The neurons themselves are connected by a number of elements entering them, called dendrites, and an element leaving them, called an axon, as shown in Figure 1.2. The neuron accepts the signals entering it via the dendrites, performs a computation on those signals, and generates a signal on the axon. These input and output sig-

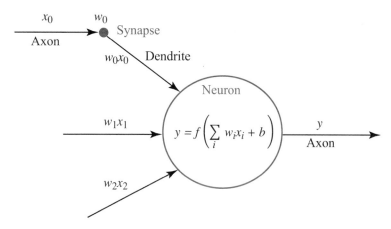

Figure 1.2: Connections to a neuron in the brain. x_i, w_i, $f(\cdot)$, and b are the activations, weights, nonlinear function, and bias, respectively. (Figure adapted from [4].)

nals are referred to as *activations*. The axon of one neuron branches out and is connected to the dendrites of many other neurons. The connections between a branch of the axon and a dendrite is called a *synapse*. There are estimated to be 10^{14} to 10^{15} synapses in the average human brain.

A key characteristic of the synapse is that it can scale the signal (x_i) crossing it, as shown in Figure 1.2. That scaling factor can be referred to as a *weight* (w_i), and the way the brain is believed to learn is through changes to the weights associated with the synapses. Thus, different weights result in different responses to an input. One aspect of learning can be thought of as the adjustment of weights in response to a learning stimulus, while the organization (what might be thought of as the program) of the brain largely does not change. This characteristic makes the brain an excellent inspiration for a machine-learning-style algorithm.

Within the brain-inspired computing paradigm, there is a subarea called spiking computing. In this subarea, inspiration is taken from the fact that the communication on the dendrites and axons are spike-like pulses and that the information being conveyed is not just based on a spike's amplitude. Instead, it also depends on the time the pulse arrives and that the computation that happens in the neuron is a function of not just a single value but the width of pulse and the timing relationship between different pulses. The IBM TrueNorth project is an example of work that was inspired by the spiking of the brain [13]. In contrast to spiking computing, another subarea of brain-inspired computing is called *neural networks*, which is the focus of this book.[2]

---

[2]Note: Recent work using TrueNorth in a stylized fashion allows it to be used to compute reduced precision neural networks [14]. These types of neural networks are discussed in Chapter 7.

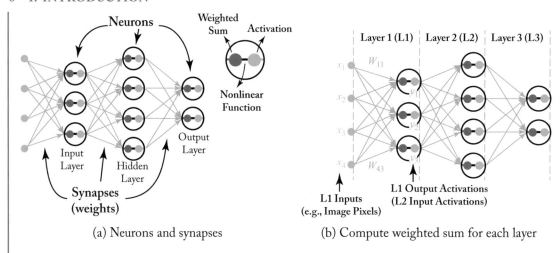

(a) Neurons and synapses   (b) Compute weighted sum for each layer

Figure 1.3: Simple neural network example and terminology. (Figure adapted from [4].)

1.1.2   NEURAL NETWORKS AND DEEP NEURAL NETWORKS

Neural networks take their inspiration from the notion that a neuron's computation involves a weighted sum of the input values. These weighted sums correspond to the value scaling performed by the synapses and the combining of those values in the neuron. Furthermore, the neuron does not directly output that weighted sum because the expressive power of the cascade of neurons involving only linear operations is just equal to that of a single neuron, which is very limited. Instead, there is a functional operation within the neuron that is performed on the combined inputs. This operation appears to be a nonlinear function that causes a neuron to generate an output only if its combined inputs cross some threshold. Thus, by analogy, neural networks apply a nonlinear function to the weighted sum of the input values.[3] These nonlinear functions are inspired by biological functions, but are not meant to emulate the brain. We look at some of those nonlinear functions in Section 2.3.3.

Figure 1.3a shows a diagram of a three-layer (non-biological) neural network. The neurons in the input layer receive some values, compute their weighted sums followed by the nonlinear function, and propagate the outputs to the neurons in the middle layer of the network, which is also frequently called a "hidden layer." A neural network can have more than one hidden layer, and the outputs from the hidden layers ultimately propagate to the output layer, which computes the final outputs of the network to the user. To align brain-inspired terminology with neural networks, the outputs of the neurons are often referred to as *activations*, and the synapses are often referred to as *weights*, as shown in Figure 1.3a. We will use the activation/weight nomenclature in this book.

---

[3]Without a nonlinear function, multiple layers could be collapsed into one.

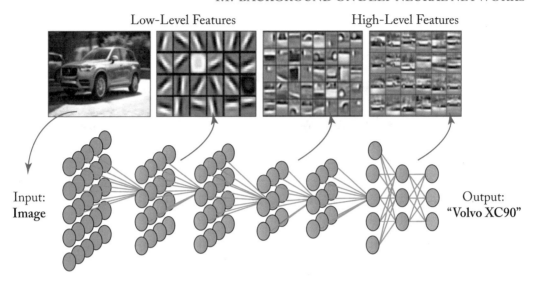

Figure 1.4: Example of image classification using deep neural networks. (Figure adapted from [15].) Note that the features go from low level to high level as we go deeper into the network.

Figure 1.3b shows an example of the computation at layer 1: $y_j = f(\sum_{i=1}^{4} W_{ij} \times x_i + b_j)$, where W_{ij}, x_i, and y_j are the weights, input activations, and output activations, respectively, and $f(\cdot)$ is a nonlinear function described in Section 2.3.3. The bias term b_j is omitted from Figure 1.3b for simplicity. In this book, we will use the color green to denote weights, blue to denote activations, and red to denote weighted sums (or partial sums, which are further accumulated to become the final weighted sums).

Within the domain of neural networks, there is an area called *deep learning*, in which the neural networks have more than three layers, i.e., more than one hidden layer. Today, the typical numbers of network layers used in deep learning range from 5 to more than a 1,000. In this book, we will generally use the terminology *deep neural networks (DNNs)* to refer to the neural networks used in deep learning.

DNNs are capable of learning high-level features with more complexity and abstraction than shallower neural networks. An example that demonstrates this point is using DNNs to process visual data, as shown in Figure 1.4. In these applications, pixels of an image are fed into the first layer of a DNN, and the outputs of that layer can be interpreted as representing the presence of different low-level features in the image, such as lines and edges. In subsequent layers, these features are then combined into a measure of the likely presence of higher-level features, e.g., lines are combined into shapes, which are further combined into sets of shapes. Finally, given all this information, the network provides a probability that these high-level fea-

Class Probabilities

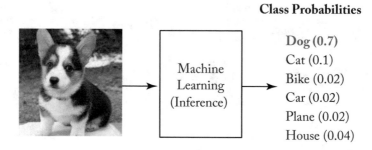

Dog (0.7)
Cat (0.1)
Bike (0.02)
Car (0.02)
Plane (0.02)
House (0.04)

Figure 1.5: Example of an image classification task. The machine learning platform takes in an image and outputs the class probabilities for a predefined set of classes.

tures comprise a particular object or scene. This deep feature hierarchy enables DNNs to achieve superior performance in many tasks.

1.2   TRAINING VERSUS INFERENCE

Since DNNs are an instance of machine learning algorithms, the basic program does not change as it learns to perform its given tasks. In the specific case of DNNs, this learning involves determining the value of the weights (and biases) in the network, and is referred to as *training* the network. Once trained, the program can perform its task by computing the output of the network using the weights determined during the training process. Running the program with these weights is referred to as *inference*.

In this section, we will use image classification, as shown in Figure 1.5, as a driving example for training and using a DNN. When we perform inference using a DNN, the input is image and the output is a vector of values representing the *class probabilities*. There is one value for each object class, and the class with the highest value indicates the most likely (predicted) class of object in the image. The overarching goal for training a DNN is to determine the weights that maximize the probability of the correct class and minimize the probabilities of the incorrect classes. The correct class is generally known, as it is often defined in the training set. The gap between the ideal correct probabilities and the probabilities computed by the DNN based on its current weights is referred to as the *loss* (L). Thus, the goal of training DNNs is to find a set of weights to minimize the average loss over a large training set.

When training a network, the weights (w_{ij}) are usually updated using a hill-climbing (hill-descending) optimization process called gradient descent. In gradient descent, a weight is updated by a scaled version of the partial derivative of the loss with respect to the weight (i.e.,

---

Sidebar: Key steps in training

Here, we will provide a very brief summary of the key steps of training and deploying a model. For more details, we recommend the reader refer to more comprehensive references such as [2]. First, we collect a labeled dataset and divide the data into subsets for training and testing. Second, we use the training set to train a model so that it can learn the weights for a given task. After achieving adequate accuracy on the training set, the ultimate quality of the model is determined by how accurately it performs on unseen data. Therefore, in the third step, we test the trained model by asking it to predict the labels for a test set that it has never seen before and compare the prediction to the ground truth labels. *Generalization* refers to how well the model maintains the accuracy between training and unseen data. If the model does not generalize well, it is often referred to as *overfitting*; this implies that the model is fitting to the noise rather than the underlying data structure that we would like it to learn. One way to combat overfitting is to have a large, diverse dataset; it has been shown that accuracy increases logarithmically as a function of the number of training examples [16]. Section 2.6.3 will discuss various popular datasets used for training. There are also other mechanisms that help with generalization including *Regularization*. It adds constraints to the model during training such as smoothness, number of parameters, size of the parameters, prior distribution or structure, or randomness in the training using dropout [17]. Further partitioning the training set into training and *validation* sets is another useful tool. Designing a DNN requires determining (tuning) a large number of hyperparameters such as the size and shape of a layer or the number of layers. Tuning the hyperparameters based on the test set may cause overfitting to the test set, which results in a misleading evaluation of the true performance on unseen data. In this circumstance, the validation set can be used instead of the test set to mitigate this problem. Finally, if the model performs sufficiently well on the test set, it can be deployed on unlabeled images.

---

updated to $w_{ij}^{t+1} = w_{ij}^t - \alpha \frac{\partial L}{\partial w_{ij}}$, where α is called the learning rate[4]). Note that this gradient indicates how the weights should change in order to reduce the loss. The process is repeated iteratively to reduce the overall loss.

An efficient way to compute the partial derivatives of the gradient is through a process called *backpropagation*. Backpropagation, which is a computation derived from the *chain rule* of

---

[4]A large learning rate increases the step size applied at each iteration, which can help speed up the training, but may also result in overshooting the minimum or cause the optimization to not converge. A small learning rate decreases the step size applied at each iteration which slows down the training, but increases likelihood of convergence. There are various methods to set the learning rate such as ADAM [18], etc. Finding the best the learning rate is one of the key challenges in training DNNs.

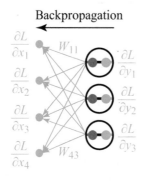

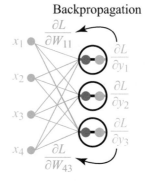

(a) Compute the gradient of the loss relative to the layer inputs ($\frac{\partial L}{\partial x_i} = \sum_j w_{ij} \frac{\partial L}{\partial y_j}$)

(b) Compute the gradient of the loss relative to the weights ($\frac{\partial L}{\partial w_{ij}} = \frac{\partial L}{\partial y_j} x_i$)

Figure 1.6: An example of backpropagation through a neural network.

calculus, operates by passing values backward through the network to compute how the loss is affected by each weight.

This backpropagation computation is, in fact, very similar in form to the computation used for inference, as shown in Figure 1.6 [19].[5] Thus, techniques for efficiently performing inference can sometimes be useful for performing training. There are, however, some important additional considerations to note. First, backpropagation requires intermediate outputs of the network to be preserved for the backward computation, thus training has increased storage requirements. Second, due to the gradients use for hill-climbing (hill-descending), the precision requirement for training is generally higher than inference. Thus, many of the reduced precision techniques discussed in Chapter 7 are limited to inference only.

A variety of techniques are used to improve the efficiency and robustness of training. For example, often, the loss from multiple inputs is computed before a single pass of weight updates is performed. This is called *batching*, which helps to speed up and stabilize the process.[6]

---

[5]To backpropagate through each layer: (1) compute the gradient of the loss relative to the weights, $\frac{\partial L}{\partial w_{ij}}$, from the layer inputs (i.e., the forward activations, x_i) and the gradients of the loss relative to the layer outputs, $\frac{\partial L}{\partial y_j}$; and (2) compute the gradient of the loss relative to the layer inputs, $\frac{\partial L}{\partial x_i}$, from the layer weights, w_{ij}, and the gradients of the loss relative to the layer outputs, $\frac{\partial L}{\partial y_j}$.

[6]There are various forms of gradient decent which differ in terms of how frequently to update the weights. *Batch Gradient Descent* updates the weights after computing the loss on the entire training set, which is computationally expensive and requires significant storage. *Stochastic Gradient Descent* update weights after computing loss on a single training example and the examples are shuffled after going through the entire training set. While it is fast, looking at a single example can be noisy and cause the weights to go in the wrong direction. Finally, *Mini-batch Gradient Descent* divides the training set into smaller sets called mini-batches, and updates weights based on the loss of each mini-batch (commonly referred to simply as "batch"); this approach is most commonly used. In general, each pass through the entire training set is referred to as an *epoch*.

There are multiple ways to train the weights. The most common approach, as described above, is called *supervised learning*, where all the training samples are labeled (e.g., with the correct class). *Unsupervised learning* is another approach, where no training samples are labeled. Essentially, the goal is to find the structure or clusters in the data. *Semi-supervised learning* falls between the two approaches, where only a small subset of the training data is labeled (e.g., use unlabeled data to define the cluster boundaries, and use the small amount of labeled data to label the clusters). Finally, *reinforcement learning* can be used to the train the weights such that given the state of the current environment, the DNN can output what action the agent should take next to maximize expected rewards; however, the rewards might not be available immediately after an action, but instead only after a series of actions (often referred to as an episode).

Another commonly used approach to determine weights is *fine-tuning*, where previously trained weights are available and are used as a starting point and then those weights are adjusted for a new dataset (e.g., transfer learning) or for a new constraint (e.g., reduced precision). This results in faster training than starting from a random starting point, and can sometimes result in better accuracy.

This book will focus on the efficient processing of DNN inference rather than training, since DNN inference is often performed on embedded devices (rather than the cloud) where resources are limited, as discussed in more details later.

1.3   DEVELOPMENT HISTORY

Although neural networks were proposed in the 1940s, the first practical application employing multiple digital neurons didn't appear until the late 1980s, with the LeNet network for hand-written digit recognition [20].[7] Such systems are widely used by ATMs for digit recognition on checks. The early 2010s have seen a blossoming of DNN-based applications, with highlights such as Microsoft's speech recognition system in 2011 [6] and the AlexNet DNN for image recognition in 2012 [7]. A brief chronology of deep learning is shown in Figure 1.7.

The deep learning successes of the early 2010s are believed to be due to a confluence of three factors. The first factor is the amount of available information to train the networks. To learn a powerful representation (rather than using a hand-crafted approach) requires a large amount of training data. For example, Facebook receives up to a billion images per day, Walmart creates 2.5 Petabytes of customer data hourly and YouTube has over 300 hours of video uploaded every minute. As a result, these and many other businesses have a huge amount of data to train their algorithms.

The second factor is the amount of compute capacity available. Semiconductor device and computer architecture advances have continued to provide increased computing capability, and we appear to have crossed a threshold where the large amount of weighted sum computation in DNNs, which is required for both inference and training, can be performed in a reasonable amount of time.

---

[7]In the early 1960s, single neuron systems built out of analog logic were used for adaptive filtering [21, 22].

> **DNN Timeline**
>
> - 1940s: Neural networks were proposed
> - 1960s: Deep neural networks were proposed
> - 1989: Neural networks for recognizing hand-written digits (LeNet)
> - 1990s: Hardware for shallow neural nets (Intel ETANN)
> - 2011: Breakthrough DNN-based speech recognition (Microsoft)
> - 2012: DNNs for vision start supplanting hand-crafted approaches (AlexNet)
> - 2014+: Rise of DNN accelerator research (Neuflow, DianNao…)

Figure 1.7: A concise history of neural networks. "Deep" refers to the number of layers in the network.

The successes of these early DNN applications opened the floodgates of algorithmic development. It has also inspired the development of several (largely open source) frameworks that make it even easier for researchers and practitioners to explore and use DNNs. Combining these efforts contributes to the third factor, which is the evolution of the algorithmic techniques that have improved accuracy significantly and broadened the domains to which DNNs are being applied.

An excellent example of the successes in deep learning can be illustrated with the ImageNet Challenge [23]. This challenge is a contest involving several different components. One of the components is an image classification task, where algorithms are given an image and they must identify what is in the image, as shown in Figure 1.5. The training set consists of 1.2 million images, each of which is labeled with one of a thousand object categories that the image contains. For the evaluation phase, the algorithm must accurately identify objects in a test set of images, which it hasn't previously seen.

Figure 1.8 shows the performance of the best entrants in the ImageNet contest over a number of years. The accuracy of the algorithms initially had an error rate of 25% or more. In 2012, a group from the University of Toronto used graphics processing units (GPUs) for their high compute capability and a DNN approach, named AlexNet, and reduced the error rate by approximately 10 percentage points [7]. Their accomplishment inspired an outpouring of deep learning algorithms that have resulted in a steady stream of improvements.

In conjunction with the trend toward using deep learning approaches for the ImageNet Challenge, there has been a corresponding increase in the number of entrants using GPUs: from 2012 when only 4 entrants used GPUs to 2014 when almost all the entrants (110) were using them. This use of GPUs reflects the almost complete switch from traditional computer vision approaches to deep learning-based approaches for the competition.

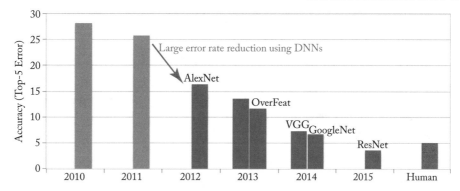

Figure 1.8: Results from the ImageNet Challenge [23].

In 2015, the ImageNet winning entry, ResNet [24], exceeded human-level accuracy with a Top-5 error rate[8] below 5%. Since then, the error rate has dropped below 3% and more focus is now being placed on more challenging components of the competition, such as object detection and localization. These successes are clearly a contributing factor to the wide range of applications to which DNNs are being applied.

1.4   APPLICATIONS OF DNNs

Many domains can benefit from DNNs, ranging from entertainment to medicine. In this section, we will provide examples of areas where DNNs are currently making an impact and highlight emerging areas where DNNs may make an impact in the future.

- **Image and Video:** Video is arguably the biggest of big data. It accounts for over 70% of today's Internet traffic [25]. For instance, over 800 million hours of video is collected daily worldwide for video surveillance [26]. Computer vision is necessary to extract meaningful information from video. DNNs have significantly improved the accuracy of many computer vision tasks such as image classification [23], object localization and detection [27], image segmentation [28], and action recognition [29].

- **Speech and Language:** DNNs have significantly improved the accuracy of speech recognition [30] as well as many related tasks such as machine translation [6], natural language processing [31], and audio generation [32].

- **Medicine and Health Care:** DNNs have played an important role in genomics to gain insight into the genetics of diseases such as autism, cancers, and spinal muscular atrophy [33–

---

[8]The Top-5 error rate is measured based on whether the correct answer appears in one of the top five categories selected by the algorithm.

36]. They have also been used in medical imaging such as detecting skin cancer [9], brain cancer [37], and breast cancer [38].

- **Game Play:** Recently, many of the grand AI challenges involving game play have been overcome using DNNs. These successes also required innovations in training techniques, and many rely on reinforcement learning [39]. DNNs have surpassed human level accuracy in playing games such as Atari [40], Go [10], and StarCraft [41], where an exhaustive search of all possibilities is not feasible due to the immense number of possible moves.

- **Robotics:** DNNs have been successful in the domain of robotic tasks such as grasping with a robotic arm [42], motion planning for ground robots [43], visual navigation [8, 44], control to stabilize a quadcopter [45], and driving strategies for autonomous vehicles [46].

DNNs are already widely used in multimedia applications today (e.g., computer vision, speech recognition). Looking forward, we expect that DNNs will likely play an increasingly important role in the medical and robotics fields, as discussed above, as well as finance (e.g., for trading, energy forecasting, and risk assessment), infrastructure (e.g., structural safety, and traffic control), weather forecasting, and event detection [47]. The myriad application domains pose new challenges to the efficient processing of DNNs; the solutions then have to be adaptive and scalable in order to handle the new and varied forms of DNNs that these applications may employ.

1.5   EMBEDDED VERSUS CLOUD

The various applications and aspects of DNN processing (i.e., training versus inference) have different computational needs. Specifically, training often requires a large dataset[9] and significant computational resources for multiple weight-update iterations. In many cases, training a DNN model still takes several hours to multiple days (or weeks or months!) and thus is typically performed in the cloud.

Inference, on the other hand, can happen either in the cloud or at the edge (e.g., Internet of Things (IoT) or mobile). In many applications, it is desirable to have the DNN inference processing at the edge near the sensor. For instance, in computer vision applications, such as measuring wait times in stores or predicting traffic patterns, it would be desirable to extract meaningful information from the video right at the image sensor rather than in the cloud, to reduce the communication cost. For other applications, such as autonomous vehicles, drone navigation, and robotics, local processing is desired since the latency and security risks of relying on the cloud are too high. However, video involves a large amount of data, which is computationally complex to process; thus, low-cost hardware to analyze video is challenging, yet critical, to enabling these applications.[10] Speech recognition allows us to seamlessly interact with electronic devices, such

---

[9]One of the major drawbacks of DNNs is their need for large datasets to prevent overfitting during training.

[10]As a reference, running a DNN on an embedded devices is estimated to consume several orders of magnitude higher energy per pixel than video compression, which is a common form of processing near image sensor [48].

as smartphones. While currently most of the processing for applications such as Apple Siri and Amazon Alexa voice services is in the cloud, it is still desirable to perform the recognition on the device itself to reduce latency. Some work have even considered partitioning the processing between the cloud and edge at a per layer basis in order to improve performance [49]. However, considerations related to dependency on connectivity, privacy, and security augur for keeping computation at the edge. Many of the embedded platforms that perform DNN inference have stringent requirements on energy consumption, compute and memory cost limitations; efficient processing of DNNs has become of prime importance under these constraints.

CHAPTER 2

Overview of Deep Neural Networks

Deep Neural Networks (DNNs) come in a wide variety of shapes and sizes depending on the application.[1] The popular shapes and sizes are also evolving rapidly to improve accuracy and efficiency. In all cases, the input to a DNN is a set of values representing the information to be analyzed by the network. For instance, these values can be pixels of an image, sampled amplitudes of an audio wave, or the numerical representation of the state of some system or game.

In this chapter, we will describe the key building blocks for DNNs. As there are *many* different types of DNNs [50], we will focus our attention on those that are most widely used. We will begin by describing the salient characteristics of commonly used DNN layers in Sections 2.1 and 2.2. We will then describe popular DNN layers and how these layers can be combined to form various types of DNNs in Section 2.3. Section 2.4 will provide a detailed discussion on convolutional neural networks (CNNs), since they are widely used and tend to provide many opportunities for efficient DNN processing. It will also highlight various popular CNN models that are often used as workloads for evaluating DNN hardware accelerators. Next, in Section 2.5, we will briefly discuss other types of DNNs and describe how they are similar to and differ from CNNs from a workload processing perspective (e.g., data dependencies, types of compute operations, etc.). Finally, in Section 2.6, we will discuss the various DNN development resources (e.g., frameworks and datasets), which researchers and practitioners have made available to help enable the rapid progress in DNN model and hardware research and development.

2.1 ATTRIBUTES OF CONNECTIONS WITHIN A LAYER

As discussed in Chapter 1, DNNs are composed of several processing layers, where in most layers the main computation is a weighted sum. There are several different types of layers, which primarily differ in terms of how the inputs and outputs are connected *within* the layers.

There are two main attributes of the connections within a layer:

1. The connection pattern between the input and output activations, as shown in Figure 2.1a: if a layer has the attribute that every input activation is connected to every output, then we

---

[1]The DNN research community often refers to the shape and size of a DNN as its "network architecture." However, to avoid confusion with the use of the word "architecture" by the hardware community, we will talk about "DNN models" and their shape and size in this book.

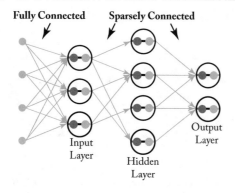

(a) Fully connected versus sparsely connected

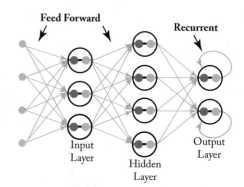

(b) Feed-forward versus feed-back (recurrent) connections

Figure 2.1: Properties of connections in DNNs (Figure adapted from [4]).

call that layer *fully connected*. On the other hand, if a layer has the attribute that only a subset of inputs are connected to the output, then we call that layer *sparsely connected*. Note that the weights associated with these connections can be zero or non-zero; if a weight happens to be zero (e.g., as a result of training), it does not mean there is no connection (i.e., the connection still exists).

For sparsely connected layers, a sub attribute is related to the structure of the connections. Input activations may connect to any output activation (i.e., global), or they may only connect to output activations in their neighborhood (i.e., local). The consequence of such local connections is that each output activation is a function of a restricted window of input activations, which is referred to as the *receptive field*.

2. The value of the weight associated with each connection: the most general case is that the weight can take on any value (e.g., each weight can have a unique value). A more restricted case is that the same value is shared by multiple weights, which is referred to as *weight sharing*.

Combinations of these attributes result in many of the common layer types. Any layer with the fully connected attribute is called a fully connected layer (FC layer). In order to distinguish the attribute from the type of layer, in this chapter, we will use the term FC layer as distinguished from the fully connected attribute. However, in subsequent chapters we will follow the common practice of using the terms interchangeably. Another widely used layer type is the convolutional (CONV) layer, which is locally, sparsely connected with weight sharing.[2] The computation in FC and CONV layers is a weighted sum. However, there are other computations that might be

---

[2]CONV layers use a specific type of weight sharing, which will be described in Section 2.4.

performed and these result in other types of layers. We will discuss FC, CONV, and these other layers in more detail in Section 2.3.

2.2   ATTRIBUTES OF CONNECTIONS BETWEEN LAYERS

Another attribute is the connections from the output of one layer to the input of *another* layer, as shown in Figure 2.1b. The output can be connected to the input of the next layer in which case the connection is referred to as *feed forward*. With feed-forward connections, all of the computation is performed as a sequence of operations on the outputs of a previous layer.[3] It has no memory and the output for an input is always the same irrespective of the sequence of inputs previously given to the network. DNNs that contain feed-forward connections are referred to as *feed-forward* networks. Examples of these types of networks include multi-layer perceptrons (MLPs), which are DNNs that are composed entirely of feed-forward FC layers and convolutional neural networks (CNNs), which are DNNs that contain both FC and CONV layers. CNNs, which are commonly used for image processing and computer vision, will be discussed in more detail in Section 2.4.

Alternatively, the output can be fed back to the input of its own layer in which case the connection is often referred to as *recurrent*. With recurrent connections, the output of a layer is a function of both the current and prior input(s) to the layer. This creates a form of memory in the DNN, which allows long-term dependencies to affect the output. DNNs that contain these connections are referred to as *recurrent* neural networks (RNNs), which are commonly used to process sequential data (e.g., speech, text), and will be discussed in more detail in Section 2.5.

2.3   POPULAR TYPES OF LAYERS IN DNNs

In this section, we will discuss the various popular layers used to form DNNs. We will begin by describing the CONV and FC layers whose main computation is a weighted sum, since that tends to dominate the computation cost in terms of both energy consumption and throughput. We will then discuss various layers that can optionally be included in a DNN and do not use weighted sums such as nonlinearity, pooling, and normalization.

These layers can be viewed as primitive layers, which can be combined to form compound layers. Compound layers are often given names as a convenience, when the same combination of primitive layer are frequently used together. In practice, people often refer to either primitive or compound layers as just layers.

2.3.1   CONV LAYER (CONVOLUTIONAL)

CONV layers are primarily composed of high-dimensional convolutions, as shown in Figure 2.2. In this computation, the input activations of a layer are structured as a 3-D *input feature map*

---

[3]Connections can come from the immediately preceding layer or an earlier layer. Furthermore, connections from a layer can go to multiple later layers.

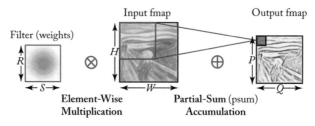

(a) 2-D convolution in traditional image processing

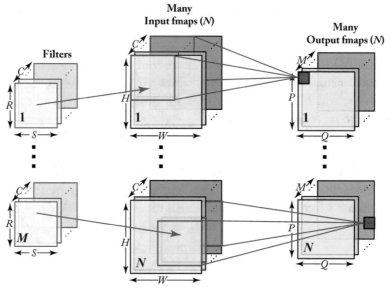

(b) High-dimensional convolutions in CNNs

Figure 2.2: Dimensionality of convolutions. (a) Shows the traditional 2-D convolution used in image processing. (b) Shows the high dimensional convolution used in CNNs, which applies a 2-D convolution on each channel.

(ifmap), where the dimensions are the height (H), width (W), and number of input channels (C). The weights of a layer are structured as a 3-D filter, where the dimensions are the height (R), width (S), and number of input channels (C). Notice that the number of channels for the input feature map and the filter are the same. For each input channel, the input feature map undergoes a 2-D convolution (see Figure 2.2a) with the corresponding channel in the filter. The results of the convolution at each point are summed across all the input channels to generate the output partial sums. In addition, a 1-D (scalar) bias can be added to the filtering results, but some recent networks [24] remove its usage from parts of the layers. The results of this

Table 2.1: Shape parameters of a CONV/FC layer

Shape Parameter	Description
N	Batch size of 3-D fmaps
M	Number of 3-D filters / number of channels of ofmap (output channels)
C	Number of channels of filter / ifmap (input channels)
H/W	Ifmap spatial height/width
R/S	Filter spatial height/width (= H/W in FC)
P/Q	Ofmap spatial height/width (= 1 in FC)

computation are the output partial sums that comprise one channel of the *output feature map* (ofmap).[4] Additional 3-D filters can be used on the same input feature map to create additional output channels (i.e., applying M filters to the input feature map generates M output channels in the output feature map). Finally, multiple input feature maps (N) may be processed together as a *batch* to potentially improve reuse of the filter weights.

Given the shape parameters in Table 2.1,[5] the computation of a CONV layer is defined as:

$$o[n][m][p][q] = (\sum_{c=0}^{C-1}\sum_{r=0}^{R-1}\sum_{s=0}^{S-1} i[n][c][Up + r][Uq + s] \times f[m][c][r][s]) + b[m],$$

$$0 \leq n < N, 0 \leq m < M, 0 \leq p < P, 0 \leq q < Q,$$
$$P = (H - R + U)/U, Q = (W - S + U)/U. \tag{2.1}$$

o, i, f, and **b** are the tensors of the ofmaps, ifmaps, filters, and biases, respectively. U is a given stride size.

Figure 2.2b shows a visualization of this computation (ignoring biases). As much as possible, we will adhere to the following coloring scheme in this book.

- **Blue**: input activations belonging to an input feature map.

- **Green**: weights belonging to a filter.

---

[4]For simplicity, in this chapter, we will refer to an array of partial sums as an output feature map. However, technically, the output feature map would be composed the values of the partial sums *after* they have gone through a nonlinear function (i.e., the output activations).

[5]In some literature, K is used rather than M to denote the number of 3-D filters (also referred to a kernels), which determines the number of output feature map channels. We opted not to use K to avoid confusion with yet other communities that use it to refer to the number of dimensions. We also have adopted the convention of using P and Q as the dimensions of the output to align with other publications and since our prior use of E and F caused an alias with the use of "F" to represent filter weights. Note that some literature also use X and Y to denote the spatial dimensions of the input rather than W and H.

Table 2.1: Shape parameters of a CONV/FC layer

Shape Parameter	Description
N	Batch size of 3-D fmaps
M	Number of 3-D filters / number of channels of ofmap (output channels)
C	Number of channels of filter / ifmap (input channels)
H/W	Ifmap spatial height/width
R/S	Filter spatial height/width (= H/W in FC)
P/Q	Ofmap spatial height/width (= 1 in FC)

computation are the output partial sums that comprise one channel of the *output feature map* (ofmap).[4] Additional 3-D filters can be used on the same input feature map to create additional output channels (i.e., applying M filters to the input feature map generates M output channels in the output feature map). Finally, multiple input feature maps (N) may be processed together as a *batch* to potentially improve reuse of the filter weights.

Given the shape parameters in Table 2.1,[5] the computation of a CONV layer is defined as:

$$o[n][m][p][q] = (\sum_{c=0}^{C-1} \sum_{r=0}^{R-1} \sum_{s=0}^{S-1} i[n][c][Up + r][Uq + s] \times f[m][c][r][s]) + b[m],$$

$$0 \leq n < N, 0 \leq m < M, 0 \leq p < P, 0 \leq q < Q,$$

$$P = (H - R + U)/U, Q = (W - S + U)/U. \tag{2.1}$$

o, **i**, **f**, and **b** are the tensors of the ofmaps, ifmaps, filters, and biases, respectively. U is a given stride size.

Figure 2.2b shows a visualization of this computation (ignoring biases). As much as possible, we will adhere to the following coloring scheme in this book.

- **Blue**: input activations belonging to an input feature map.

- **Green**: weights belonging to a filter.

---

[4]For simplicity, in this chapter, we will refer to an array of partial sums as an output feature map. However, technically, the output feature map would be composed the values of the partial sums *after* they have gone through a nonlinear function (i.e., the output activations).

[5]In some literature, K is used rather than M to denote the number of 3-D filters (also referred to a kernels), which determines the number of output feature map channels. We opted not to use K to avoid confusion with yet other communities that use it to refer to the number of dimensions. We also have adopted the convention of using P and Q as the dimensions of the output to align with other publications and since our prior use of E and F caused an alias with the use of "F" to represent filter weights. Note that some literature also use X and Y to denote the spatial dimensions of the input rather than W and H.

- **Red**: partial sums—Note: since there is no formal term for an array of partial sums, we will sometimes label an array of partial sums as an output feature map and color it red (even though, technically, output feature maps are composed of activations derived from partial sums that have passed through a nonlinear function and therefore should be blue).

Returning to the CONV layer calculation in Equation (2.1), one notes that the operands (i.e., the ofmaps, ifmaps, and filters) have many dimensions. Therefore, these operands can be viewed as *tensors* (i.e., high-dimension arrays) and the computation can be treated as a tensor algebra computation where the computation involves performing binary operations (e.g., multiplications and additions forming dot products) between tensors to produce new tensors. Since the CONV layer can be viewed as a tensor algebra operation, it is worth noting that an alternative representation for a CONV layer can be created using the *tensor index notation* found in [51], which describes a compiler for sparse tensor algebra computations.[6] The tensor index notation provides a compact way to describe a kernel's functionality. For example, in this notation matrix multiply $Z = AB$ can be written as:

$$\mathbf{Z}_{ij} = \sum\nolimits_{k} \mathbf{A}_{ik} \mathbf{B}_{kj}. \tag{2.2}$$

That is, the output point (i, j) is formed by taking a dot product of k values along the i-th row of A and the j-th column of B.[7] Extending this notation to express computation on the index variables (by putting those calculations in parenthesis) allows a CONV layer in tensor index notation to be represented quite concisely as:

$$\mathbf{O}_{nmpq} = \left(\sum\nolimits_{crs} \mathbf{I}_{nc(Up+r)(Uq+s)} \mathbf{F}_{mcrs}\right) + \mathbf{b}_{m}. \tag{2.3}$$

In this calculation, each output at a point (n, m, p, q) is calculated as a dot product taken across the index variables c, r, and s of the specified elements of the input activation and filter weight tensors. Note that this notation attaches no significance to the order of the index variables in the summation. The relevance of this will become apparent in the discussion of dataflows (Chapter 5) and mapping computations onto a DNN accelerator (Chapter 6).

Finally, to align the terminology of CNNs with the generic DNN,

- filters are composed of weights (i.e., synapses), and

- input and output feature maps (ifmaps, ofmaps) are composed of input and output activations (partial sums after application of a nonlinear function) (i.e., input and output neurons).

---

[6]Note that many of the values in the CONV layer tensors are zero, making the tensors *sparse*. The origins of this sparsity, and approaches for performing the resulting sparse tensor algebra, are presented in Chapter 8.

[7]Note that Albert Einstein popularized a similar notation for tensor algebra which omits any explicit specification of the summation variable.

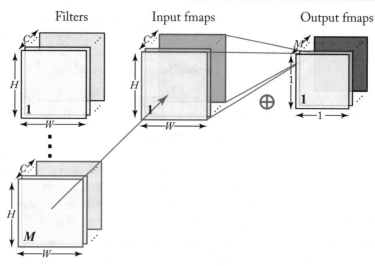

Figure 2.3: Fully connected layer from convolution point of view with $H = R$, $W = S$, $P = Q = 1$, and $U = 1$.

2.3.2   FC LAYER (FULLY CONNECTED)

In an FC layer, every value in the output feature map is a weighted sum of every input value in the input feature map (i.e., it is fully connected). Furthermore, FC layers typically do not exhibit weight sharing and as a result the computation tends to be memory-bound. FC layers are often processed in the form of a matrix multiplication, which will be explained in Chapter 4. This is the reason while matrix multiplication is often associated with DNN processing.

An FC layer can also be viewed as a special case of a CONV layer. Specifically, a CONV layer where the filters are of the same size as the input feature maps. Therefore, it does not have the local, sparsely connected with weight sharing property of CONV layers. Therefore, Equation (2.1) still holds for the computation of FC layers with a few additional constraints on the shape parameters: $H = R$, $W = S$, $P = Q = 1$, and $U = 1$. Figure 2.3 shows a visualization of this computation and in the tensor index notation from Section 2.3.1 it is:

$$\mathbf{O}_{nm} = \sum_{chw} \mathbf{I}_{nchw} \mathbf{F}_{mchw}. \tag{2.4}$$

2.3.3   NONLINEARITY

A nonlinear activation function is typically applied after each CONV or FC layer. Various nonlinear functions are used to introduce nonlinearity into the DNN, as shown in Figure 2.4. These include historically conventional nonlinear functions such as sigmoid or hyperbolic tangent. These were popular because they facilitate mathematical analysis/proofs. The rectified linear unit

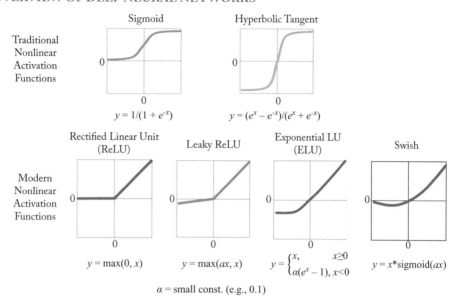

Figure 2.4: Various forms of nonlinear activation functions. (Figure adapted from [62].)

(ReLU) [52] has become popular in recent years due to its simplicity and its ability to enable fast training, while achieving comparable accuracy.[8] Variations of ReLU, such as leaky ReLU [53], parametric ReLU [54], exponential LU [55], and Swish [56] have also been explored for improved accuracy. Finally, a nonlinearity called maxout, which takes the maximum value of two intersecting linear functions, has shown to be effective in speech recognition tasks [57, 58].

2.3.4 POOLING AND UNPOOLING

There are a variety of computations that can be used to change the spatial resolution (i.e., H and W or P and Q) of the feature map depending on the application. For applications such as image classification, the goal is to summarize the entire image into one label; therefore, reducing the spatial resolution may be desirable. Networks that reduce input into a sparse output are often referred to as *encoder networks*. For applications such as semantic segmentation, the goal is to assign a label to each pixel in the image;[9] as a result, increasing the spatial resolution may be desirable. Networks that expand input into a dense output are often referred to as *decoder networks*.

Reducing the spatial resolution of a feature map is referred to as *pooling* or more generically downsampling. Pooling, which is applied to each channel separately, enables the network to be

---

[8]In addition to being simple to implement, ReLU also increases the sparsity of the output activations, which can be exploited by a DNN accelerator to increase throughput, reduce energy consumption and reduce storage cost, as described in Section 8.1.1.

[9]In the literature, this is often referred to *dense* prediction.

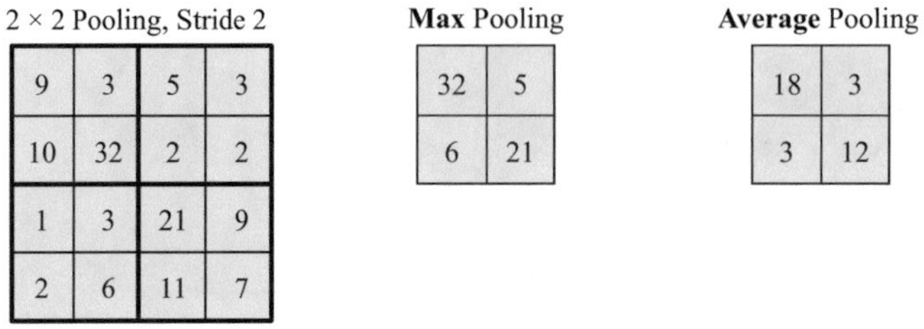

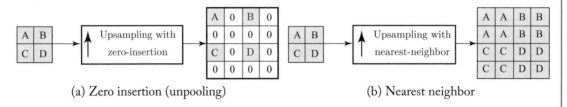

Figure 2.5: Various forms of pooling.

(a) Zero insertion (unpooling)   (b) Nearest neighbor

Figure 2.6: Various forms of unpooling/upsampling. (Figures adapted from [64].)

robust and invariant to small shifts and distortions. Pooling combines, or *pools*, a set of values in its *receptive field* into a smaller number of values. Pooling can be parameterized based on the size of its receptive field (e.g., 2×2) and pooling operation (e.g., max or average), as shown in Figure 2.5. Typically, pooling occurs on non-overlapping blocks (i.e., the stride is equal to the size of the pooling). Usually a stride of greater than one is used such that there is a reduction in the spatial resolution of the representation (i.e., feature map). Pooling is usually performed *after* the nonlinearity.

   Increasing the spatial resolution of a feature map is referred to as *unpooling* or more generically as upsampling. Commonly used forms of upsampling include inserting zeros between the activations, as shown in Figure 2.6a (this type of upsampling is commonly referred to as unpooling[10]), interpolation using nearest neighbors [63, 64], as shown in Figure 2.6b, and interpolation with bilinear or bicubic filtering [65]. Upsampling is usually performed *before* the CONV or FC layer. Upsampling can introduce structured sparsity in the input feature map that can be exploited for improved energy efficiency and throughput, as described in Section 8.1.1.

2.3.5   NORMALIZATION

Controlling the input distribution across layers can help to significantly speed up training and improve accuracy. Accordingly, the distribution of the layer input activations (σ, μ) are normal-

---

[10]There are two versions of unpooling: (1) zero insertion is applied in a regular pattern, as shown in Figure 2.6a [60]—this is most commonly used; and (2) unpooling is paired with a max pooling layer, where the location of the max value during pooling is stored, and during unpooling the location of the non-zero value is placed in the location of the max value before pooling [61].

ized such that it has a zero mean and a unit standard deviation. In batch normalization (BN), the normalized value is further scaled and shifted, as shown in Equation (2.5), where the parameters (γ, β) are learned from training [66]:[11,12]

$$y = \frac{x - \mu}{\sqrt{\sigma^2 + \epsilon}}\gamma + \beta,\tag{2.5}$$

where ϵ is a small constant to avoid numerical problems.

Prior to the wide adoption of BN, local response normalization (LRN) [7] was used, which was inspired by lateral inhibition in neurobiology where excited neurons (i.e., high value activations) should subdue its neighbors (i.e., cause low value activations); however, BN is now considered standard practice in the design of CNNs while LRN is mostly deprecated. Note that while LRN is usually performed after the nonlinear function, BN is usually performed between the CONV or FC layer and the nonlinear function. If BN is performed immediately after the CONV or FC layer, its computation can be folded into the weights of the CONV or FC layer resulting in no additional computation for inference.

2.3.6   COMPOUND LAYERS

The above primitive layers can be combined to form compound layers. For instance, *attention* layers are composed of matrix multiplications and feed-forward, fully connected layers [68]. Attention layers have become popular for processing a wide range of data including language and images and are commonly used in a type of DNNs called Transformers. We will discuss transformers in more detail in Section 2.5. Another example of a compound layer is the *up-convolution* layer [60], which performs zero-insertion (unpooling) on the input and then applies a convolutional layer.[13] Up-convolution layers are typically used in DNNs such as General Adversarial Networks (GANs) and Auto Encoders (AEs) that process image data. We will discuss GANs and AEs in more detail in Section 2.5.

2.4   CONVOLUTIONAL NEURAL NETWORKS (CNNs)

CNNs are a common form of DNNs that are composed of multiple CONV layers, as shown in Figure 2.7. In such networks, each layer generates a successively higher-level abstraction of

---

[11]It has been recently reported that the reason batch normalization enables faster and more stable training is due to the fact that it makes the optimization landscape smoother resulting in more predictive and stable behavior of the gradient [67]; this is in contrast to the popular belief that batch normalization stabilizes the distribution of the input across layers. Nonetheless, batch normalization continues to be widely used for training and thus needs to be supported during inference.

[12]During training, parameters σ and μ are computed per batch, and γ and β are updated per batch based on the gradient; therefore, training for different batch sizes will result in different σ and μ parameters, which can impact accuracy. Note that each channel has its own set of σ, μ, γ, and β parameters. During inference, all parameters are fixed, where σ and μ are computed from the entire training set. To avoid performing an extra pass over the entire training set to compute σ and μ, σ and μ are usually implemented as the running average of the per batch σ and μ computed during training.

[13]Note variants of the up CONV layer with different types of upsampling include deconvolution layer, sub-pixel or fractional convolutional layer, transposed convolutional layer, and backward convolution layer [69].

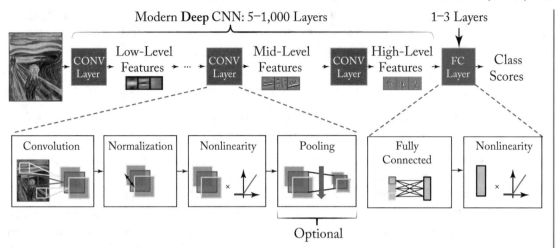

Figure 2.7: Convolutional Neural Networks.

the input data, called a *feature map* (fmap), which preserves essential yet unique information. Modern CNNs are able to achieve superior performance by employing a very deep hierarchy of layers. CNNs are widely used in a variety of applications including image understanding [7], speech recognition [70], game play [10], robotics [42], etc. This book will focus on its use in image processing, specifically for the task of image classification [7]. Modern CNN models for image classification typically have 5 [7] to more than a 1,000 [24] CONV layers. A small number, e.g., 1 to 3, of FC layers are typically applied after the CONV layers for classification purposes.

2.4.1 POPULAR CNN MODELS

Many CNN models have been developed over the past two decades. Each of these models are different in terms of number of layers, layer types, layer shapes (i.e., filter size, number of channels and filters), and connections between layers. Understanding these variations and trends is important for incorporating the right flexibility in any efficient DNN accelerator, as discussed in Chapter 3.

In this section, we will give an overview of various popular CNNs such as LeNet [71] as well as those that competed in and/or won the ImageNet Challenge [23], as shown in Figure 1.8, most of whose models with pre-trained weights are publicly available for download; the CNN models are summarized in Table 2.2. Two results for Top-5 error are reported. In the first row, the accuracy is boosted by using multiple crops from the image and an ensemble of multiple trained models (i.e., the CNN needs to be run several times); these results were used to compete in the ImageNet Challenge. The second row reports the accuracy if only a single crop was used

Table 2.2: Summary of popular CNNs [7, 24, 71, 73, 74]. †Accuracy is measured based on Top-5 error on ImageNet [23] using multiple crops. ‡This version of LeNet-5 has 431k weights for the filters and requires 2.3M MACs per image, and uses ReLU rather than sigmoid.

Metrics	LeNet 5	AlexNet	Overfeat Fast	VGG 16	GoogLeNet V1	ResNet 50
Top-5 error†	n/a	16.4	14.2	7.4	6.7	5.3
Top-5 error (single crop)†	n/a	19.8	17.0	8.8	10.7	7.0
Input size	28×28	227×227	231×231	224×224	224×224	224×224
Number of CONV layers	2	5	5	13	57	53
Depth in number of CONV layers	2	5	5	13	21	49
Filter sizes	5	3, 5, 11	2, 5, 11	3	1, 3, 5, 7	1, 3, 7
Number of channels	1, 20	3–256	3–1,024	3–512	3–832	3–2,048
Number of filters	20, 50	96–384	96–1,024	64–512	16–384	64–2,048
Stride	1	1, 4	1, 4	1	1, 2	1, 2
Weights	2.6 k	2.3 M	16 M	14.7 M	6.0 M	23.5 M
MACs	283 k	666 M	2.67 G	15.3 G	1.43 G	3.86 G
Number of FC layers	2	3	3	3	1	1
Filter sizes	1, 4	1, 6	1, 6, 12	1, 7	1	1
Number of channels	50, 500	256–4,096	1,024–4,096	512–4,096	1,024	2,048
Number of filters	10, 500	1,000–4,096	1,000–4,096	1,000–4,096	1,000	1,000
Weights	58 k	58.6 M	130 M	124 M	1 M	2 M
MACS	58 K	58.6 M	130 M	124 M	1 M	2 M
Total weights	60 k	61 M	146 M	138 M	7 M	25.5 M
Total MACs	341 k	724 M	2.8 G	15.5 G	1.43 G	3.9 G
Pretrained model website	[77]‡	[78, 79]	n/a	[78, 79, 80]	[78, 79, 80]	[78, 79, 80]

(i.e., the CNN is run only once), which is more consistent with what would likely be deployed in real-time and/or energy-constrained applications.

LeNet [20] was one of the first CNN approaches introduced in 1989. It was designed for the task of digit classification in grayscale images of size 28×28. The most well known version, LeNet-5, contains two CONV layers followed by two FC layers [71]. Each CONV layer uses filters of size 5×5 (1 channel per filter) with 6 filters in the first layer and 16 filters in the second layer. Average pooling of 2×2 is used after each convolution and a sigmoid is used for the non-linearity. In total, LeNet requires 60k weights and 341k multiply-and-accumulates (MACs) per image. LeNet led to CNNs' first commercial success, as it was deployed in ATMs to recognize digits for check deposits.

AlexNet [7] was the first CNN to win the ImageNet Challenge in 2012. It consists of five CONV layers followed by three FC layers. Within each CONV layer, there are 96 to 384 filters and the filter size ranges from 3×3 to 11×11, with 3 to 256 channels each. In the first layer, the three channels of the filter correspond to the red, green, and blue components of the input image. A ReLU nonlinearity is used in each layer. Max pooling of 3×3 is applied to the outputs of layers 1, 2, and 5. To reduce computation, a stride of 4 is used at the first layer of the network. AlexNet introduced the use of LRN in layers 1 and 2 before the max pooling, though LRN is no longer popular in later CNN models. One important factor that differentiates AlexNet from LeNet is that the number of weights is much larger and the shapes vary from layer to layer. To reduce the amount of weights and computation in the second CONV layer, the 96 output channels of the first layer are split into two groups of 48 input channels for the second layer, such that the filters in the second layer only have 48 channels. This approach is referred to as "grouped convolution" and illustrated in Figure 2.8.[14] Similarly, the weights in fourth and fifth layer are also split into two groups. In total, AlexNet requires 61M weights and 724M MACs to process one 227×227 input image.

Overfeat [72] has a very similar architecture to AlexNet with five CONV layers followed by three FC layers. The main differences are that the number of filters is increased for layers 3 (384 to 512), 4 (384 to 1024), and 5 (256 to 1024), layer 2 is not split into two groups, the first FC layer only has 3072 channels rather than 4096, and the input size is 231×231 rather than 227×227. As a result, the number of weights grows to 146M and the number of MACs grows to 2.8G per image. Overfeat has two different models: fast (described here) and accurate. The accurate model used in the ImageNet Challenge gives a 0.65% lower Top-5 error rate than the fast model at the cost of 1.9× more MACs.

VGG-16 [73] goes deeper to 16 layers consisting of 13 CONV layers followed by 3 FC layers. In order to balance out the cost of going deeper, larger filters (e.g., 5×5) are built from multiple smaller filters (e.g., 3×3), which have fewer weights, to achieve the same effective receptive fields, as shown in Figure 2.9a. As a result, all CONV layers have the same filter size of 3×3. In total, VGG-16 requires 138M weights and 15.5G MACs to process one 224×224 input image. VGG has two different models: VGG-16 (described here) and VGG-19. VGG-19 gives a 0.1% lower Top-5 error rate than VGG-16 at the cost of 1.27× more MACs.

GoogLeNet [74] goes even deeper with 22 layers. It introduced an inception module, shown in Figure 2.10, whose input is distributed through multiple feed-forward connections to several parallel layers. These parallel layers contain different sized filters (i.e., 1×1, 3×3, 5×5), along with 3×3 max-pooling, and their outputs are concatenated for the module output. Using multiple filter sizes has the effect of processing the input at multiple scales. For improved train-

---

[14]This grouped convolution approach is applied more aggressively when performing co-design of algorithms and hardware to reduce complexity, which will be discussed in Chapter 9.

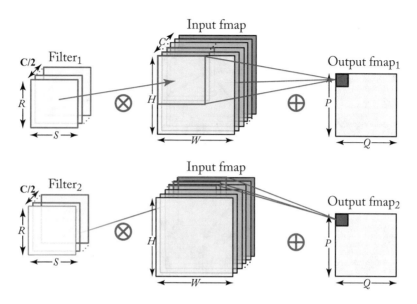

Figure 2.8: An example of dividing feature map into two *grouped convolutions*. Each filter requires 2× fewer weights and multiplications.

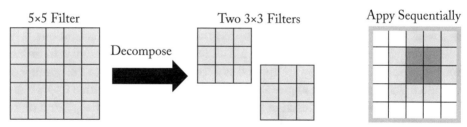

(a) Constructing a 5×5 support from 3×3 filters. Used in VGG-16.

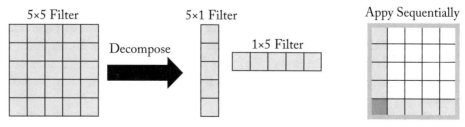

(b) Constructing a 5×5 support from 1×5 and 5×1 filter. Used in GoogLeNet/Inception v3 and v4.

Figure 2.9: Decomposing larger filters into smaller filters.

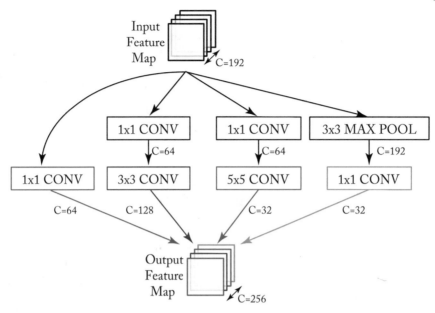

Figure 2.10: Inception module from GoogLeNet [74] with example channel lengths. Note that each CONV layer is followed by a ReLU (not drawn).

ing speed, GoogLeNet is designed such that the weights and the activations, which are stored for backpropagation during training, could all fit into the GPU memory. In order to reduce the number of weights, 1×1 filters are applied as a "bottleneck" to reduce the number of channels for each filter [75], as shown in Figure 2.11. The 22 layers consist of three CONV layers, followed by nine inceptions modules (each of which are two CONV layers deep), and one FC layer. The number of FC layers was reduce from three to one using a global average pooling layer, which summarizes the large feature map from the CONV layers into one value; global pooling will be discussed in more detail in Section 9.1.2. Since its introduction in 2014, GoogLeNet (also referred to as Inception) has multiple versions: v1 (described here), v3,[15] and v4. Inception-v3 decomposes the convolutions by using smaller 1-D filters, as shown in Figure 2.9b, to reduce number of MACs and weights in order to go deeper to 42 layers. In conjunction with batch normalization [66], v3 achieves over 3% lower Top-5 error than v1 with 2.5× more MACs [76]. Inception-v4 uses residual connections [77], described in the next section, for a 0.4% reduction in error.

ResNet [24], also known as Residual Net, uses feed-forward connections that connects to layers beyond the immediate next layer (often referred to as *residual*, *skip* or *identity* connections); these connections enable a DNN with many layers (e.g., 34 or more) to be trainable. It was

---
[15]v2 is very similar to v3.

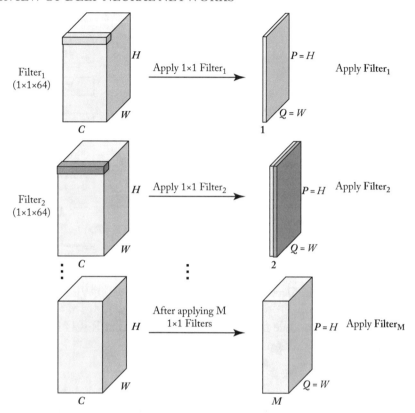

Figure 2.11: Apply 1×1×C filter (usually referred to as 1×1) to capture cross-channel correlation, but no spatial correlation. This bottleneck approach reduces the number of channels in next layer assuming the number of filters applied (M) is less than the original number of channels (C).

the first entry CNN in ImageNet Challenge that exceeded human-level accuracy with a Top-5 error rate below 5%. One of the challenges with deep networks is the vanishing gradient during training [78]; as the error backpropagates through the network the gradient shrinks, which affects the ability to update the weights in the earlier layers for very deep networks. ResNet introduces a "shortcut" module which contains an identity connection such that the weight layers (i.e., CONV layers) can be skipped, as shown in Figure 2.12. Rather than learning the function for the weight layers $F(x)$, the shortcut module learns the residual mapping ($F(x) = H(x) - x$). Initially, $F(x)$ is zero and the identity connection is taken; then gradually during training, the actual forward connection through the weight layer is used. ResNet also uses the "bottleneck" approach of using 1×1 filters to reduce the number of weights. As a result, the two layers in the shortcut module are replaced by three layers (1×1, 3×3, 1×1) where the first 1×1 layer reduces the number of activations and thus weights in the 3×3 layer, the last 1×1 layer restores

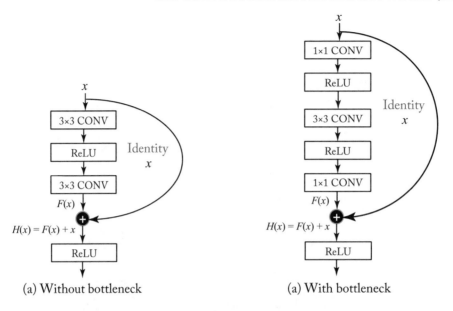

(a) Without bottleneck   (a) With bottleneck

Figure 2.12: Shortcut module from ResNet [24]. Note that ReLU following last CONV layer in shortcut is *after* the addition.

the number of activations in the output of the third layer. ResNet-50 consists of one CONV layer, followed by 16 shortcut layers (each of which are 3 CONV layers deep), and 1 FC layer; it requires 25.5M weights and 3.9G MACs per image. There are various versions of ResNet with multiple depths (e.g., *without bottleneck:* 18, 34; *with bottleneck:* 50, 101, 152). The ResNet with 152 layers was the winner of the ImageNet Challenge requiring 11.3G MACs and 60M weights. Compared to ResNet-50, it reduces the Top-5 error by around 1% at the cost of 2.9× more MACs and 2.5× more weights.

Several trends can be observed in the popular CNNs shown in Table 2.2. Increasing the depth of the network tends to provide higher accuracy. Controlling for number of weights, a deeper network can support a wider range of nonlinear functions that are more discriminative and also provides more levels of hierarchy in the learned representation [24, 73, 74, 79]. The number of filter shapes continues to vary across layers, thus flexibility is still important. Furthermore, most of the computation has been placed on CONV layers rather than FC layers. In addition, the number of weights in the FC layers is reduced and in most recent networks (since GoogLeNet) the CONV layers also dominate in terms of weights. Thus, the focus of hardware implementations targeted at CNNs should be on addressing the efficiency of the CONV layers, which in many domains are increasingly important.

Since ResNet, there have been several other notable networks that have been proposed to increase accuracy. *DenseNet* [84] extends the concept of skip connections by adding skip con-

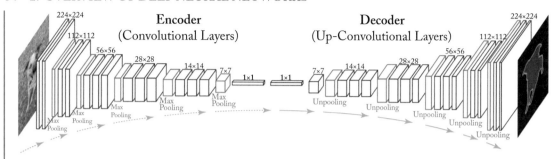

Figure 2.13: Auto Encoder network for semantic segmentation. Feature maps along with pooling and upsampling layers are shown. (Figure adapted from [92].)

nection from *multiple* previous layers to strengthen feature map propagation and feature reuse. This concept, commonly referred to as *feature aggregation*, continues to be widely explored. *WideNet* [85] proposes increasing the width (i.e., the number of filters) rather than depth of network, which has the added benefit that increasing width is more parallel-friendly than increasing depth. *ResNeXt* [86] proposes increasing the number of convolution groups (referred to as cardinality) instead of depth and width of network and was used as part of the winning entry for ImageNet in 2017. Finally, *EfficientNet* [87] proposes uniformly scaling all dimensions including depth, width, and resolution rather than focusing on a single dimension since there is an interplay between the different dimensions (e.g., to support higher input image resolution, the DNN needs higher depth to increase the receptive field and higher width to capture more fine-grained patterns). WideNet, ResNeXt, and EfficientNet demonstrate that there exists methods beyond increasing depth for increasing accuracy, and thus highlights that there remains much to be explored and understood about the relationship between layer shape, number of layers, and accuracy.

2.5   OTHER DNNs

There are other types of DNNs beyond CNNs including Recurrent Neural Networks (RNNs) [88, 89], Transformers [68], Auto Encoders (AEs) [90], and General Adversarial Networks (GANs) [91]. The diverse types of DNNs allow them to handle a wide range of inputs for a wide range of tasks. For instance, RNNs and Transformers are often used to handle sequential data that can have variable length (e.g., audio for speech recognition, or text for natural language processing). AEs and GANs can be used to generate dense output predictions by combining encoder and decoder networks. Example applications that use AEs include predicting pixel-wise depth values for depth estimation [64] and assigning pixel-wise class labels for semantic segmentation [92], as shown in Figure 2.13. Example applications that use GANs to generate images with the same statistics as the training set include image synthesis [93] and style transfer [94].

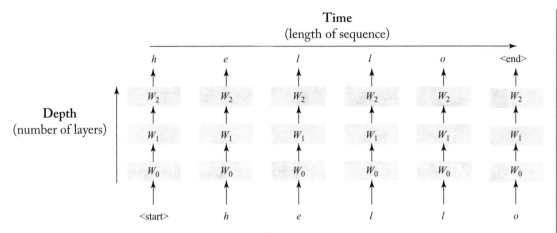

Figure 2.14: Dependencies in RNN are in both the time and depth dimension. The same weights (W_i) are used across time, while different weights are used across depth. (Figure adapted from [4].)

While their applications may differ from the CNNs described in Section 2.4, many of the building blocks and primitive layers are similar. For instance, RNNs and transformers heavily rely on matrix multiplications, which means that they have similar challenges as FC layers (e.g., they are memory bound due to lack of data reuse); thus, many of the techniques used to accelerate FC layers can also be used to accelerate RNNs and transformers (e.g., tiling discussed in Chapter 4, network pruning discussed in Chapter 8, etc.). Similarly, the decoder network of GANs and AEs for image processing use *up-convolution layers*, which involves upsampling the input feature map using zero insertion (unpooling) before applying a convolution; thus, many of the techniques used to accelerate CONV layers can also be used to accelerate the decoder network of GANs and AEs for image processing (e.g., exploit input activation sparsity discussed in Chapter 8).

While the dominant compute aspect of these DNNs are similar to CNNs, they do often require some other forms of compute. For instance, RNNs, particularly Long Short-Term Memory networks (LSTMs) [95], require support of element-wise multiplications as well a variety of nonlinear functions (sigmoid, tanh), unlike CNNs which typically only use ReLU. However, these operations do not tend to dominate run-time or energy consumption; they can be computed in software [96] or the nonlinear functions can be approximated by piecewise linear look up tables [97]. For GANs and AEs, additional support is required for upsampling.

Finally, RNNs have additional dependencies since the output of a layer is fed back to its input, as shown in Figure 2.14. For instance, the inputs to layer i at time t depends on the output of layer $i - 1$ at time t and layer i at time $t - 1$. This is similar to the dependency across layers, in that the output of layer i is the input to layer $i + 1$. These dependencies limit what inputs can be processed in parallel (e.g., within the same batch). For DNNs with feed-forward layers, *any* inputs can be processed at the same time (i.e., batch size greater than one); however, multiple layers of the *same* input cannot be processed at the same time (e.g., layers i and $i + 1$). In contrast, RNNs can only process multiple inputs at the same time if the inputs are *not sequentially dependent*; in other words, RNNs can process two separate sequences at the same time, but not multiple elements within the sequence (e.g., inputs t and $t + 1$ of the same sequence) and not multiple layers of the same input (which is similar to feed-forward networks).

2.6   DNN DEVELOPMENT RESOURCES

One of the key factors that has enabled the rapid development of DNNs is the set of development resources that have been made available by the research community and industry. These resources are also key to the development of DNN accelerators by providing characterizations of the workloads and facilitating the exploration of trade-offs in model complexity and accuracy. This section will describe these resources such that those who are interested in this field can quickly get started.

2.6.1   FRAMEWORKS

For ease of DNN development and to enable the sharing of trained networks, several deep learning frameworks have been developed from various sources. These open-source libraries contain software libraries for DNNs. Caffe was made available in 2014 from UC Berkeley [59]. It supports C, C++, Python, and MATLAB. Tensorflow [98] was released by Google in 2015, and supports C++ and Python; it also supports multiple CPUs and GPUs and has more flexibility than Caffe, with the computation expressed as dataflow graphs to manage the "tensors" (multidimensional arrays). Another popular framework is Torch, which was developed by Facebook and NYU and supports C, C++, and Lua; PyTorch [99] is its successor and is built in Python. There are several other frameworks such as Theano, MXNet, CNTK, which are described in [100]. There are also higher-level libraries that can run on top of the aforementioned frameworks to provide a more universal experience and faster development. One example of such libraries is Keras, which is written in Python and supports Tensorflow, CNTK, and Theano.

The existence of such frameworks are not only a convenient aid for DNN researchers and application designers, but they are also invaluable for engineering high performance or more efficient DNN computation engines. In particular, because the frameworks make heavy use of a set of primitive operations, such as the processing of a CONV layer, they can incorporate use of optimized software or hardware accelerators. This acceleration is transparent to the user of the framework. Thus, for example, most frameworks can use Nvidia's cuDNN library for rapid execution on Nvidia GPUs. Similarly, transparent incorporation of dedicated hardware accelerators can be achieved as was done with the Eyeriss chip using Caffe [101].

Finally, these frameworks are a valuable source of workloads for hardware researchers. They can be used to drive experimental designs for different workloads, for profiling different workloads and for exploring hardware-algorithm trade-offs.

MNIST   ImageNet

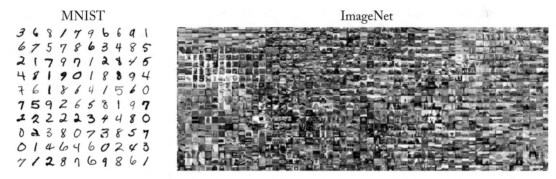

Figure 2.15: MNIST (10 classes, 60k training, 10k testing) [103] versus ImageNet (1000 classes, 1.3M training, 100k testing) [23] dataset.

2.6.2   MODELS

Pretrained DNN models can be downloaded from various websites [80–83] for the various different frameworks. It should be noted that even for the same DNN (e.g., AlexNet) the accuracy of these models can vary by around 1 to 2% depending on how the model was trained and tested, and thus the results do not always exactly match the original publication.

These pre-trained models often are tied to a given framework. In order to facilitate easier exchange between different networks, Open Neural Network Exchange (ONNX) has been established as an open ecosystem for interchangeable DNN models [102]; the current participants include Amazon, Facebook, and Microsoft.

2.6.3   POPULAR DATASETS FOR CLASSIFICATION

It is important to factor in the difficulty of the task when comparing different DNN models. For instance, the task of classifying handwritten digits from the MNIST dataset [103] is much simpler than classifying an object into one of 1000 classes as is required for the ImageNet dataset [23] (Figure 2.15). It is expected that the size of the DNNs (i.e., number of weights) and the number of MACs will be larger for the more difficult task than the simpler task and thus require more energy and have lower throughput. For instance, LeNet-5[71] is designed for digit classification, while AlexNet[7], VGG-16[73], GoogLeNet[74], and ResNet[24] are designed for the 1000-class image classification.

There are many AI tasks that come with publicly available datasets in order to evaluate the accuracy of a given DNN. Public datasets are important for comparing the accuracy of different approaches. The simplest and most common task in computer vision is image classification, which involves being given an entire image, and selecting 1 of N classes that the image most likely belongs to. There is no localization or detection.

MNIST is a widely used dataset for digit classification that was introduced in 1998 [103]. It consists of 28×28 pixel grayscale images of handwritten digits. There are 10 classes (for 10 digits) and 60,000 training images and 10,000 test images. LeNet-5 was able to achieve an accuracy of 99.05% when MNIST was first introduced. Since then the accuracy has increased to 99.79% using regularization of neural networks with dropconnect [104]. Thus, MNIST is now considered a fairly easy dataset.

CIFAR is a dataset that consists of 32×32 pixel colored images of various objects, which was released in 2009 [105]. CIFAR is a subset of the 80 million Tiny Image dataset [106]. CIFAR-10 is composed of 10 mutually exclusive classes. There are 50,000 training images (5000 per class) and 10,000 test images (1000 per class). A two-layer convolutional deep belief network was able to achieve 64.84% accuracy on CIFAR-10 when it was first introduced [107]. Since then the accuracy has increased to 96.53% using fractional max pooling [108].

ImageNet is a large-scale image dataset that was first introduced in 2010; the dataset stabilized in 2012 [23]. It contains images of 256×256 pixel in color with 1000 classes. The classes are defined using the WordNet as a backbone to handle ambiguous word meanings and to combine together synonyms into the same object category. In other words, there is a hierarchy for the ImageNet categories. The 1000 classes were selected such that there is no overlap in the ImageNet hierarchy. The ImageNet dataset contains many fine-grained categories including 120 different breeds of dogs. There are 1.3M training images (732 to 1300 per class), 100,000 testing images (100 per class) and 50,000 validation images (50 per class).

The accuracy for the image classification task in the ImageNet Challenge are reported using two metrics: Top-5 and Top-1 accuracy.[16] Top-5 accuracy means that if any of the top five scoring categories are the correct category, it is counted as a correct classification. Top-1 accuracy requires that the top scoring category be correct. In 2012, the winner of the ImageNet Challenge (AlexNet) was able to achieve an accuracy of 83.6% for the Top-5 (which is substantially better than the 73.8% which was second place that year that did not use DNNs); it achieved 61.9% on the Top-1 of the validation set. In 2019, the state-of-the-art DNNs achieve accuracy above 97% for the Top-5 and above 84% for the Top-1 [87].

In summary of the various image classification datasets, it is clear that MNIST is a fairly easy dataset, while ImageNet is a more challenging one with a wider coverage of classes. Thus, in terms of evaluating the accuracy of a given DNN, it is important to consider that dataset upon which the accuracy is measured.

2.6.4   DATASETS FOR OTHER TASKS

Since the accuracy of the state-of-the-art DNNs are performing better than human-level accuracy on image classification tasks, the ImageNet Challenge has started to focus on more difficult tasks such as single-object localization and object detection. For single-object localization, the target object must be localized and classified (out of 1000 classes). The DNN outputs the top five categories and top five bounding box locations. There is no penalty for identifying an object that is in the image but not included in the ground truth. For object detection, all objects in the

---

[16]Note that in some parts of the book we use Top-1 and Top-5 *error*. The error can be computed as 100% minus accuracy.

image must be localized and classified (out of 200 classes). The bounding box for all objects in these categories must be labeled. Objects that are not labeled are penalized as well as duplicated detections.

Beyond ImageNet, there are also other popular image datasets for computer vision tasks. For object detection, there is the PASCAL VOC (2005-2012) dataset that contains 11k images representing 20 classes (27k object instances, 7k of which have detailed segmentation) [109]. For object detection, segmentation, and recognition in context, there is the M.S. COCO dataset with 2.5M labeled instances in 328k images (91 object categories) [110]; compared to ImageNet, COCO has fewer categories but more instances per category, which is useful for precise 2-D localization. COCO also has more labeled instances per image to potentially help with contextual information.

Most recently, even larger scale datasets have been made available. For instance, Google has an Open Images dataset with over 9M images [111], spanning 6000 categories. There is also a YouTube dataset with 8M videos (0.5M hours of video) covering 4800 classes [112]. Google also released an audio dataset comprised of 632 audio event classes and a collection of 2M human-labeled 10-second sound clips [113]. These large datasets will be evermore important as DNNs become deeper with more weights to train. In addition, it has been shown that accuracy increases logarithmically based on the amount of training data [16].[17]

Undoubtedly, both larger datasets and datasets for new domains will serve as important resources for profiling and exploring the efficiency of future DNN engines.

2.6.5   SUMMARY

The development resources presented in this section enable us to evaluate hardware using the appropriate DNN model and dataset. In particular, it's important to realize that difficult tasks typically require larger models; for instance, LeNet would not apply to the ImageNet Challenge. In addition, different datasets are required for different tasks; for instance, self-driving cars require high-definition video, and thus a network trained on the low resolution ImageNet dataset may not be sufficient. To address these requirements, the number of datasets continues to grow at a rapid pace.

---

[17]This was demonstrated on Google's internal JFT-300M dataset with 300M images and 18,291 classes, which is two orders of magnitude larger than ImageNet. However, performing four iterations across the entire training set using 50 K-80 GPUs required two months of training, which further emphasizes that compute is one of the main bottlenecks in the advancement of DNN research.

PART II

Design of Hardware for Processing DNNs

CHAPTER 3

Key Metrics and Design Objectives

Over the past few years, there has been a significant amount of research on efficient processing of DNNs. Accordingly, it is important to discuss the key metrics that one should consider when comparing and evaluating the strengths and weaknesses of different designs and proposed techniques and that should be incorporated into design considerations. While efficiency is often only associated with the number of operations per second per Watt (e.g., floating-point operations per second per Watt as FLOPS/W or tera-operations per second per Watt as TOPS/W), it is actually composed of many more metrics including accuracy, throughput, latency, energy consumption, power consumption, cost, flexibility, and scalability. Reporting a comprehensive set of these metrics is important in order to provide a complete picture of the trade-offs made by a proposed design or technique.

In this chapter, we will

- discuss the importance of each of these metrics;

- breakdown the factors that affect each metric. When feasible, present equations that describe the relationship between the factors and the metrics;

- describe how these metrics can be incorporated into design considerations for both the DNN hardware and the DNN model (i.e., workload); and

- specify what should be reported for a given metric to enable proper evaluation.

Finally, we will provide a case study on how one might bring all these metrics together for a holistic evaluation of a given approach. But first, we will discuss each of the metrics.

3.1 ACCURACY

Accuracy is used to indicate the quality of the result for a given task. The fact that DNNs can achieve state-of-the-art accuracy on a wide range of tasks is one of the key reasons driving the popularity and wide use of DNNs today. The units used to measure accuracy depend on the task. For instance, for image classification, accuracy is reported as the percentage of correctly classified images, while for object detection, accuracy is reported as the mean average precision (mAP), which is related to the trade off between the true positives, false positives, and false negatives.

Factors that affect accuracy include the difficulty of the task and dataset.[1] For instance, classification on ImageNet is much more difficult than on MNIST, and object detection is usually more difficult than classification. As a result, a DNN model that performs well on MNIST may not necessarily perform well on ImageNet.

Achieving high accuracy on difficult tasks or datasets typically requires more complex DNN models (e.g., a larger number of MAC operations and more distinct weights, increased diversity in layer shapes, etc.), which can impact how efficiently the hardware can process the DNN model.

Accuracy should therefore be interpreted in the context of the difficulty of the task and dataset.[2] Evaluating hardware using well-studied, widely used DNN models, tasks, and datasets can allow one to better interpret the significance of the accuracy metric. Recently, motivated by the impact of the SPEC benchmarks for general purpose computing [114], several industry and academic organizations have put together a broad suite of DNN models, called *MLPerf*, to serve as a common set of well-studied DNN models to evaluate the performance and enable fair comparison of various software frameworks, hardware accelerators, and cloud platforms for both training and inference of DNNs [115].[3] The suite includes various types of DNNs (e.g., CNN, RNN, etc.) for a variety of tasks including image classification, object identification, translation, speech-to-text, recommendation, sentiment analysis, and reinforcement learning.

3.2   THROUGHPUT AND LATENCY

Throughput is used to indicate the amount of data that can be processed or the number of executions of a task that can be completed in a given time period. High throughput is often critical to an application. For instance, processing video at 30 frames per second is often necessary for delivering real-time performance. For data analytics, high throughput means that more data can be analyzed in a given amount of time. As the amount of visual data is growing exponentially, high-throughput big data analytics becomes increasingly important, particularly if an action needs to be taken based on the analysis (e.g., security or terrorist prevention; medical diagnosis or drug discovery). Throughput is often generically reported as the number of operations per second. In the case of inference, throughput is reported as inferences per second.

Latency measures the time between when the input data arrives to a system and when the result is generated. Low latency is necessary for real-time interactive applications, such as augmented reality, autonomous navigation, and robotics. Latency is typically reported in seconds per inference.

---

[1]Ideally, robustness and fairness should be considered in conjunction with accuracy, as there is also an interplay between these factors; however, these are areas of on-going research and beyond the scope of this book.

[2]As an analogy, getting 9 out of 10 answers correct on a high school exam is different than 9 out of 10 answers correct on a college-level exam. One must look beyond the score and consider the difficulty of the exam.

[3]Earlier DNN benchmarking efforts including DeepBench [116] and Fathom [117] have now been subsumed by MLPerf.

Throughput and latency are often assumed to be directly derivable from one another. However, they are actually quite distinct. A prime example of this is the well-known approach of batching input data (e.g., batching multiple images or frames together for processing) to increase throughput since it amortizes overhead, such as loading the weights; however, batching also increases latency (e.g., at 30 frames per second and a batch of 100 frames, some frames will experience at least 3.3 second delay), which is not acceptable for real-time applications, such as high-speed navigation where it would reduce the time available for course correction. Thus, achieving low latency and high throughput simultaneously can sometimes be at odds depending on the approach and both should be reported.[4]

There are several factors that affect throughput and latency. In terms of throughput, the number of inferences per second is affected by

$$\frac{\text{inferences}}{\text{second}} = \frac{\text{operations}}{\text{second}} \times \frac{1}{\frac{\text{operations}}{\text{inference}}}, \tag{3.1}$$

where the number of *operations per second* is dictated by both the DNN hardware and DNN model, while the number of *operations per inference* is dictated by the DNN model.

When considering a system comprised of multiple processing elements (PEs), where a PE corresponds to a simple or primitive core that performs a single MAC operation, the number of operations per second can be further decomposed as follows:

$$\frac{\text{operations}}{\text{second}} = \underbrace{\left(\frac{1}{\frac{\text{cycles}}{\text{operation}}} \times \frac{\text{cycles}}{\text{second}} \right)}_{\text{for a single PE}} \times \text{number of PEs} \times \text{utilization of PEs}. \tag{3.2}$$

The first term reflects the peak throughput of a single PE, the second term reflects the amount of parallelism, while the last term reflects degradation due to the inability of the architecture to effectively utilize the PEs.

Since the main operation for processing DNNs is a MAC, we will use number of operations and number of MAC operations interchangeably.

One can increase the peak throughput of a single PE by increasing the number of *cycles per second*, which corresponds to a higher clock frequency achieved by reducing the critical path

---

[4]The phenomenon described here can also be understood using Little's Law [118] from queuing theory, where the relationship between average throughput and average latency are related by the average number of tasks in flight, as defined by

$$\overline{\text{throughput}} = \frac{\overline{\text{tasks-in-flight}}}{\overline{\text{latency}}}.$$

A DNN-centric version of Little's Law would have throughput measured in inferences per second, latency measured in seconds, and inferences-in-flight, as the tasks-in-flight equivalent, measured in the number of images in a batch being processed simultaneously. This helps to explain why increasing the number of inferences in flight to increase throughput may be counterproductive because some techniques that increase the number of inferences in flight (e.g., batching) also increase latency.

at the circuit or micro-architectural level; alternatively, one can also reduce the number of *cycles per operations*, which can be affected by the design of the MAC (e.g., a bit-serial, multi-cycle MAC would have more cycles per operation).

While the above approaches increase the throughput of a single PE, the overall throughput can be increased by increasing the *number of PEs*, and thus the maximum number of MAC operations that can be performed in parallel. The number of PEs is dictated by the area of the PE and the area cost of the system. If the area cost of the system is fixed, then increasing the number of PEs requires either reducing the area per PE or trading off on-chip storage area for more PEs. Reducing on-chip storage, however, can affect the utilization of the PEs, which we will discuss next.

Reducing the area per PE can also be achieved by reducing the logic associated with delivering operands to a MAC. This can be achieved by controlling multiple MACs with a single piece of logic. This is analogous to the situation in instruction-based systems such as CPUs and GPUs that reduce instruction bookkeeping overhead by using large aggregate instructions (e.g., single-instruction, multiple-data (SIMD), vector instructions, single-instruction, multiple-threads (SIMT), or tensor instructions, where a single instruction can be used to initiate multiple operations.

The number of PEs and the peak throughput of a single PE only indicate the theoretical maximum throughput (i.e., peak performance) when all PEs are performing computation (100% utilization). In reality, the achievable throughput depends on the actual utilization of those PEs, which is affected by several factors as follows:

$$\text{utilization of PEs} = \frac{\text{number of active PEs}}{\text{number of PEs}} \times \text{utilization of active PEs}. \quad (3.3)$$

The first term reflects the ability to distribute the workload to PEs, while the second term reflects how efficiently those active PEs are processing the workload.

The *number of active PEs* is the number of PEs that receive work; therefore, it is desirable to distribute the workload to as many PEs as possible. The ability to distribute the workload is determined by the flexibility of the architecture, for instance the on-chip network, to support the different layer shapes in the DNN model.

Within the constraints of the on-chip network, the *number of active PEs* is also determined by the specific allocation of work to PEs by the mapping process. The mapping process involves the placement and scheduling in space and time of every MAC operation (including the delivery of the appropriate operands) onto the PEs. The mapper can be thought of as a compiler for the DNN hardware. The design of on-chip networks and mappings are discussed in Chapters 5 and 6.

The *utilization of the active PEs* is largely dictated by the timely delivery of work to the PEs such that the active PEs do not become idle while waiting for the data to arrive. This can be affected by the bandwidth and latency of the (on-chip and off-chip) memory and network. The bandwidth requirements can be affected by the amount of data reuse available in the

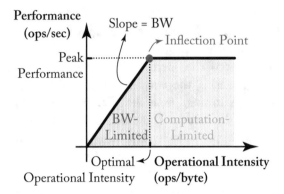

Figure 3.1: The roofline model. The peak *operations per second* is indicated by the bold line; when the operation intensity, which dictates by amount of compute per byte of data, is low, the *operations per second* is limited by the data delivery. The design goal is to operate as close as possible to the peak *operations per second* for the operation intensity of a given workload.

DNN model and the amount of data reuse that can be exploited by the memory hierarchy and dataflow. The dataflow determines the order of operations and where data is stored and reused. The amount of data reuse can also be increased using a larger batch size, which is one of the reasons why increasing batch size can increase throughput. The challenge of data delivery and memory bandwidth are discussed in Chapters 5 and 6. The *utilization of the active PEs* can also be affected by the imbalance of work allocated across PEs, which can occur when exploiting sparsity (i.e., avoiding unnecessary work associated with multiplications by zero); PEs with less work become idle and thus have lower utilization.

There is also an interplay between the number of PEs and the utilization of PEs. For instance, one way to reduce the likelihood that a PE needs to wait for data is to store some data locally near or within the PE. However, this requires increasing the chip area allocated to on-chip storage, which, given a fixed chip area, would reduce the number of PEs. Therefore, a key design consideration is the area allocation between compute (which increases the number of PEs) versus on-chip storage (which increases the utilization of PEs).

The impact of these factors can be captured using Eyexam, which is a systematic way of understanding the performance limits for DNN processors as a function of specific characteristics of the DNN model and accelerator design. Eyexam includes and extends the well-known roofline model [119]. The roofline model, as illustrated in Figure 3.1, relates average bandwidth demand and peak computational ability to performance. Eyexam is described in Chapter 6.

While the number of *operations per inference* in Equation (3.1) depends on the DNN model, the *operations per second* depends on both the DNN model and the hardware. For example, designing DNN models with efficient layer shapes (also referred to efficient network architectures), as described in Chapter 9, can reduce the number of MAC operations in the

DNN model and consequently the number of *operations per inference*. However, such DNN models can result in a wide range of layer shapes, some of which may have poor utilization of PEs and therefore reduce the overall *operations per second*, as shown in Equation (3.2).

A deeper consideration of the *operations per second*, is that all operations are not created equal and therefore *cycles per operation* may not be a constant. For example, if we consider the fact that anything multiplied by zero is zero, some MAC operations are ineffectual (i.e., they do not change the accumulated value). The number of ineffectual operations is a function of both the DNN model and the input data. These ineffectual MAC operations can require fewer cycles or no cycles at all. Conversely, we only need to process effectual (or non-zero) MAC operations, where both inputs are non-zero; this is referred to as exploiting sparsity, which is discussed in Chapter 8.

Processing only effectual MAC operations can increase the *(total) operations per second* by increasing the *(total) operations per cycle*.[5] Ideally, the hardware would skip all ineffectual operations; however, in practice, designing hardware to skip all ineffectual operations can be challenging and result in increased hardware complexity and overhead, as discussed in Chapter 8. For instance, it might be easier to design hardware that only recognizes zeros in one of the operands (e.g., weights) rather than both. Therefore, the ineffectual operations can be further divided into those that are exploited by the hardware (i.e., skipped) and those that are unexploited by the hardware (i.e., not skipped). The number of operations actually performed by the hardware is therefore *effectual operations plus unexploited ineffectual operations*.

Equation (3.4) shows how *operations per cycle* can be decomposed into

1. the number of *effectual operations plus unexploited ineffectual operations per cycle*, which remains somewhat constant for a given hardware accelerator design;

2. the ratio of *effectual operations* over *effectual operations plus unexploited ineffectual operations*, which refers to the ability of the hardware to exploit ineffectual operations (ideally unexploited ineffectual operations should be zero, and this ratio should be one); and

3. the number of *effectual operations out of (total) operations*, which is related to the amount of sparsity and depends on the DNN model.

As the amount of sparsity increases (i.e., the number of *effectual operations out of (total) operations* decreases), the *operations per cycle* increases, as shown in Equation (3.4); this subsequently increases *operations per second*, as shown in Equation (3.2):

$$\frac{\text{operations}}{\text{cycle}} = \frac{\text{effectual operations + unexploited ineffectual operations}}{\text{cycle}}$$
$$\times \frac{\text{effectual operations}}{\text{effectual operations + unexploited ineffectual operations}} \times \frac{1}{\frac{\text{effectual operations}}{\text{operations}}}.$$

$$(3.4)$$

---

[5]By *total* operations we mean both effectual and ineffectual operations.

Table 3.1: Classification of factors that affect inferences per second

Factor	Hardware	DNN Model	Input Data
Operations per inference		✓	
Operations per cycle	✓		
Cycles per second	✓		
Number of PEs	✓		
Number of active PEs	✓	✓	
Utilization of active PEs	✓	✓	
Effectual operations out of (total) operations		✓	✓
Effectual operations plus unexploited ineffectual operations per cycle	✓		

However, exploiting sparsity requires additional hardware to identify when inputs to the MAC are zero to avoid performing unnecessary MAC operations. The additional hardware can increase the critical path, which decreases cycles per second, and also increase the area of the PE, which reduces the number of PEs for a given area. Both of these factors can reduce the *operations per second*, as shown in Equation (3.2). Therefore, the complexity of the additional hardware can result in a trade off between reducing the number of *unexploited ineffectual operations* and increasing critical path or reducing the number of PEs.

Finally, designing hardware and DNN models that support reduced precision (i.e., fewer bits per operand and per operations), which is discussed in Chapter 7, can also increase the number of *operations per second*. Fewer bits per operand means that the memory bandwidth required to support a given operation is reduced, which can increase the utilization of active PEs since they are less likely to be starved for data. In addition, the area of each PE can be reduced, which can increase the number of PEs for a given area. Both of these factors can increase the *operations per second*, as shown in Equation (3.2). Note, however, that if *multiple* levels of precision need to be supported, additional hardware is required, which can, once again, increase the critical path and also increase the area of the PE, both of which can reduce the *operations per second*, as shown in Equation (3.2).

In this section, we discussed multiple factors that affect the number of inferences per second. Table 3.1 classifies whether the factors are dictated by the hardware, by the DNN model or both.

In summary, the number of MAC operations in the DNN model alone is not sufficient for evaluating the throughput and latency. While the DNN model can affect the number of MAC operations per inference based on the network architecture (i.e., layer shapes) and the sparsity of the weights and activations, the overall impact that the DNN model has on throughput and

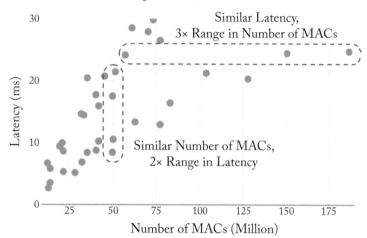

Figure 3.2: The number of MAC operations in various DNN models versus latency measured on Pixel phone. Clearly, the number of MAC operations is not a good predictor of latency. (Figure from [120].)

latency depends on the ability of the hardware to add support to recognize these approaches without significantly reducing utilization of PEs, number of PEs, or cycles per second. This is why the number of MAC operations is not necessarily a good proxy for throughput and latency (see Figure 3.2), and it is often more effective to design efficient DNN models with hardware in the loop. Techniques for designing DNN models with hardware in the loop are discussed in Chapter 9.

Similarly, the number of PEs in the hardware and their peak throughput are not sufficient for evaluating the throughput and latency. It is critical to report actual runtime of the DNN models on hardware to account for other effects such as utilization of PEs, as highlighted in Equation (3.2). Ideally, this evaluation should be performed on clearly specified DNN models, for instance those that are part of the MLPerf benchmarking suite. In addition, batch size should be reported in conjunction with the throughput in order to evaluate latency.

3.3   ENERGY EFFICIENCY AND POWER CONSUMPTION

Energy efficiency is used to indicate the amount of data that can be processed or the number of executions of a task that can be completed for a given unit of energy. High energy efficiency is important when processing DNNs at the edge in embedded devices with limited battery capacity (e.g., smartphones, smart sensors, robots, and wearables). Edge processing may be preferred over the cloud for certain applications due to latency, privacy, or communication bandwidth limitations. Energy efficiency is often generically reported as the number of operations per joule. In the case of inference, energy efficiency is reported as inferences per joule and energy consumption is reported as joules per inference.

Power consumption is used to indicate the amount of energy consumed per unit time. Increased power consumption results in increased heat dissipation; accordingly, the maximum power consumption is dictated by a design criterion typically called the thermal design power (TDP), which is the power that the cooling system is designed to dissipate. Power consumption is important when processing DNNs in the cloud as data centers have stringent power ceilings due to cooling costs; similarly, handheld and wearable devices also have tight power constraints since the user is often quite sensitive to heat and the form factor of the device limits the cooling mechanisms (e.g., no fans). Power consumption is typically reported in watts or joules per second.

Power consumption in conjunction with energy efficiency limits the throughput as follows:

$$\frac{\text{inferences}}{\text{second}} \leq \text{Max}\left(\frac{\text{joules}}{\text{second}}\right) \times \frac{\text{inferences}}{\text{joule}}. \tag{3.5}$$

Therefore, if we can improve energy efficiency by increasing the number of *inferences per joule*, we can increase the number of *inferences per second* and thus throughput of the system.

There are several factors that affect the energy efficiency. The number of inferences per joule can be decomposed into

$$\frac{\text{inferences}}{\text{joule}} = \frac{\text{operations}}{\text{joule}} \times \frac{1}{\frac{\text{operations}}{\text{inference}}}, \tag{3.6}$$

where the number of operations per joule is dictated by both the hardware and DNN model, while the number of operations per inference is dictated by the DNN model.

There are various design considerations for the hardware that will affect the energy per operation (i.e., joules per operation). The energy per operation can be broken down into the energy required to move the input and output data, and the energy required to perform the MAC computation

$$\text{Energy}_{total} = \text{Energy}_{data} + \text{Energy}_{MAC}. \tag{3.7}$$

For each component the joules per operation[6] is computed as

$$\frac{\text{joules}}{\text{operation}} = \alpha \times C \times V_{DD}^{2}, \tag{3.8}$$

where C is the total switching capacitance, V_{DD} is the supply voltage, and α is the switching activity, which indicates how often the capacitance is charged.

The energy consumption is dominated by the data movement as the capacitance of data movement tends to be much higher that the capacitance for arithmetic operations such as a

---

[6]Here, an operation can be a MAC operation or a data movement.

Operation:	Energy (pJ)	Relative Energy Cost
8b Add	0.03	
16b Add	0.05	
32b Add	0.1	
16b FP Add	0.4	
32b FP Add	0.9	
8b Multiply	0.2	
32b Multiply	3.1	
16b FP Multiply	1.1	
32b FP Multiply	3.7	
32b SRAM Read (8KB)	5	
32b DRAM Read	640	

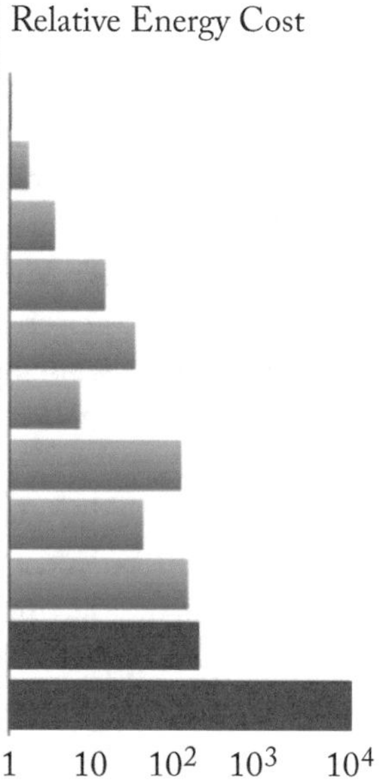

1   10 10^2 10^3 10^4

Figure 3.3: The energy consumption for various arithmetic operations and memory accesses in a 45 nm process. The relative energy cost (computed relative to the 8b add) is shown on a log scale. The energy consumption of data movement (red) is significantly higher than arithmetic operations (blue). (Figure adapted from [121].)

MAC (Figure 3.3). Furthermore, the switching capacitance increases with the distance the data needs to travel to reach the PE, which consists of the distance to get out of the memory where the data is stored and the distance to cross the network between the memory and the PE. Accordingly, larger memories and longer interconnects (e.g., off-chip) tend to consume more energy than smaller and closer memories due to the capacitance of the long wires employed. In order to reduce the energy consumption of data movement, we can exploit data reuse where the data is moved once from a distant large memory (e.g., off-chip DRAM) and reused for multiple operations from a local smaller memory (e.g., on-chip buffer or scratchpad within the PE). Optimizing data movement is a major consideration in the design of DNN accelerators; the design of the dataflow, which defines the processing order, to increase data reuse within the memory hierarchy is discussed in Chapter 5. In addition, advanced device and memory technologies can be used to reduce the switching capacitance between compute and memory, as described in Chapter 10.

This raises the issue of the appropriate scope over which energy efficiency and power consumption should be reported. Including the entire system (out to the fans and power supplies) is beyond the scope of this book. Conversely, ignoring off-chip memory accesses, which can vary greatly between chip designs, can easily result in a misleading perception of the efficiency of the system. Therefore, it is critical to not only report the energy efficiency and power consumption of the chip, but also the energy efficiency and power consumption of the off-chip memory (e.g., DRAM) or the amount of off-chip accesses (e.g., DRAM accesses) if no specific memory technology is specified; for the latter, it can be reported in terms of the total amount of data that is read and written off-chip per inference.

Reducing the joules per MAC operation itself can be achieved by reducing the switching activity and/or capacitance at a circuit level or micro-architecture level. This can also be achieved by reducing precision (e.g., reducing the bit width of the MAC operation), as shown in Figure 3.3 and discussed in Chapter 7. Note that the impact of reducing precision on accuracy must also be considered.

For instruction-based systems such as CPUs and GPUs, this can also be achieved by reducing instruction bookkeeping overhead. For example, using large aggregate instructions (e.g., single-instruction, multiple-data (SIMD), vector instructions, single-instruction, multiple-threads (SIMT), tensor instructions), a single instruction can be used to initiate multiple operations.

Similar to the throughput metric discussed in Section 3.2, the number of *operations per inference* depends on the DNN model, however the *operations per joules* may be a function of the ability of the hardware to exploit sparsity to avoid performing ineffectual MAC operations. Equation (3.9) shows how *operations per joule* can be decomposed into:

1. the number of *effectual operations plus unexploited ineffectual operations per joule*, which remains somewhat constant for a given hardware architecture design;

2. the ratio of *effectual operations* over *effectual operations plus unexploited ineffectual operations*, which refers to the ability of the hardware to exploit ineffectual operations (ideally unexploited ineffectual operations should be zero, and this ratio should be one); and

3. the number of *effectual operations out of (total) operations*, which is related to the amount of sparsity and depends on the DNN model.

$$\frac{\text{operations}}{\text{joule}} = \frac{\text{effectual operations} + \text{unexploited ineffectual operations}}{\text{joule}}$$
$$\times \frac{\text{effectual operations}}{\text{effectual operations} + \text{unexploited ineffectual operations}}$$
$$\times \frac{1}{\frac{\text{effectual operations}}{\text{operations}}}. \qquad (3.9)$$

For hardware that can exploit sparsity, increasing the amount of sparsity (i.e., decreasing the number of *effectual operations out of (total) operations*) can increase the number of *operations per joule*, which subsequently increases *inferences per joule*, as shown in Equation (3.6). While exploiting sparsity has the potential of increasing the number of *(total) operations per joule*, the additional hardware will decrease the *effectual operations plus unexploited ineffectual operations per joule*. In order to achieve a net benefit, the decrease in *effectual operations plus unexploited ineffectual operations per joule* must be more than offset by the decrease of *effectual operations out of (total) operations*.

In summary, we want to emphasize that the number of MAC operations and weights in the DNN model are not sufficient for evaluating energy efficiency. From an energy perspective, all MAC operations or weights are not created equal. This is because the number of MAC operations and weights do not reflect where the data is accessed and how much the data is reused, both of which have a significant impact on the *operations per joule*. Therefore, the number of MAC operations and weights is not necessarily a good proxy for energy consumption and it is often more effective to design efficient DNN models with hardware in the loop. Techniques for designing DNN models with hardware in the loop are discussed in Chapter 9.

In order to evaluate the energy efficiency and power consumption of the entire system, it is critical to not only report the energy efficiency and power consumption of the chip, but also the energy efficiency and power consumption of the off-chip memory (e.g., DRAM) or the amount of off-chip accesses (e.g., DRAM accesses) if no specific memory technology is specified; for the latter, it can be reported in terms of the total amount of data that is read and written off-chip per inference. As with throughput and latency, the evaluation should be performed on clearly specified, ideally widely used, DNN models.

3.4   HARDWARE COST

In order to evaluate the desirability of a given architecture or technique, it is also important to consider the *hardware cost* of the design. Hardware cost is used to indicate the monetary cost to build a system.[7] This is important from both an industry and a research perspective as it dictates whether a system is financially viable. From an industry perspective, the cost constraints are related to volume and market; for instance, embedded processors have a much more stringent cost limitations than processors in the cloud.

One of the key factors that affect cost is the chip area (e.g., square millimeters, mm^2) in conjunction with the process technology (e.g., 45 nm CMOS), which constrains the amount of on-chip storage and amount of compute (e.g., the number of PEs for DNN accelerators, the number of cores for CPUs and GPUs, the number of digital signal processing (DSP) engines for FPGAs, etc.). To report information related to area without specifying a specific process

---

[7]There is also cost associated with operating a system, such as the electricity bill and the cooling cost, which are primarily dictated by the energy efficiency and power consumption, respectively. In addition, there is cost associated with designing the system. The operating cost is covered by the section on energy efficiency and power consumption and we limited our coverage of design cost to the fact that custom DNN accelerators have a higher design cost (after amortization) than off-the-shelf CPUs and GPUs. We consider anything beyond this, e.g., the economics of the semiconductor business, including how to price platforms, is outside the scope of this book.

technology, one can report the amount of on-chip memory (e.g, storage capacity of the global buffer) and compute (e.g., number of PEs) as a proxy for area.

Another important factor is the amount of off-chip bandwidth, which dictates the cost and complexity of the packaging and printed circuit board (PCB) design (e.g., High Bandwidth Memory (HBM) [122] to connect to off-chip DRAM, NVLink to connect to other GPUs, etc.), as well as whether additional chip area is required for a transceiver to handle signal integrity at high speeds. The off-chip bandwidth, which is typically reported in gigabits per second (Gbps) and the number of I/O ports, can be used as a proxy for packaging and PCB cost.

There is also an interplay between the costs attributable to the chip area and off-chip bandwidth. For instance, increasing on-chip storage, which increases chip area, can reduce off-chip bandwidth. Accordingly, both metrics should be reported in order to provide perspective on the total cost of the system.

Of course reducing cost alone is not the only objective. The design objective is invariably to maximize the throughput or energy efficiency for a given cost, specifically, to maximize *inferences per second per cost* (e.g., $) and/or *inferences per joule per cost*. This is closely related to the previously discussed property of utilization; to be cost efficient, the design should aim to utilize every PE to increase inferences per second, since each PE increases the area and thus the cost of the chip; similarly, the design should aim to effectively utilize all the on-chip storage to reduce off-chip bandwidth, or increase operations per off-chip memory access as expressed by the roofline model (see Figure 3.1), as each byte of on-chip memory also increases cost.

3.5 FLEXIBILITY

The merit of a DNN accelerator is also a function of its *flexibility*. Flexibility refers to the range of DNN models that can be supported on the DNN processor and the ability of the software environment (e.g., the mapper) to maximally exploit the capabilities of the hardware for any desired DNN model. Given the fast-moving pace of DNN research and deployment, it is increasingly important that DNN processors support a wide range of DNN models and tasks.

We can define *support* in two tiers: the first tier requires that the hardware only needs to be able to *functionally* support different DNN models (i.e., the DNN model can run on the hardware). The second tier requires that the hardware should also *maintain efficiency* (i.e., high throughput and energy efficiency) across different DNN models.

To maintain efficiency, the hardware should not rely on certain properties of the DNN models to achieve efficiency, as the properties of DNN models are diverse and evolving rapidly. For instance, a DNN accelerator that can efficiently support the case where the entire DNN model (i.e., all the weights) fits on-chip may perform extremely poorly when the DNN model grows larger, which is likely given that the size of DNN models continue to increase over time, as discussed in Section 2.4.1; a more flexible processor would be able to efficiently handle a wide range of DNN models, even those that exceed on-chip memory.

The degree of flexibility provided by a DNN accelerator presents a complex trade-off with accelerator cost. Specifically, additional hardware usually needs to be added in order to flexibly support a wider range of workloads and/or improve their throughput and energy efficiency. Thus, the design objective is to reduce the overhead (e.g., area cost and energy consumption) of supporting flexibility while maintaining efficiency across the wide range of DNN models. Thus, evaluating flexibility would entail ensuring that the extra hardware is a net benefit across multiple workloads.

Flexibility has become increasingly important when we factor in the many techniques that are being applied to the DNN models with the promise to make them more efficient, since they increase the diversity of workloads that need to be supported. These techniques include DNNs with different network architectures (i.e., different layer shapes, which impacts the amount of required storage and compute and the available data reuse that can be exploited), as described in Chapter 9, different levels of precision (i.e., different number of bits across layers and data types), as described in Chapter 7, and different degrees of sparsity (i.e., number of zeros in the data), as described in Chapter 8. There are also different types of DNN layers and computations beyond MAC operations (e.g., activation functions) that need to be supported.

Actually getting a performance or efficiency benefit from these techniques invariably requires additional hardware. Again, it is important that the overhead of the additional hardware does not exceed the benefits of these techniques. This encourages a hardware and DNN model *co-design* approach.

To date, exploiting the flexibility of DNN hardware has relied on mapping processes that act like static per-layer compilers. As the field moves to DNN models that change dynamically, mapping processes will need to dynamically adapt at runtime to changes in the DNN model or input data, while still maximally exploiting the flexibility of the hardware to improve efficiency.

In summary, to assess the flexibility of DNN processors, its efficiency (e.g., inferences per second, inferences per joule) should be evaluated on a wide range of DNN models. The MLPerf benchmarking workloads are a good start; however, additional workloads may be needed to represent efficient techniques such as efficient network architectures, reduced precision and sparsity. The workloads should match the desired application. Ideally, since there can be many possible combinations, it would also be beneficial to define the range and limits of DNN models that can be *efficiently* supported on a given platform (e.g., maximum number of weights per filter or DNN model, minimum amount of sparsity, required structure of the sparsity, levels of precision such as 8-bit, 4-bit, 2-bit, or 1-bit, types of layers and activation functions, etc.).

3.6   SCALABILITY

Scalability has become increasingly important due to the wide use cases for DNNs. This is demonstrated by emerging technologies used for scaling up not just the size of the chip, but also building systems with multiple chips (often referred to as chiplets) [123] or even wafer-scale chips [124]. Scalability refers to how well a design can be scaled up to achieve higher performance (i.e., latency and throughput) and energy efficiency when increasing the amount of resources (e.g., the number of PEs and on-chip storage). This evaluation is done under the assumption that the system does not have to be significantly redesigned (e.g., the design only needs to be replicated) since major design changes can be expensive in terms of time and cost. Ideally, a scalable design can be used for low-cost embedded devices and high-performance devices in the cloud simply by scaling up the resources.

Ideally, the performance would scale linearly and proportionally with the number of PEs. When the problem size (e.g., the batch size) is held constant, this is referred to as strong scaling, and is the more challenging type of scaling. On the other hand, scaling performance while allowing the problem size to increase (e.g., by increasing batch size) is called weak scaling and is also an important objective in some situations. Similarly, the energy efficiency would also improve with more on-chip storage, however, this would be likely be nonlinear (e.g., increasing the on-chip storage such that the entire DNN model fits on chip would result in an abrupt improvement in energy efficiency). In practice, this is often challenging due to factors such as the reduced utilization of PEs and the increased cost of data movement due to long distance interconnects.

Scalability can be connected with cost efficiency by considering how *inferences per second per cost* and *inferences per joule per cost* changes with scale. For instance, if throughput increases linearly with number of PEs (with proportional scaling of all storage), then the *inferences per second per cost* could be constant. It is also possible for the *inferences per second per cost* to improve super-linearly with increasing number of PEs, due to increased sharing of data across PEs. On the other hand, inferences per joule per cost might remain constant or even improve as a consequence of more sharing of data by multiple PEs.

In summary, to understand the scalability of a DNN accelerator design, it is important to report its performance and efficiency metrics as the number of PEs and storage capacity increases. This may include how well the design might handle technologies used for scaling up, such as inter-chip interconnect.

3.7   INTERPLAY BETWEEN DIFFERENT METRICS

It is important that all metrics are accounted for in order to fairly evaluate the design trade-offs. For instance, without the accuracy given for a specific dataset and task, one could run a simple DNN model and easily claim low power, high throughput, and low cost—however, the processor might not be usable for a meaningful task; alternatively, without reporting the off-chip bandwidth, one could build a processor with only MACs and easily claim low cost, high throughput, high accuracy, and low *chip* power—however, when evaluating *system* power, the off-chip memory access would be substantial. Finally, the test setup should also be reported, including whether the results are measured or obtained from simulation[8] and how many images were tested.

In summary, the evaluation process for whether a DNN system is a viable solution for a given application might go as follows:

1. the accuracy determines if it can perform the given task;

2. the latency and throughput determine if it can run fast enough and in real time;

3. the energy and power consumption will primarily dictate the form factor of the device where the processing can operate;

4. the cost, which is primarily dictated by the chip area and external memory bandwidth requirements, determines how much one would pay for this solution;

5. flexibility determines the range of tasks it can support; and

6. the scalability determines whether the same design effort can be amortized for deployment in multiple domains, (e.g., in the cloud and at the edge), and if the system can efficiently be scaled with DNN model size.

---

[8]If obtained from simulation, it should be clarified whether it was after synthesis or post place-and-route and what library corner (e.g., process corner, supply voltage, temperature) was used.

CHAPTER 4

Kernel Computation

The fundamental computation of both CONV and FC layers described in Chapter 2 are multiply-and-accumulate (MAC) operations. Because there are negligible dependencies between these operations and the accumulations are commutative, there is considerable flexibility in the order in which MACs can be scheduled and these computations can be easily parallelized. Therefore, in order to achieve high performance for DNNs, highly parallel compute paradigms are very commonly used. These architectural paradigms can be categorized as being either temporal or spatial, as shown in Figure 4.1.

Temporal architectures use centralized control for a large number of arithmetic logic units (ALUs). These ALUs typically can only fetch data from the memory hierarchy and cannot communicate directly with each other. Such architectures, which appear mostly in CPUs or GPUs, employ a variety of techniques to improve parallelism such as vector instructions (e.g., single-instruction-multiple-data, SIMD, instructions) or parallel threads (e.g., single-instruction-multiple-thread, SIMT, architectures). In contrast, spatial architectures allow for communication between ALUs, and use dataflow processing (i.e., the ALUs form a processing chain so that they can pass data from one to another directly). Sometimes each ALU can have its own control logic and local memory, called a scratchpad or register file. We refer to an ALU with its own local memory as a processing engine (PE). Spatial architectures are commonly used for processing DNNs in ASIC- and FPGA-based designs.

With the rise in popularity of DNNs, many programmable temporal systems (i.e., CPUs and GPUs) started adding features that target DNN processing. For instance, the Intel Knights Landing CPU featured special vector instructions for deep learning that performed multiple fused multiply accumulate operations; the Nvidia PASCAL GP100 GPU featured 16-bit floating point (fp16) arithmetic support to perform two fp16 operations on a single precision core for faster deep learning computation. As will be described in Section 4.1, DNN calculations can often be cast as matrix multiplications. As a result, the Nvidia VOLTA GV100 GPU featured a special compute unit for performing matrix multiplication and accumulation. Activity on that unit is invoked with individual instructions that perform many MAC operations.

We also have been seeing systems built specifically for DNN processing such as Facebook's Big Basin custom DNN server [125] and Nvidia's DGX-1. DNN inference also started

Temporal Architecture
(SIMD/SIMT)

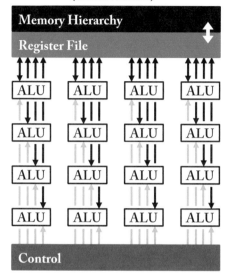

Spatial Architecture
(Dataflow Processing)

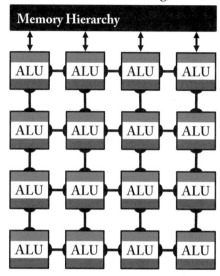

Figure 4.1: Highly parallel compute paradigms.

appearing on various embedded System-on-Chips (SoC[1]) such as Apple's A, Nvidia's Tegra, and Samsung's Exynos.

In this chapter and Chapter 5, we will discuss the different design strategies for efficient processing on these different platforms, without any impact on accuracy (i.e., all approaches in this chapter produce bit-wise identical results[2]); specifically,

- for *temporal* architectures such as CPUs and GPUs, we will discuss how DNN algorithms can be mapped and optimized on these platforms and how *computational transforms* on the kernel can reduce the number of multiplications to *increase throughput* and how the computation (e.g., MACs) can be ordered (i.e., *tiled*) to improve memory subsystem behavior (this chapter); and

- for *spatial* architectures used in accelerators, we will discuss how *dataflows* can increase data reuse from low cost memories in the memory hierarchy to *reduce energy consumption* and how other architectural features can help optimize data movement (Chapter 5).

---

[1]A system-on-chip (SoC) refers to when a CPU and accelerators such as GPUs or application specific processing modules for video compression engines and baseband communications, are all integrated on a chip.

[2]There may be some minor mismatch due to order of operations for floating point operations, however this does not affect accuracy.

4.1   MATRIX MULTIPLICATION WITH TOEPLITZ

CPUs and GPUs use hardware parallelizaton techniques such as SIMD or SIMT to perform the MACs in parallel. All the ALUs share the same control and memory (register file). Such a scheme is naturally amenable to perform the many regular parallel multiplications found in matrix-matrix or matrix-vector multiplication.[3] Therefore, the kernel computation of both the FC and CONV layers are often mapped to matrix multiplication. Figure 4.2 shows how a matrix multiplication is used for the FC layer. The height of the filter matrix is the number of 3-D filters (M) and the width is the number of weights per 3-D filter (input channels (C) × height (H) × width (W), since the filter height (R) equals H and the filter width (S) equals W in the FC layer); the height of the input feature maps matrix is the number of activations per 3-D input feature map ($C \times H \times W$), and the width is the number of 3-D input feature maps, also referred to as the batch size (one in Figure 4.2a and N in Figure 4.2b); finally, the height of the output feature map matrix is the number of channels in the output feature maps (M), and the width is the number of 3-D output feature maps/batch size (N), where each output feature map of the FC layer has the dimension of 1×1×number of output channels (M).

The CONV layer in a DNN can also be mapped to a matrix multiplication using a relaxed form of the Toeplitz matrix, as shown in Figure 4.3. In this form, the input activations in the input feature map are replicated to correspond to the input activation convolutional reuse. The downside of using matrix multiplication for the CONV layers is that there is redundant data in the input feature map matrix, as highlighted in Figure 4.3a. This can lead to either inefficiency in storage, or a complex memory access pattern.

Since convolving an input by a filter is mathematically equivalent to convolving the filter by the input, one can also convert convolution into a matrix multiply by replicating the filter weights to correspond to the filter weight convolutional reuse. Such a transformation is illustrated in Figure 4.4. As was the case when replicating input activations the downside is the redundant data in the filter matrix. Furthermore, since the filter size is typically much smaller than the size of the input feature map, this transformation results in a sparse matrix. The combination of these factors can again lead to either inefficiency in storage or a complex memory access pattern.

4.2   TILING FOR OPTIMIZING PERFORMANCE

As described in the previous section, many DNN computations can be formulated as a matrix multiplication. To efficiently perform these computations, there are many software libraries designed for CPUs (e.g., OpenBLAS, Intel MKL, etc.) and GPUs (e.g., cuBLAS, cuDNN, etc.) that have optimized implementations of matrix multiplication. A key characteristic of these libraries is that they strive to optimize memory subsystem behavior. In specific, given the conventional memory subsystem organization that places increasingly smaller, faster, and lower energy

---

[3]These operations are often implemented in generalized matrix-matrix multiplication (GEMM) or generalized matrix-vector multiplication (GEMV) libraries.

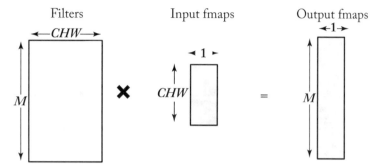

(a) Matrix vector multiplication is used when computing a single output feature map from a single input feature map.

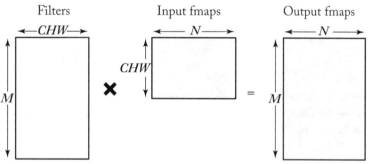

(b) Matrix multiplications is used when computing *N* output feature maps from *N* input feature maps.

Figure 4.2: Mapping to matrix multiplication for FC layers (where filter size is equal to input feature map size, such that $R = H$ and $S = W$); in other words, CHW in the figure is the same as CRS. A batch size of $N = 1$ results in a matrix-vector multiplication, while a batch size greater than 1 ($N > 1$) results in a matrix-matrix multiplication.

consuming memories closer to the compute units, the libraries attempt to maximize reuse of the values held in the smaller, faster, and more energy-efficient memories.

To understand how one might maximize reuse of values in the memories closest to the compute units, consider a naive implementation of matrix multiplication used in a fully connected computation, as illustrated in Figure 4.5. The figure shows how rows of the filter weight matrix are combined with columns of the input feature map. This combination involves the element-wise multiplication of the values from the filter row and the input feature map column and final summing of all the elements of the resulting vector (i.e., doing an *inner product*). Each such inner product produces the value for one element of the output feature map matrix. This computation style is referred to as the *inner-product approach* to matrix multiplication.

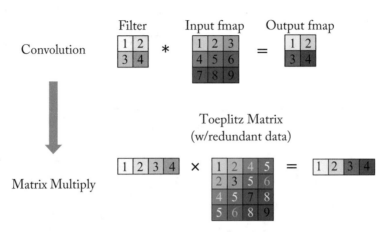

(a) Mapping convolution to matrix multiplication. Convolution is performed via matrix multiplication after flattening the filter weights and output fmap into vectors and expanding the input fmap values through replication into the *Toeplitz matrix*.

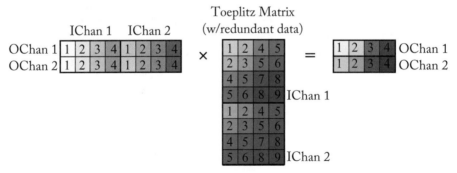

(b) Extend Toeplitz matrix to multiple input and output channels. Additional rows and columns corresponding to additional input and output channels are added to the filter weights, output fmap, and Toeplitz matrix to perform multiple input channel/multiple output channel convolution via matrix multiplication.

Figure 4.3: Mapping to matrix multiplication for CONV layers. (*Continues.*)

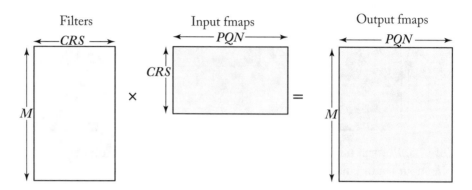

(c) Dimensions of matrix multiplication, where $P = (H - R + U)/U$ and $Q = (W - S + U)/U$, as defined in Equation (2.1) and Table 2.1.

Figure 4.3: (*Continued.*) Mapping to matrix multiplication for convolutional layers.

Convolution:

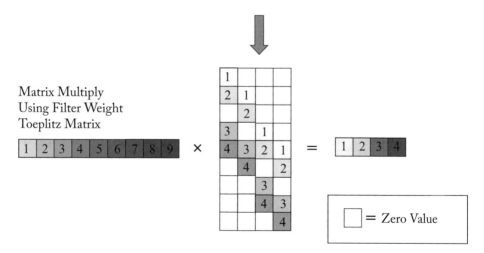

Figure 4.4: Mapping to matrix multiplication for CONV layers by weight replication.

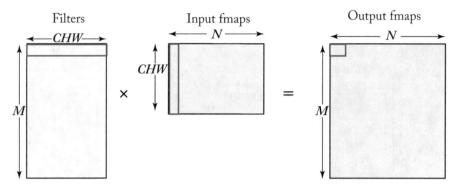

Figure 4.5: Illustration of dot product operation on a row of the filter weight matrix and column of the input feature map (fmap).

If the inner-product computation is ordered such that a single row of the filter matrix is combined successively with each of the columns of the input feature map, then there is apparently good reuse of the elements of the row of the filter matrix. However, often the size of a row in the filter matrix[4] is larger than the memory closest to the compute units, e.g., cache. This results in poor reuse, because values from the filter matrix in the small memory cannot be held long enough that they are available to be reused by the computation on the next column of the input feature map matrix. Therefore, they must be reloaded from the next level of the memory hierarchy resulting in significant inefficiency.

To ameliorate the memory inefficiencies that result from calculating the inner products on full rows and columns of a matrix multiply, libraries will invariably partition or *tile* the computation to fit in the various levels of the memory hierarchy. The principle behind tiling is illustrated in Figure 4.6, where the inner products are done on a 2-D partition, or tile, of the full matrices. For each pair of tiles in the matrices, the same inner-product approach can be employed on the partial rows of filter weights and partial columns in the input feature map matrix to create a tile of partial results in the output feature map. As computations for all the pairs of tiles are done, the subsequent partial results are added to the partial results in the output feature map from previous partial result computations. If a single tile of filter weights is used repeatedly to create a series of partial results, and if the tile is small enough to be held in the memory closest to the compute units, then reuse in that memory will be higher.

Tiling of matrix multiply can be applied recursively to improve efficiency at each level of the memory hierarchy. Tiling can also be applied to parallelize the computation across multiple CPUs or the many threads of a GPU. Therefore, CPU and GPU libraries for matrix multiplication have been optimized by tiling the computations appropriately for every architecture of

---

[4]Note that each row of the filter matrix represents a filter, and the size of the row is the size of the filter, which for fully connected is $C \times H \times W$.

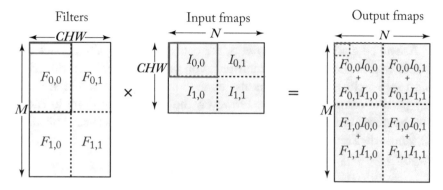

Figure 4.6: Illustration of dot product operation on a tiled row of a filter weight matrix and a tiled column of the input feature map (fmap). Note that the output is highlighted with a dotted line since only a partial value has been computed (from tiles $F_{0,0}$ and $I_{0,0}$). To complete the computation of the output, one needs to perform the dot product operation on tiles $F_{0,1}$ and $I_{1,0}$ and accumulate with the previous result.

interest based on characteristics like sizes of the cache at each level in the memory hierarchy and the topology of parallel computation units.

Tiling algorithms require considerable sophistication. For instance, an additional complication arises because set associative caches have a policy to determine which data is retained when new data needs to be added (i.e., on a cache miss). These policies are implemented with sophisticated hardware-based replacement algorithms (e.g., least recently used (LRU) or dynamic re-reference interval prediction (DRRIP) [126]). Therefore, the libraries also have to try to account for exactly how tiles will be retained to achieve optimal performance. To handle the wide variety of hardware platforms and layer shapes, these libraries often have different implementations of an algorithm, like matrix multiply, that are optimized for a particular hardware platform and layer shape. This optimization invariably includes the tiling strategy. So, when the library is called, it will dynamically select and run the most appropriate implementation.

In addition to the pre-existing libraries that dynamically pick from a menu of implementations, there has been considerable work on compilers that optimize a user-written program for tiling, e.g., [127]. One segment of this work relies on creating a polyhedral model of the computation and using a boolean satisfiability (SAT) solver to optimally tile and schedule the program [128]. Such techniques have been added to the popular GCC and LLVM compiler frameworks. Halide [129] is an example of another approach that decouples the basic expression of the algorithm from user-provided annotations that describe the desired scheduling and tiling of the algorithm. Finally, TVM [130] is a compiler that exposes graph-level and operator-level optimizations for DNN workloads across diverse hardware back-ends. In addition to tiling for hiding memory latency, TVM performs optimization such as high-level operator fusion (e.g., performing a CONV layer and ReLU together with one pass through memory), and mapping to arbitrary hardware primitives.

As will be discussed in Chapter 5, the concept of tiling also applies to specialized DNN architectures that perform matrix multiplication as well as those that directly perform convolutions. In such architectures, tiling is also done to maximize data reuse in the memory hierarchy and by parallel computation units. However, the tiling is not done in caches (i.e., implicitly data orchestrated units) but rather in more specialized buffers (i.e., explicitly data orchestrated units), whose data retention characteristics are more directly under program control, as described in Section 5.8.

4.3 COMPUTATION TRANSFORM OPTIMIZATIONS

DNN calculations can sometimes be further sped up by applying computational transforms to the data to reduce the number of (typically expensive) multiplications, while still giving the same bit-wise result. The objective is to improve performance or reduce energy consumption, although this can come at a cost of more intermediate results, an increased number of additions, and a more irregular data access pattern.

4.3.1 GAUSS' COMPLEX MULTIPLICATION TRANSFORM

One way to view these transforms is as more sophisticated versions of Gauss' technique for multiplying complex numbers. In the standard approach for complex multiplication the computation $(a + bi)(c + di)$ is performed with a full set of cross-term multiplications as follows:

$$(ac - bd) + (bc - ad)i.$$

In this case, the computation requires *4 multiplications and 3 additions*. However, the computation can be transformed by a re-association of operations into the computation of three intermediate terms and then the real and imaginary parts of the result are computed via a sum and difference of the intermediate terms (k_j) as follows:

$$k_1 = c(a + b)$$
$$k_2 = a(d - c)$$
$$k_3 = b(c + d)$$
$$\text{Real part} = k_1 - k_3$$
$$\text{Imaginary part} = k_1 + k_2.$$

In the transformed form, the computation is reduced to *3 multiplications and 5 additions*. In the following sections, we will discuss more sophisticated re-association of computations for matrix multiplication and direct convolution that can result in reducing the number of multiplications in those computations.

4.3.2   STRASSEN'S MATRIX MULTIPLICATION TRANSFORM

For matrix multiplication, a well-known re-association technique is Strassen's algorithm, which has been explored for reducing the number of multiplications in DNNs [131]. Strassen's algorithm rearranges the computations of a matrix multiplication in a recursive manner to reduce the number of multiplications from $O(N^3)$ to $O(N^{2.807})$. An illustration of the application of the Strassen algorithm to a 2×2 matrix multiplication is shown as follows:

$$A = \begin{bmatrix} a & b \\ c & d \end{bmatrix}$$

$$B = \begin{bmatrix} e & f \\ g & h \end{bmatrix}$$

$$AB = \begin{bmatrix} ae + bg & af + bh \\ ce + dg & cf + dh \end{bmatrix}.$$

In this example, the *8 multiplications and 4 additions* are converted into *7 multiplications and 18 additions* along with the creation of 7 intermediate values (k_j) as follows:

$$k_1 = a(f - h)$$
$$k_2 = (a + b)h$$
$$k_3 = (c + d)e$$
$$k_4 = d(g - e)$$
$$k_5 = (a + d)(e + h)$$
$$k_6 = (b - d)(g + h)$$
$$k_7 = (a - c)(e + f).$$

The final output AB is constructed from the intermediate (k_i) values as follows:

$$AB = \begin{bmatrix} k_5 + k_4 - k_2 + k_6 & k_1 + k_2 \\ k_3 + k_4 & k_1 + k_5 - k_3 - k_7 \end{bmatrix}.$$

Note that, when used for DNN calculations, one of the matrices will contain filter weights, which will be constant across different inputs. In this example, pre-calculation using a constant B matrix reduces the number of additions to 13. In summary, Strassen can be used to reduce the number of multiplications, but its benefits come at the cost of increased storage requirements for intermediate results and sometimes reduced numerical stability [132]. Furthermore, the benefit of Strassen is primarily manifest for matrices larger than those typically used in DNNs.

4.3.3   WINOGRAD TRANSFORM

Winograd's algorithm [133, 134] applies a re-association of the arithmetic operations on the feature map and filter to reduce the number of multiplications required specifically for *convolution*, as opposed to a generic matrix multiply handled by the previously described transforms.

The Winograd transform allows for the efficient computation of multiple convolutions using the same filter weights. For example, the computation of two 1×3 convolutions of input activations (i_j) and filter weights (f_j), normally takes *6 multiplications and 4 additions* as follows:

$$\begin{bmatrix} i_0 & i_1 & i_2 \\ i_1 & i_2 & i_3 \end{bmatrix} \begin{bmatrix} f_0 \\ f_1 \\ f_2 \end{bmatrix} = \begin{bmatrix} o_0 \\ o_1 \end{bmatrix}.$$

However, using the Winograd re-association, this reduces to *4 multiplications and 12 additions* along with 2 shifts (to implement divide by 2) using 4 intermediate values (k_j) as follows:

$$k_1 = (i_0 - i_2) f_0$$
$$k_2 = (i_1 + i_2) \frac{f_0 + f_1 + f_2}{2}$$
$$k_3 = (i_2 - i_1) \frac{f_0 - f_1 + f_2}{2}$$
$$k_4 = (i_1 - i_3) f_2.$$

The final outputs (o_j) are constructed from the intermediate (k_i) values as follows:

$$\begin{bmatrix} o_0 \\ o_1 \end{bmatrix} = \begin{bmatrix} k_1 + k_2 + k_3 \\ k_2 - k_3 - k_4 \end{bmatrix}.$$

With constant filter weights, this reduces to *4 multiplications and 8 additions*. Note that each application of the Winograd transform only does a convolution on a small number of input activations (i.e., a tile of input activations). Therefore, to do the entire set of convolutions for an input feature map requires the application of Winograd on a tile-by-tile basis; as a result, a series of separate tiles of output are generated using a sliding window of input activations (i.e., inputs are reused across output tiles). Winograd transforms can apply to 2-D convolutions by repeated application of the transform.

The reduction in multiplications that Winograd achieves varies based on the filter and tile size. A larger tile size results in a larger reduction in multiplications at the cost of higher complexity transforms. A particularly attractive filter size is 3×3, which can reduce the number of multiplications by 2.25× when computing a tile of 2×2 outputs. Note that Winograd requires specialized processing depending on the size of the filter and tile, so Winograd hardware typically support only specific tile and filter sizes.[5]

---

[5]For instance, NVDLA only supports 3×3 filters [135].

A matrix linear algebraic formulation of Winograd is shown as follows:

$$B^T = \begin{bmatrix} 1 & 0 & -1 & 0 \\ 0 & 1 & 1 & 0 \\ 0 & -1 & 1 & 0 \\ 0 & 1 & 0 & -1 \end{bmatrix}$$

$$G = \begin{bmatrix} 1 & 0 & 0 \\ \frac{1}{2} & \frac{1}{2} & \frac{1}{2} \\ \frac{1}{2} & -\frac{1}{2} & \frac{1}{2} \\ 0 & 0 & 1 \end{bmatrix}$$

$$A^T = \begin{bmatrix} 1 & 1 & 1 & 0 \\ 0 & 1 & -1 & -1 \end{bmatrix}$$

$$f = \begin{bmatrix} f_0 & f_1 & f_2 \end{bmatrix}^T$$

$$i = \begin{bmatrix} i_0 & i_1 & i_2 & i_3 \end{bmatrix}^T.$$

In this formulation the first steps of the computation are transformations of both the filter weights (f_j) and input activations (i_j) by sandwiching those matrices in a chain of matrix multiplies by constant matrices, GfG^T and $B^T i B$, respectively. The resulting values can be considered as existing in a "Winograd" space, where a convolution can be performed by combining those matrices with element-wise multiplication, which is computationally efficient, as follows:

$$[GfG^T] \odot [B^T i B].$$

Finally, a reverse transformation out of the "Winograd" space is performed by, again, sandwiching the result of the element-wise multiplication in a chain of matrix multiplies by constant matrices A^T and A.

$$Y = A^T \Big[[GfG^T] \odot [B^T i B] \Big] A.$$

Note that since the filter weights are constant across many applications of the tiled convolution, the transformation of the filter weights into the "Winograd" space, GfG^T, only needs to be performed once.

4.3.4 FAST FOURIER TRANSFORM

The Fast Fourier Transform (FFT) [19, 136] follows a similar pattern to the Winograd transform to convert a convolution into a new space where convolution is more computationally efficient. This well-known approach, shown in Figure 4.7, reduces the number of multiplications for each input channel from $O(RSPQ)$ to $O(PQ \log_2 PQ)$, where the output size is $P \times Q$ and the filter size is $R \times S$.[6] To perform the convolution, we take the FFT of the filter and input

---

[6]Note that convolutions for DNNs do not assume any zero padding, which differ from the traditional forms of convolutions associated with FFTs. For more on this, please see [137].

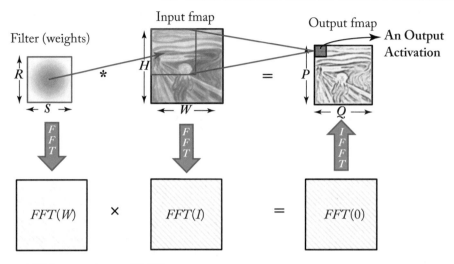

Figure 4.7: FFT to accelerate DNN.

feature map, and then perform the element-wise multiplication in the frequency domain; we then apply an inverse FFT to the resulting product to recover the output feature map in the spatial domain. However, there are several drawbacks to using FFT: (1) the benefits of FFTs decrease with filter size;[7] (2) the size of the FFT is dictated by the output feature map size which is often much larger than the filter; and (3) the coefficients in the frequency domain are complex. As a result, while FFT reduces computation, it requires larger storage capacity and bandwidth. Finally, a popular approach for reducing complexity is to make the weights sparse, which will be discussed in Section 8.1.2; using FFTs makes it difficult for this sparsity to be exploited.

Several optimizations can be performed on FFT to make it more effective for DNNs. To reduce the number of operations, the FFT of the filter can be precomputed and stored. In addition, the FFT of the input feature map can be computed once and used to generate multiple channels in the output feature map. Finally, since an image contains only real values, its Fourier Transform is symmetric and this can be exploited to reduce storage and computation cost.

4.3.5   SELECTING A TRANSFORM

In practice, different algorithms might be used for different layer shapes and sizes (e.g., FFT for filters greater than 5×5, and Winograd for filters 3×3 and below). Existing platform libraries, such as MKL and cuDNN, dynamically choose the appropriate algorithm for a given shape and size [138, 139].

---

[7]The benefits of FFT depends on the ratio between the filter size $R \times S$ and the output size $P \times Q$. Specifically, after accounting for constant terms, one needs $RS > \log_2 PQ$ for there to be a benefit.

4.4   SUMMARY

In this chapter, we discussed different approaches for achieving efficient processing on temporal platforms, such as CPUs and GPUs. The objective was to restructure the computations to improve efficiency without any impact on accuracy (i.e., all approaches in this chapter produced near bit-wise identical results). These approaches can target reducing memory bandwidth (e.g., via tiling); reshaping the computation to be more efficient; or reducing high-cost operations. One significant example of reshaping the computation was the Toeplitz transformation, which is widely used in CPUs and GPUs. The Toeplitz transformation converts a convolution into a matrix multiply by replicating values, which allows application of routines from any of the highly optimized matrix multiply libraries. Finally, a variety of other transformations are aimed at reducing the number of high-cost operations (i.e., multiplies) through algebraic re-association including the Strassen, Winograd, and FFT transforms.

CHAPTER 5

Designing DNN Accelerators

In Chapter 4, we discussed how DNN processing can undergo transforms to leverage optimized libraries or reduce the number of operations, specifically multiplications, in order to achieve higher performance (i.e., higher throughput and/or lower latency) on off-the-shelf general-purpose processors such as CPUs and GPUs. In this chapter, we will focus on optimizing the processing of DNNs directly by designing specialized hardware.

A major motivation for designing specialized hardware instead of just trying to improve general-purpose processors is described by John Hennessy and Dave Patterson in their 2018 Turing Award Lecture [11]. In that lecture, they argue that with the end of Moore's law [140] there is a need to employ domain-specific hardware/software co-design (e.g., domain-specific languages such as TensorFlow) in computing systems to continue to improve performance and energy efficiency for important computational domains.

An architectural recipe for designing such specialized systems is outlined in Leiserson et al. [141]. That article describes how improving performance and/or energy efficiency can be achieved by: (1) identifying opportunities for significant parallelism and data locality in workloads in the domain of interest/importance; (2) design a hardware organization that exploits that parallelism and data locality; and (3) *streamline* that hardware to maximize efficiency, possibly through hardware-software as well as hardware-algorithm co-design. Here, we distinguish between these two forms of co-design: hardware-software co-design refers to the development of new software and languages which improves ease of use; furthermore, the compiler can map such workloads better to domain-specific hardware to enable improvements in performance and energy efficiency. Hardware-algorithm co-design refers to modifying the algorithm, and thus its workloads, in conjunction with the hardware for improvements in performance and energy efficiency that could not achieved by each approach individually. Note that in this book we primarily focus on hardware-algorithm co-design.[1]

The domain of DNN acceleration fits perfectly into the paradigm just described. First, the computational domain is important and admits of considerable parallelism and data locality (i.e., reuse opportunities). Thus, the goal is to design specialized DNN hardware to further improve key metrics, such as performance and energy efficiency, over general-purpose processors across a wide domain of DNN computations. So, second, we will explore (in this chapter) hardware organizations that can exploit both the parallelism and data locality in DNN computations to

---

[1]Sometimes in the literature hardware-software co-design is used to encompass both hardware-software and hardware-algorithm co-design. However, in this book, we want to make these forms of co-design distinct.

achieve those goals. Third, in Chapters 7, 8, and 9, we will explore how to co-design the hardware and DNN algorithms to further improve efficiency.

When considering the hardware organizations for DNN acceleration, the design space for specialized DNN hardware is quite large. This is due to the fact that there are no constraints on the execution order of the MAC operations within a DNN layer. As a result, the hardware designer has significant flexibility to choose the execution order of operations and optimize the hardware for the target metrics, under some given resource constraints (e.g., number of compute datapaths and amount of storage capacity).

In order to approach the hardware design in a large design space, we will discuss several key design decisions and how they affect performance and energy efficiency, and then show how these design decisions can be formally described using a *loop nest*, which is commonly used to describe the processing of DNNs. We will then discuss several design patterns that are commonly used in real-world architectures for DNN acceleration using loop nests. In the final sections of this chapter, we will discuss design patterns for efficient and flexible data storage and data movement via a flexible *network on chip* (NoC).

5.1   EVALUATION METRICS AND DESIGN OBJECTIVES

As discussed in Chapter 3, energy consumption and performance are two of the principal driving metrics for the design of specialized hardware. In modern compute systems, energy consumption is often dominated by data movement, especially memory access [121]. This is because accessing data from memory, especially off-chip storage such as DRAM, can consume orders of magnitude more energy than the actual computation of operands (e.g., MACs). Even for the on-chip memories, accessing data from the larger memory (e.g., caches) is also much more energy consuming than accessing data from the smaller memory (e.g., registers). Thus, in order to reduce energy consumption, one objective is to design hardware that reduces data movement by:

- reducing the number of times values are moved from sources that have a high energy cost, such as DRAM or large on-chip buffers; and

- reducing the cost of moving each value. For example, by reducing the data's bit-width, which we will discuss in Chapter 7.

Performance in terms of throughput and latency, on the other hand, is largely dictated by the number of processing elements (PEs) and more specifically the number of multipliers that can operate in parallel.[2] Therefore, another objective is to design hardware that:

- allocates work to as many PEs as possible so that they can operate in parallel; and

---

[2]The throughput of each PE also affects performance, but we consider that as an orthogonal micro-architectural design decision.

- minimizes the number of idle cycles per PE by ensuring that there is sufficient memory bandwidth to deliver the data that needs to be processed, the data is delivered before it is needed, and workload imbalance among the parallel PEs is minimized.

Note that energy consumption and hardware performance can also be improved by designing hardware that only performs the necessary operations; for instance, when the data is sparse (i.e., has many zeros), the hardware should only perform MACs on the non-zero data, and it can also use representations that exploit zeros to save space and data movement. Exploiting sparsity will be discussed in Chapter 8. Finally, improving energy consumption and performance might be improved with new technologies, which are described in Chapter 10.

5.2   KEY PROPERTIES OF DNN TO LEVERAGE

There are several properties of DNNs that can be leveraged by the hardware to optimize for the design objectives discussed in Section 5.1, thus improving the hardware performance and energy efficiency. First, as previously mentioned, the key computation of DNNs involves many MACs that have no restriction on their execution order within a layer. Therefore, although DNNs require a significant number of MACs per layer, the hardware can still achieve higher throughput and lower latency by exploiting high compute parallelism. However, the challenge of reducing the energy consumption of moving data to the parallel PEs remains.

Thankfully, the data movement cost can be addressed by another important property of DNNs, which is that the same piece of data is often used for multiple MAC operations. This property results in three forms of data reuse, as shown in Figure 5.1.

- *Input feature map reuse*: Different filters (from dimension M) are applied to the same input feature map in order to generate multiple output feature maps; therefore, each input activation is reused M times.

- *Filter reuse*: When processing a batch (of size N) of input feature maps, the same filter is applied to all inputs in the batch; therefore, each filter weight is reused N times.

- *Convolutional reuse*: For convolutional layers, the size of a filter ($R \times S$) is smaller than the size of an input feature map ($H \times W$), and the filter slides across different positions (often overlapping with each other) in the input feature map to generate an output feature map. As a result, each weight and input activation are further reused $P \times Q$ and $R \times S$ times,[3] respectively, to generate different output activations.

Data reuse can translate to reduced energy consumption for data movement through reading data once from a large but expensive (in terms of energy cost) memory, and either:

---

[3]For simplicity, we ignore the halo effect at the edges of input feature maps that results in less reuse for input activations and assume a stride of 1.

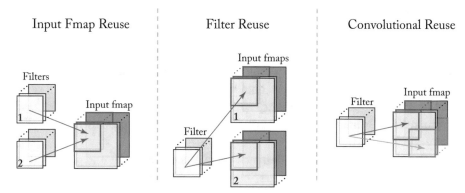

Figure 5.1: Data reuse opportunities in a CONV or FC layers of DNNs.

- store the data in a relatively small, but cheap memory, and (temporally) reuse that data multiple times at that cheap memory; and

- send the same data to multiple PEs and (spatially) use the data at those distinct PEs.

Both methods save accesses to the expensive memory, and therefore reduce the overall energy cost. While the maximum amount of data reuse for a DNN layer is defined by its shape and size (e.g., number of channels, filter size, etc.), the corresponding energy cost savings is determined by the amount of data reuse that is actually harnessed by the specialized hardware through these methods. We will discuss how to apply these methods in the hardware in Sections 5.4 and 5.5.

In addition to exploiting data reuse for input activations and filter weights, specialized hardware can also reduce the energy cost of data movement by properly orchestrating the movement of partial sums, which are the intermediate results from the multiplications. If partial sums can be temporally accumulated in a small buffer (e.g., registers) the energy cost is less than reading and updating a value from a larger buffer. Alternatively, accumulating the partial sums in the same cycle through an adder tree (i.e., as a *spatial sum*) can reduce the required storage capacity [135, 142]. In the best case, all $C \times R \times S$ partial sums for each output activation can be accumulated in one cycle, and therefore partial sums never need to be stored in memory.

Employing a spatial sum can also reduce the energy and latency compared to the same capability implemented with individual two-input adders [143]; specifically, rather than performing a carry propagation after each addition, using a redundant binary representation for the adder tree allows the carry propagation to be deferred to a single adder at the bottom of the tree. The magnitude of the benefit of the spatial sum over temporal sum depends on the reduction factor [144].

Although the cost of data movement for each data type can be minimized individually, it is not possible to do so for all three data types (i.e., input activations, weights, and partial sums) at the same time. This is due to the fact that different MACs can only reuse data for at most one data type at a time. For example, MACs that reuse the same weight require different input activations and the generated partial sums cannot be further accumulated together. Thus, the methods mentioned above can only be applied to reduce the energy cost of one data type at a time and will increase the cost of data movement for the other two data types. Therefore, in order to minimize the overall energy consumption, it is important to balance the cost of data movement for all three data types through an optimization process instead of just minimizing for any specific one of them.

The optimization process depends on the specific shape and size of the DNN layer in addition to the hardware constraints. However, while the hardware constraints stay fixed, the shape and size of a DNN layer can vary dramatically across different DNNs and also across different layers within a DNN. Therefore, the optimization of data movement has to be performed for each DNN layer separately,[4] and the hardware needs to be able to support different configurations of data movement based on the results of the optimization. This *flexibility* requirement raises several considerations that are critical to the design of specialized DNN hardware, which we will discuss in the next section.

5.3 DNN HARDWARE DESIGN CONSIDERATIONS

The challenges of designing specialized DNN hardware involve designing a flexible architecture and then finding the best ways to configure the architecture to get optimal hardware performance and energy efficiency for different DNN layers. These two aspects are tightly correlated, since the chance of finding the optimal configuration depends on the flexibility of the hardware, while higher flexibility often implies a loss of efficiency since additional hardware is required. Therefore, it often takes an iterative process to distill down to the best design. This is in contrast to the approach taken in Chapter 4, where the hardware architecture is already fixed and the focus is on adapting the computation to fit in the compute paradigm of the hardware.

Given a particular DNN model, it is necessary to be able to configure the hardware to minimize the overall energy consumption while maintaining high performance. This process involves finding an optimal *mapping*, where a mapping defines: (1) the execution order of the MAC operations, both temporally (i.e., serial order on the same PE) and spatially (i.e., across many parallel PEs); and (2) how to tile and move data across the different levels of the memory hierarchy to carry out the computation in accord with that execution order. For a given DNN layer, there often exists a large number of possible mappings. As a result, it is very crucial to be able to find the best mappings for the desired metrics. In Section 5.4, we will show examples of different mappings and discuss their impact on the performance of the hardware. Chapter 6 extends that discussion with more details of the mapping optimization process.

---

[4]Most of the focus in this chapter and the next is on per layer processing. Section 5.7.6 has a brief discussion of cross-layer optimization.

Given the sheer number of possible spatio-temporal orderings of MAC operations for a DNN layer, it is generally very unlikely for a hardware architecture to support the execution of all these orderings. Therefore, practical hardware will support a more restricted set of orderings, which consequently reduces the number of legal mappings. Therefore, a very critical design consideration for DNN hardware is to select a subset of mappings to support. In other words, the hardware architecture has to set up rules that narrow down the range of considered mappings in the optimization. An important set of rules that determine the subset of supported mappings is called a *dataflow*. Dataflow is one of the most important attributes that define the architecture of a DNN hardware, as the subset of supported mappings directly impact the quality of the optimization. In Section 5.6, we will formally define dataflow and mapping using a loop nest-based representation.

An additional design consideration is that the shape and size of a DNN layer, as described in Chapter 2, can vary dramatically across different DNNs and also across different layers within a DNN. Furthermore, the number of layers that are found in emerging DNNs continues to grow. Therefore, given this rapidly growing and changing field, it is increasingly important to design DNN hardware that is sufficiently flexible and scalable to address these varying needs. Specifically, the hardware should avoid making assumptions about the layer shape and size; for instance, since DNNs have tended to grow in size, one cannot assume the entire model can always be stored on-chip. The degree of flexibility needed in the hardware to efficiently support a wide variety of DNNs has become one of the main challenges in the design of DNN accelerators.

To summarize, the design and use of specialized DNN hardware involves multiple steps.

- **At design time**, an architecture is specified with a set of attributes. These attributes include: (1) the dataflow or dataflows supported; (2) the number of PEs and the number of multipliers and adders per PE;[5] (3) the memory hierarchy, including the number of storage levels and the storage capacity at each level; and (4) the allowed patterns of data delivery for the NoC within the memory hierarchy and between the memory and PEs. Note, these attributes set certain limitations on the legal mappings of an architecture. For example, the amount of storage capacity at each level of the memory hierarchy impacts how much data reuse can be exploited in the hardware, which limits the set of supported mappings. We will discuss these limitations in Section 5.5.

- **At mapping time**, given a DNN model, a mapping that optimizes the desired operational metrics is selected from among all the mappings supported by the accelerator. In Chapter 6, we will discuss the process of finding optimal mappings for a DNN accelerator.

- **At configuration time**, a configuration derived from the selected mapping is loaded into the accelerator.

---

[5]Without losing generality, we will assume that each PE contains a single multiplier and adder in the rest of the chapter unless explicitly specified.

- **At runtime**, the desired DNN input data is loaded and processed according to the loaded configuration. However, depending on the capabilities of the accelerator (e.g., how many layers can it process per configuration) it might need to iterate between the configuration step and the run step multiple times.

In the next section, we will explore how to exploit data reuse.

5.4 ARCHITECTURAL TECHNIQUES FOR EXPLOITING DATA REUSE

As hinted in Section 5.2, there are two methods for exploiting data reuse to reduce the energy cost of data movement. These architectural techniques are called *temporal reuse* and *spatial reuse*. In this section, we will formally define them and describe how to apply these techniques in the hardware.

5.4.1 TEMPORAL REUSE

Temporal reuse occurs when the same data value is used more than once by the same consumer (e.g., a PE). It can be exploited by adding an intermediate memory level to the memory hierarchy of the hardware, where the intermediate memory level has a smaller storage capacity than the level that acts as the original source of the data; an example is shown in Figure 5.2d. Since smaller memories consume less energy to access than larger memories, the data value is transferred once from the source level (i.e., larger memory) to the intermediate level (i.e., smaller memory), and used multiple times at the intermediate level, which reduces the overall energy cost.[6]

Since the intermediate memory level has a smaller storage capacity, it cannot fit all data from the source level at the same time. As a result, data in the intermediate level may be replaced by new data from the source level and lose the chance to further exploit temporal reuse. Whether data in the intermediate level will be replaced or not depends on the *reuse distance*. For exploiting temporal reuse, the reuse distance is defined as the number of data accesses required by the consumer in between the accesses to the same data value, which is a function of the ordering of operations. Figure 5.2 shows an example to illustrate this phenomenon. For the example 1-D convolution workload shown in Figure 5.2a, two different operation orderings are shown in Figures 5.2b and 5.2c: the former has a temporal reuse distance for weights of 4, while the latter has a temporal reuse distance for weights of 1. Given the memory hierarchy shown in Figure 5.2d, in which we are only showing the memory space allocated to weights and it only has 1 slot in the intermediate memory level ($L1$) allocated to weights, the ordering in Figure 5.2b will have to keep swapping the weights in $L1$, while the ordering in Figure 5.2c can read each weight once from the source into $L1$ and reuse it for 4 times. Therefore, if the reuse distance for a data type is smaller than or equal to the storage capacity of the intermediate memory level,

---

[6]The assumption is that the ratio of energy per access between the two levels is large enough so that it is worthwhile to move data from the large memory to the smaller memory for reuse.

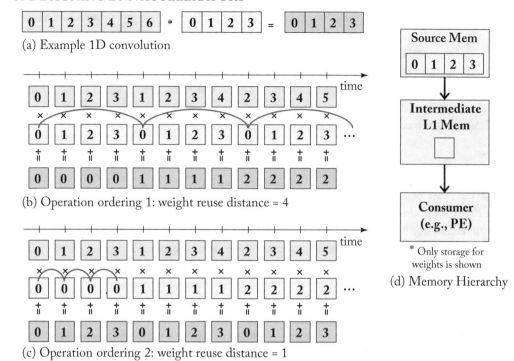

(a) Example 1D convolution

(b) Operation ordering 1: weight reuse distance = 4

(c) Operation ordering 2: weight reuse distance = 1

(d) Memory Hierarchy

* Only storage for weights is shown

Figure 5.2: To run the example 1-D convolution in (a), two possible operation orderings are shown in (b) and (c). (b) has a reuse distance of 4 for weights, while (c) has a reuse distance of 1. Given the memory hierarchy shown in (d), which has only 1 slot in the intermediate memory allocated to weights, the ordering in (c) can fully exploit temporal reuse while the ordering in (b) cannot, since it has a reuse distance larger than the storage capacity of the intermediate memory. Note that the numbers shown here are the indices of the values in each of the data vector.

temporal reuse can be exploited for all values of that data type. However, if the reuse distance is larger, then a part or all of the data values that were stored in the intermediate level would be replaced before the reuse opportunities are fully exploited. In other words, the storage capacity of the intermediate memory level limits the maximum reuse distance where temporal reuse can be exploited.

Reducing the reuse distance of one data type often comes at the cost of increasing the reuse distance of other data types, which should be taken into account at the same time. Although it is possible to increase the storage capacity of the intermediate memory level to exploit temporal reuse on larger reuse distances, it has the counter effect that the energy cost per memory access also goes up, which increases the average energy cost of data movement for all reuse distances. In order to keep the memory small, an alternative solution is to reduce the reuse distance by changing the processing order of MACs. Various techniques for reducing reuse distance will be discussed in Section 5.5.

Temporal reuse can be exploited at multiple levels of the memory hierarchy. By treating the intermediate memory level as the new consumer of data from the source level, additional levels of memory can be added to further exploit temporal reuse. However, adding more memory levels requires more area and results in reduced ratio of energy cost between different levels, which diminishes the effectiveness of exploiting temporal reuse.

5.4.2   SPATIAL REUSE

Spatial reuse occurs when the same data value is used by more than one consumer (e.g., a group of PEs) at different spatial locations of the hardware. It can be exploited by reading the data once from the source memory level and multicasting it to all of the consumers. Exploiting spatial reuse has the benefits of (1) reducing the number of accesses to the source memory level, which reduces the overall energy cost, and (2) reducing the bandwidth required from the source memory level, which helps to keep the PEs busy and therefore increases performance.

If a group of consumers of a data value have no storage capacity, spatial reuse can only be exploited by the subset of consumers that can process the data in the same cycle as the multicast of the data; the other subset of consumers that requires the same data value, but cannot process it in the same cycle, needs to be sent the data again by the source memory level. In contrast, if each consumer in the group has some storage capacity, then a certain time span (dictated by the storage capacity) in which the same data value is processed by multiple consumers can be tolerated when exploiting spatial reuse with multicast. In other words, whether a consumer in the group that uses the same data value can exploit spatial reuse or not also depends on the *reuse distance*. For exploiting spatial reuse, the reuse distance is defined as the maximum number of data accesses in between any pair of consumers that access the same data value, which is again a function of the ordering of operations.

An example is shown in Figure 5.3. For the same 1-D convolution workload in Figure 5.2, we run it on a new architecture with the memory hierarchy shown in Figure 5.3a, which has four consumers, $C0$ to $C3$. Each consumer also has 1 slot of local storage for weights. Figures 5.3b, 5.3c, and 5.3d show three different orderings of operations (with only the weights being shown for simplicity). In addition to the time axis, the orderings also have the space axis to indicate the ordering at each parallel consumer. The spatial reuse distance for weights are 0, 1, and 3 for the orderings in Figure 5.3b, 5.3c, and 5.3d, respectively. Since the reuse distance of the orderings in Figure 5.3b and 5.3c are smaller than or equal to the storage capacity at the consumer level, only one single multicast from the source memory to all consumers is needed for each weight. However, for the ordering in Figure 5.3d, multiple reads from the source memory for each weight is needed since its reuse distance is larger than the storage capacity at the consumers.

In addition to the reuse distance, since spatial reuse involves routing data to multiple destinations in the hardware, the NoC that distributes data also plays an important role for

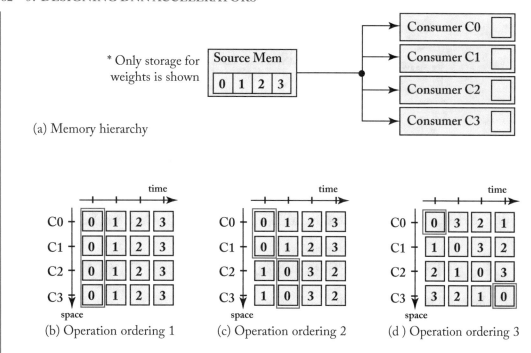

(a) Memory hierarchy

(b) Operation ordering 1

(c) Operation ordering 2

(d) Operation ordering 3

Figure 5.3: To run the same 1-D convolution, as shown in Figure 5.2a, on the memory hierarchy shown in (a), three possible operation orderings are shown in (b), (c), and (d), which have a spatial reuse distance of 0, 1, and 3, respectively. The distance between the red boxes in the figures shows the reuse distance. Since each consumer has 1 slot of local storage for storing weights, the orderings in (b) and (c) can fully exploit spatial reuse of weights since their reuse distances are smaller than or equal to the storage capacity at the consumers, while the ordering in (d) has to read each weight multiple times from the source memory level.

achieving spatial reuse. Specifically, the NoC has to support the data distribution patterns as derived from the ordering of operations. For example, as shown in Figure 5.4, if a data value is used by all consumers in the same cycle (e.g., Figure 5.3b), but the memory level $L1$ is banked in a way that each bank only connects to a subset of consumers, then spatial reuse needs to be first exploited from a higher memory level $L2$, where data is multicast from $L2$ to all banks in $L1$ with each bank serving as a single consumer, and then multicast again from the $L1$ banks to all of the consumers. Exploiting spatial reuse at higher levels of the memory hierarchy creates more duplicated data in the hardware, which is not desirable. However, supporting multicast to all consumers in a single level at a large scale can also be expensive. Therefore, it is a design trade-off to determine where and how to exploit spatial reuse in the hardware. We will discuss the design of NoCs for DNN accelerators in Section 5.9.

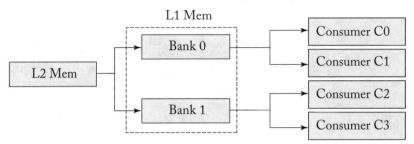

Figure 5.4: The physical connectivity of the NoC between the $L1$ memory and the consumers limits the multicast from any banks in $L1$ to all consumers. Therefore, data needs to be first multicast from the $L2$ memory to all banks in $L1$, and then each $L1$ bank further multicast the data to the associated consumers. In this case, data is being duplicated twice in $L1$.

5.5   TECHNIQUES TO REDUCE REUSE DISTANCE

As discussed in Section 5.4, the order of operations determines the reuse distance, which impacts the effectiveness of exploiting either temporal reuse or spatial reuse. In this section, we will discuss the various methods for manipulating the order of operations to reduce the reuse distance.

First, since it is not feasible to minimize the reuse distance for all data types simultaneously, as described in Section 5.2, the order of operations has to prioritize reducing the reuse distance of a certain data type over the others. Figures 5.2b and 5.2c are two examples of operation ordering that prioritize reducing the reuse distance of partial sums and weights, respectively, for temporal reuse. Each data value of the prioritized data type appears to stay *stationary* over time in the sequence of operations (i.e., the same value gets accessed many times from memory consecutively). Later on, when we discuss loop nests in Section 5.6, we will show that the stationariness of different data types is controlled by the ordering of the nested loops, which determines their priority in reuse distance reduction.

Data tiling, also referred to as blocking, is another technique that is commonly used to reduce the reuse distance, such as tiling for CPUs or GPUs, as described in Section 4.2. Data is partitioned into smaller *tiles* through tiling, and the processing only focuses on a tile at a time for each data type. There are many ways to tile the data for processing. In addition to the size of each tile, deciding along which dimensions the data (which is 4-D for each data type in DNNs) is being tiled is another design decision. The goal is to tile the data so that the reuse distance becomes smaller. For example, in the operation ordering in Figure 5.2c, where weights are prioritized (i.e., each weight stays stationary across multiple cycles) the temporal reuse distance of partial sums is large, and is the same as the length of the entire output activation vector, which is 4 in this case. Tiling can be applied to reduce the reuse distance of partial sums, as shown in Figure 5.5, by only working on half of the output activations at a time, thus cutting

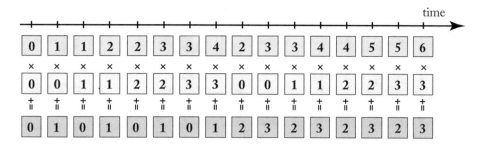

Figure 5.5: A temporally tiled version of the operation ordering in Figure 5.2c.

the reuse distance of partial sums by two times down to 2. However, this also increases the average reuse distance of weights, though still being the most prioritized data type.

Tiling can be employed to exploit either temporal reuse or spatial reuse. Tiling for temporal reuse, or *temporal tiling*, focuses on reducing the reuse distance of specific data types to make it smaller than the storage capacity of a certain memory level in the memory hierarchy. For example, assuming the intermediate memory level $L1$ in Figure 5.2d has a total storage capacity of three data values,[7] the un-tiled ordering in Figure 5.2c can only exploit temporal reuse of weights, since the temporal reuse distance of partial sums is so large that they have to be stored and fetched from higher memory levels. However, with the tiled ordering in Figure 5.5, one weight value and a tile of two partial sum values can fit into $L1$ to exploit temporal reuse. Note that, however, each weight value that is written into $L1$ can only exploit temporal reuse for two times in this case, and needs to be read twice from a higher memory level into $L1$ instead of once as in the un-tiled ordering.

On the other hand, tiling for spatial reuse, or *spatial tiling*, focuses on (1) reusing the same data value by as many consumers as possible, and (2) reducing the reuse distance so that one multicast can serve as many consumers as possible given a fixed amount of storage capacity at each consumer. For example, Figures 5.6a, 5.6b, and 5.6c show the impact of three different spatial tilings that result in operation orderings with different degrees of spatial reuse. The ordering in Figure 5.6a has no spatial reuse of weights as each weight is only used by a single consumer. The ordering in Figure 5.6c has the highest degree of spatial reuse as each weight is used by all consumers. Spatial tiling effectively improves the amount of spatial reuse, saving the number of reads from the source memory.

The goal of both temporal and spatial tiling is to exactly match the reuse distance of a data type to the available storage capacity at a certain memory level to maximally exploit data reuse at that level. Overly reducing the reuse distance does not bring additional benefits but only increases the reuse distance of other data types. However, exactly matching the reuse distance

---

[7]We will ignore the differences in bit-width between different data types for now. In general, increasing the bit-width of a data type would increase its impact on energy consumption, and thus the dataflow would be more sensitive to that data type.

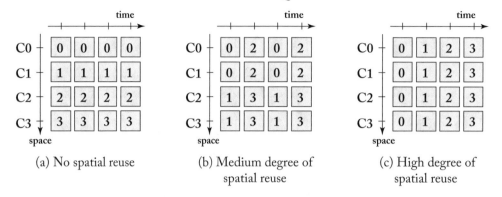

Figure 5.6: Operation orderings with different degrees of spatial reuse for the memory hierarchy shown in Figure 5.3.

to the storage capacity is not always feasible since the shapes and sizes of data vary across different layers and different DNNs, while the hardware storage capacity remains fixed. This causes fragmentation in mapping, in which case the workload cannot be evenly divided to run on the hardware and therefore would result in under-utilization of the hardware.

In addition, for spatial tiling, we would like to fully utilize all PEs to achieve maximum performance while exploiting spatial reuse, which is also not always possible since there might not be sufficient spatial reuse to exploit. For example, a common way to distribute data across PEs to exploit parallelism is shown in Figure 5.7: data from different output channels (M) are sent to different PEs vertically, while data from different input channels (C) are sent to different PEs horizontally. Therefore, each PE gets a unique weight, and each input activation is reused spatially across a column of PEs with weights from different output channels. However, if M is smaller than the number of PEs in a column and/or C is smaller than the number of PEs in a row, only a portion of the PEs are utilized (e.g., the colored ones in Figure 5.7). To improve the utilization in such cases, multiple tiles of data can run at the same time while not getting more spatial reuse, as discussed in Section 5.7.4. This technique, however, is only feasible if there is sufficient flexibility of data delivery and the NoC provides sufficient bandwidth.

Both data prioritization and tiling can be applied independently at each level of the memory hierarchy. Different data types can be prioritized at different levels of the memory hierarchy, which helps to balance the energy cost of accessing all types of data. For example, while weights are prioritized for access from memory level $L1$ to the PEs, partial sums can be prioritized for access from memory level $L2$ to $L1$. Also, each level of the memory hierarchy can perform either temporal or spatial tiling, or both at the same time. There are no direct interactions between the tiling decisions for spatial and temporal; instead, they can be interleaved at different levels of the storage hierarchy. For example, in Figure 5.4, temporal tiling and spatial tiling are applied at both level $L1$ and $L2$ of the storage hierarchy. The consumers at the last level can have their

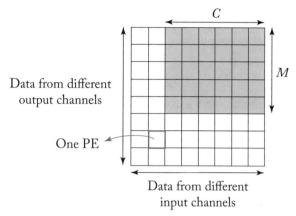

Figure 5.7: Spatial tiling can result in under-utilization of the parallelism if there is insufficient spatial reuse to exploit. For example, if the number of output channels (M) or input channels (C) is smaller than the number of PEs in a column or row, respectively, the specific mapping shown here will only have a portion of PEs utilized for processing (e.g., the colored ones).

local storage for temporal tiling; however, if the consumers at the last level have no local storage, they will operate like a vector-based architecture as mentioned in Chapter 4. Note that, unlike many vector architectures, the consumers can communicate data with each other, which is often used to do spatial accumulation of the partial sums across PEs.

Finally, in addition to the reuse distance, the bandwidth requirement should also be taken into account when performing operation reordering or data tiling. Specifically, certain operation orderings can have higher peak bandwidth requirement than others even though the average bandwidth is the same, which often happens during the ramp-up or ramp-down of the computation. For example, when multiple output activations are generated by parallel PEs in the same cycle, it will require a high peak bandwidth in order to store them to memory immediately. Also, if techniques such as double buffering are used to prefetch data and hide the data access latency, the effective storage capacity of the memory levels becomes smaller under a fixed budget of total storage capacity, which should be accounted for when calculating the required reuse distance.

While the techniques introduced in this section can be applied in many different ways, there are a few common approaches in term of how they are applied in existing designs, which forms a taxonomy of dataflows. In the next section, we will go through these dataflows in detail by formally introducing a loop nest-based representation for dataflows.

5.6   DATAFLOWS AND LOOP NESTS

Using the tensor index notation from Chapter 2, the calculation for a CONV layer (with unit stride) is:

$$\mathbf{O}_{nmpq} = \left(\sum_{crs} \mathbf{I}_{nc(p+r)(q+s)} \mathbf{F}_{mcrs}\right) + \mathbf{b}_m. \qquad (5.1)$$

That calculation, which can serve as a specification of a desired computation, imposes no ordering or notion of parallelism on the individual calculations, but those characteristics are important to the performance and efficiency of the computation when run on hardware. Specifying an ordering, and which calculations run in parallel, is called a *dataflow*.

A dataflow defines some of the specific rules for controlling activity on an accelerator. These include the ordering of the operations, and how to prioritize the use and transfers of data temporally and spatially across the memory hierarchy and compute datapaths. It also dictates which mappings are legal and directly impacts the performance and energy efficiency of the DNN accelerator. Therefore, it is very crucial to be able to precisely describe a dataflow. In this section, we will formally introduce *loop nests*, which is a powerful tool for this purpose.

Loop nests are a compact way to describe various properties of a given DNN accelerator design and, in specific, its dataflow. Figure 5.8a shows the loop nest that can represent the operation ordering of the example 1-D convolution in Figure 5.2c. The loop with variable s, which indexes filter weights, is placed as the outermost loop. That loop traverses the open range $[0, S)$, where S is the number of filter weights and is defined as part of the *shape* of the convolution.[8] Another component of the shape of the convolution is Q (the number of partial sums), which is traversed over the open range $[0, Q)$ by the inner loop using the variable q. The final component of the shape is W (the number of input activations), which is traversed via a simple computation on variables s and q. More discussion on such computations are included in Section 6.3.

Since in Figure 5.8a the weights are traversed in the outermost loop, the ordering prioritizes activity to reduce the reuse distance of filter weights. We call the dataflow constructed this way a *weight-stationary* (WS) dataflow. While we are illustrating the idea with an 1-D example, it can be generalized to data of higher dimensions: it is a WS dataflow as long as the loops that go through different weights are placed above all the other loops.

Using the space-time diagram in Figure 5.9, we can see the reference activity to each data array in each cycle (i.e., execution of the body of the loop nest). In that figure, we can clearly see the horizontal sequence of green •s, which represent the same filter weight being reused multiple times before processing moves to a new weight. The partial sums are accessed repeatedly after a long interval and the input activations are accessed in a large sliding window pattern. In both cases, the reuse interval is Q.

Different dataflows can be created by reordering the loops as convolution imposes no ordering constraints. Therefore, we can also create an *output-stationary* (OS) dataflow, as shown

---

[8]By convention in this book, we will typically use a capital letter to represent a constant value that specifies a part of the shape of a computation (e.g., M, C, H, W, R, and S for a CONV layer) and the associated small letter (e.g., m, c, h, w, r, s) to represent a variable that accesses a data structure along a dimension associated with the corresponding capital letter.

```
# i[W] - input activations
# f[S] - filter weights
# o[Q] - output activations

for s in range(S):
  for q in range(Q):
    w = q+s
    o[q] += i[w] * f[s]
```
(a) Weight stationary

```
# i[W] - input activations
# f[S] - filter weights
# o[Q] - output activations

for q in range(Q):
  for s in range(S):
    w = q+s
    o[q] += i[w] * f[s]
```
(b) Output stationary

```
# i[W] - input activations
# f[S] - filter weights
# o[Q] - output activations

for w in range(W):
  for s in range(S):
    q = w-s
    o[q] += i[w] * f[s]
```
(c) Input stationary

Figure 5.8: Three un-tiled loop nests for the 1-D convolution workload in Figure 5.2a. In this case, $W = 6$, $S = 4$, and $Q = 4$. (a) is a weight-stationary dataflow that generates the ordering in Figure 5.2c, while (b) is an output-stationary dataflow that generates the ordering in Figure 5.2b. (c) is an input-stationary dataflow.

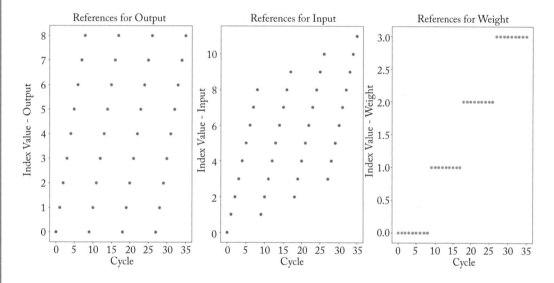

Figure 5.9: Space-time diagrams of partial sums (red), input activations (blue), and filter weights (green) for the 1-D weight-stationary convolution dataflow from Figure 5.8a with workload shape, $S = 4$, $Q = 9$, and $W = 12$. On each plot, each step on the x-axis is a cycle (time) and the y-axis represents an offset into the corresponding data array (space). Thus, a point at $(x, y) = (20, 1)$ represents an access to element 1 during cycle 20.

in Figure 5.8b. In this case, the loops that go through different partial sums are placed as the outermost loop.

Figure 5.10a shows the space-time diagrams for this dataflow. The reference pattern for the outputs show that each partial sum is referenced multiple times in succession and the partial

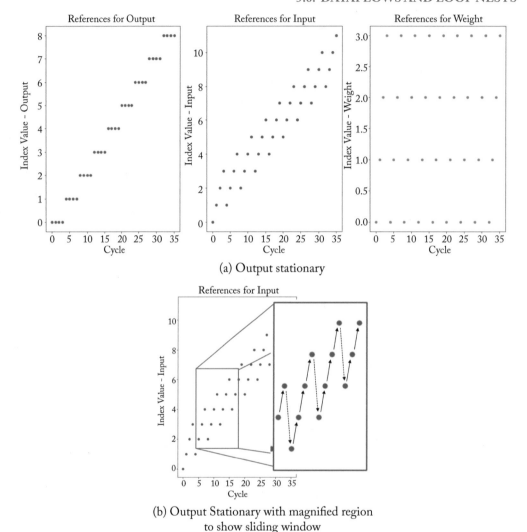

(a) Output stationary

(b) Output Stationary with magnified region
to show sliding window

Figure 5.10: Space-time diagrams of partial sums (red), input activations (blue), and filter weights (green) for the 1-D output-stationary convolution dataflow from Figure 5.8b with workload shape, $S = 4$, $Q = 9$, and $W = 12$.

sum is completed before processing moves to the next partial sum. The reference pattern for filter weights shows that all the weights are used repeatedly through the run with reuse distance S. Finally, the blowup of the reference pattern for input activations in Figure 5.10b illustrates a sliding window references of size S through the input activations.

```
# i[W] - input activations
# f[S] - filter weights
# o[Q] - output activations

for q1 in range(Q1):
    for s in range(S):
        for q0 in range(Q0):
            q = q1*Q0 + q0
            w = q+s
            o[q] += i[w] * f[s]
```

Figure 5.11: A tiled loop nest for the weight-stationary dataflow in Figure 5.8a, which can generate the ordering, as shown in Figure 5.5, by setting $Q0 = 2$ and $Q1 = 2$.

It is also possible to create an *input-stationary* (IS) dataflow; however, we have to use w and s as the loop variables and calculate the index variable q, as shown in Figure 5.8c. The characteristics and associated architecture designs for each of these dataflows will be discussed in Section 5.7.

The loop nests we have introduced so far are not tiled (i.e., the loops in these loop nests go through the entire dimension of the data). Tiling the data of a specific data type involves picking a specific dimension of the corresponding data type and breaking it up into multiple loops. Figure 5.11 shows an example loop nest of the tiled ordering shown in Figure 5.5. The loop that originally goes through dimension Q is now being divided up into two loops with new loop bounds $Q1$ and $Q0$. The inner bound $Q0$ defines the size of a tile, while the outer bound $Q1$ defines the number of tiles. The same process can be repeated to break it up into more loops, which creates multi-level tiles that can be put into different storage levels in the memory hierarchy. Tiling can also be applied to multiple data dimensions simultaneously by breaking up loops for different dimensions.

Parallel processing can also be described by the loop nest by introducing `parallel-for` loops in addition to the conventional `for` loops. The different operations that are iterated through in the `parallel-for` loops will then run on parallel consumers, e.g., PEs. For example, we can make a parallel weight-stationary dataflow by making the loop with variable $s0$ a `parallel-for` in a filter weight tiled version of the loop nest of Figure 5.8a. The new loop nest is shown in Figure 5.12.

A space-time diagram of the parallel weight-stationary dataflow is shown in Figure 5.13. Here we see that two weights are used repeatedly—one in each PE. Furthermore, both PEs access the same partial sum, thus providing an opportunity for a spatial sum (see Section 5.2). With respect to the input activations, the two PEs use the same input activation in successive cycles providing an opportunity for inter-PE communication or multicast if the PE buffer can hold more than one input activation (i.e., the data will arrive at the PE marked by "•", and wait a cycle until it is used).

```
# i[W]  - input activations
# f[S]  - filter weights
# o[Q]  - output activations

for s1 in range(S1):
  for q in range(Q):
    parallel-for s0 in range(S0):
      s = s1*S0 + s0
      w = q+s
      o[q] += i[w] * f[s]
```

Figure 5.12: A loop nest for the weight-stationary dataflow in Figure 5.8a with parallel processing for the a tile of filter weights (data dimension $S0$).

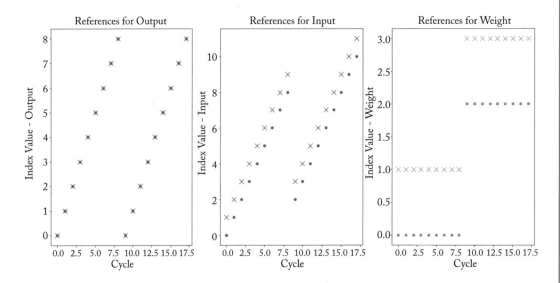

Figure 5.13: Space-time diagrams of partial sums (red), input activations (blue), and filter weights (green) for the 1-D weight-stationary convolution dataflow from Figure 5.12 with two parallel PEs. Activity in one PE is marked with a • and the other an ×. The workload shape is $S = 4$, $Q = 9$, $W = 12$, and S is tiled into $S1 = 2$, $S0 = 2$.

Figure 5.13 also illustrates the impact of tiling on reference patterns. Because the S dimension is tiled as $S1 = 2$ and $S0 = 2$, the weights are divided into two tiles of two weights each. In this case, the first tile is used in the first nine (Q) cycles and the second in the second nine (Q) cycles. Both tiles also contribute to the same partial sums, so partial sums have a long tile-related reuse distance of nine cycles (Q). Finally, most of the input activations are used by both tiles, so they also are used with a long reuse distance of $Q - 1$ cycles.

Table 5.1: Classification of recent work by dataflow

Dataflow	Recent Work
Weight Stationary (Section 5.7.1)	NVDLA [132], TPU [142], neuFlow [143], Sankaradas et al. [144], Park et al. [145], Chakradhar et al. [146], Sriram et al. [147], Origami [148]
Output Stationary (Section 5.7.2)	DaDianNao [149], DianNao [150], Zhang et al. [151], Moons et al. [152], ShiDianNao [153], Gupta et al. [154], Peeman et al. [155]
Input Stationary (Section 5.7.3)	SCNN [156]
Row Stationary (Section 5.7.4)	Eyeriss v1 [157, 139], Eyeriss v2 [158]

A dataflow only defines the following aspects of a loop nest: (1) the specific order of the loops to prioritize the data types; (2) the number of loops for each data dimension to describe the tiling; and (3) whether each of the loops is temporal (`for`) or spatial (`parallel-for`). The maximum number of loops that each data dimension can have is capped by the number of storage levels in the hierarchy that the specific data type can utilize.

The specific loop bounds in the loop nest, e.g., $S0$ and $S1$ in Figure 5.11, are not defined by the dataflow. However, the maximum value of each loop bound can be limited by a variety of factors, including: the storage capacity for the temporal loops, by the number of reachable consumers through the multicast network for the spatial loops (i.e., `parallel-for`), or by the size of the data dimension. Determination of the specific values of the loop bounds to use for a particular workload are determined by the optimization process that finds the optimal mapping as will be discussed in more depth in Chapter 6.

5.7   DATAFLOW TAXONOMY

In Sections 5.4 and 5.5, we have introduced several techniques to exploit data reuse. While there are many ways to apply these techniques, there are several commonly used design patterns that can be categorized into a taxonomy of dataflows: Weight Stationary (WS), Output Stationary (OS), Input Stationary (IS), and Row Stationary (RS). These dataflows can be seen in many recent works of DNN accelerator design, as shown in Table 5.1. In this section, we will use a generic architecture to describe how these dataflows are used in the recent works. As shown in Figure 5.14, this architecture consists of an array of PEs, with each PE having some local storage called a register file (RF), and the array of PEs shares a common storage level called the global buffer.

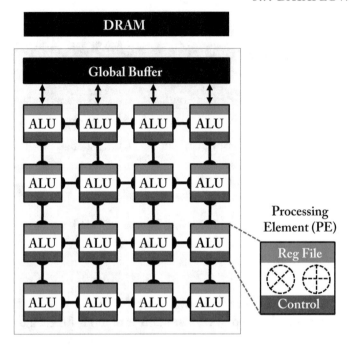

Figure 5.14: A generic DNN accelerator architecture.

5.7.1   WEIGHT STATIONARY (WS)

The weight-stationary dataflow is designed to minimize the energy consumption of reading weights by maximizing the reuse of weights from the register file (RF) at each PE (Figure 5.15a). The reuse distance of weights is minimized; therefore, each weight is read from DRAM into the RF of each PE and stays stationary for further accesses. The processing runs as many MACs that use the same weight as possible while the weight is present in the RF; it maximizes convolutional and filter reuse of weights. The inputs and partial sums must move through the spatial array and global buffer. The input feature map activations are broadcast to all PEs and then the partial sums are spatially accumulated across the PE array.

One example of previous work that implements a weight-stationary dataflow is nn-X (also called neuFlow) [146], which uses eight 2-D convolution engines, each of which is capable of processing a 2-D filter up to 10×10 in size. There are a total of 100 MAC units (i.e., PEs) per engine with each PE having a weight that stays stationary for processing. Figure 5.16 shows one 2-D convolution engine that can process 3×3 filters as a simplified example. The input feature map activations are broadcast to all MAC units and the partial sums are accumulated across the MAC units. In order to accumulate the partial sums correctly, additional delay storage elements are required, which are counted into the required size of local storage.

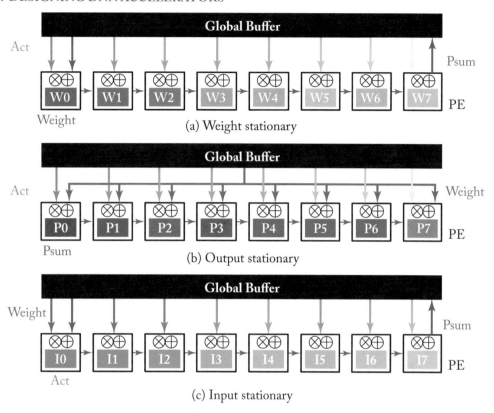

Figure 5.15: The taxonomy of commonly seen dataflows for DNN processing. *Act* means input activation. The color gradient is used to note different values of the same data type.

Another example architecture that implements the weight-stationary dataflow is Nvidia's Deep Learning Accelerator (NVDLA) [135], as shown in Figure 5.17. While weights stay stationary in each PE, the way input activations and partial sums are orchestrated through the PE array is different from the nn-X example, and is the same as the example shown in Figure 5.7. Note the opportunity for a spatial sum (as described in Section 5.2) vertically along the column of PEs. The loop nest representation of a simplified version of NVDLA is shown in Figure 5.18, which illustrates the parallelism over the input and output channels and the stationarity of the weights.

Google's TPU is another design that features a weight-stationary dataflow [145]. A major difference between TPU and NVDLA is that TPU utilizes a systolic array to share input activations and accumulate partial sums across the PEs. Other weight-stationary examples are found in [147–151].

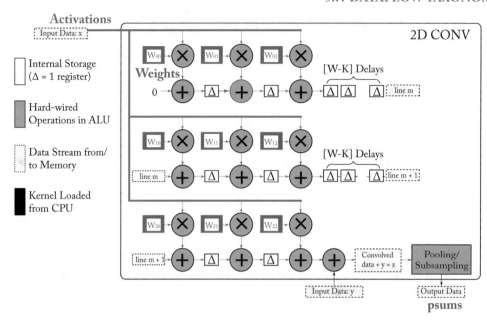

Figure 5.16: WS dataflow as implemented in nn-X or neuFlow [146].

5.7.2 OUTPUT STATIONARY (OS)

The output-stationary dataflow is designed to minimize the energy consumption of reading and writing the partial sums (Figure 5.15b). Through minimizing the reuse distance of partial sums, it keeps the accumulation of partial sums for the same output activation value local in the RF. In order to keep the accumulation of partial sums stationary in the RF, one common implementation is to stream the input activations across the PE array and broadcast the weights to all PEs in the array from the global buffer.

Figure 5.19 shows one example that implements an output-stationary dataflow presented by ShiDianNao [156], where each PE handles the processing for each output activation value by fetching the corresponding input activations from neighboring PEs. The PE array implements dedicated NoCs to pass data horizontally and vertically. Each PE also has data delay registers to keep data around for the required number of cycles. At the system level, the global buffer streams the input activations and broadcasts the weights into the PE array. The partial sums are accumulated inside each PE and then get streamed out back to the global buffer. Another example can be seen from the work by Moons et al. [155]. As shown in Figure 5.20, each input activation is reused across all PEs in the same column, while each weight is reused across all PEs in the same row. Other examples of output stationary are found in [157, 158].

There are multiple possible variants of output stationary, as shown in Figure 5.21, since the output activations that get processed at the same time can come from different dimensions.

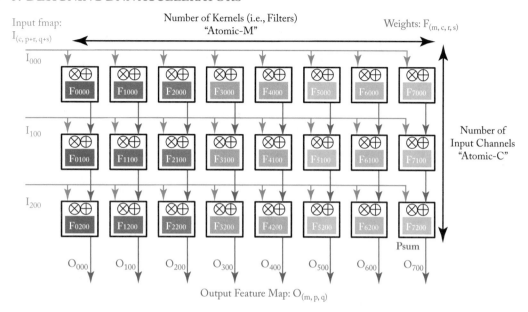

Figure 5.17: Simplified version of a WS dataflow as implemented in NVDLA [135], which processes a weight from each of three input channels (c) and eight output channels (m) each cycle. The figure shows the first time step of computation. Subsequent time steps will increment the values of the indices of the output activations (p and q) and then more slowly the current weights (r and s).

For example, the variant OS_A targets the processing of CONV layers, and therefore focuses on the processing of output activations from the same channel at a time in order to maximize convolutional data reuse opportunities. The variant OS_C targets the processing of FC layers, and focuses on generating output activations from all different channels, since each channel only has one output activation. The variant OS_B is something in between OS_A and OS_C. Example of variants OS_A, OS_B, and OS_C are [156], [155], and [158], respectively.

5.7.3 INPUT STATIONARY (IS)

Similar to the previous two dataflows, the input-stationary dataflow is designed to minimize the energy consumption of reading input activations (Figure 5.15c). With minimized reuse distance, each input activation is read from DRAM and put into the RF of each PE and stays stationary for further access. Then, it runs through as many MACs as possible in the PE to reuse the same input activation. It maximizes the convolutional and input feature map reuse of input activations. While each input activation stays stationary in the RF, unique filter weights are uni-cast into the

```
# i[C,H,W]     -   Input activations
# f[M,C,R,S]   -   Filter weights
# o[M,P,Q];    -   Output activations

parallel-for m in range(M):
  parallel-for c in range(C):
    for r in range(R):
      for s in range(S):
        for p in range(P):
          for q in range(Q):
            o[m,p,q] += i[c,p+r,q+s] * f[m,c,r,s]
```

Figure 5.18: Loop nest for simplified variant of the WS dataflow implemented in NVDLA [135]. To handle larger numbers of input channels (C) or output channels (M), additional temporal **for** loops would need to be added. Additional parallelism could also be added, for example, over the weight indices (r and s).

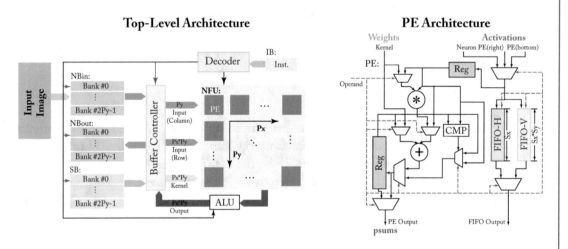

Figure 5.19: OS dataflow as implemented in ShiDianNao [156].

PEs at each cycle, while the partial sums are spatially accumulated across the PEs to generate the final output activation.

One example that implements the input-stationary dataflow is SCNN [159], where each PE, as shown in Figure 5.22, can process four stationary input activations in parallel, with each input activation being processed by a SIMD lane of width four. Therefore, each PE has a par-

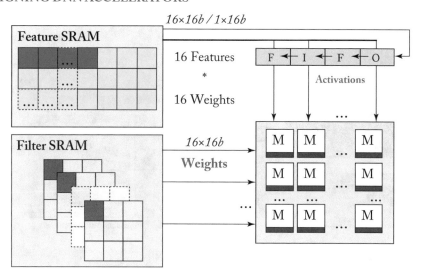

Figure 5.20: OS dataflow as implemented in Moons et al. [155].

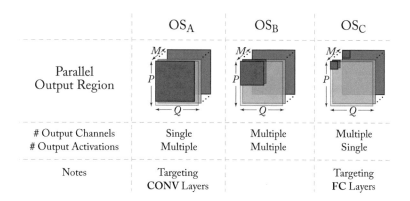

Figure 5.21: Variations of output stationary [142].

allelism of 16 MACs. SCNN takes advantage of the Cartesian product: any input activation in a plane of $H \times W$ feature map (i.e., a single input channel) can be reused across $R \times S \times M$ filter weights, and vice versa. Therefore, the PE first fetches 4 input activations out of the input feature map of size $H \times W$, goes through 4 out of the $R \times S \times M$ weights each cycle until all weights are looped through, and then switches to the next 4 input activations. Each cycle the PE takes 4 input activations and 4 weights, and processes 16 MACs. The partial sums for each output activation are spatially accumulated 4 times each cycle, and are put into a partial sum RF to be further accumulated across cycles. SCNN also supports processing sparse data, which will be discussed in Chapter 8.

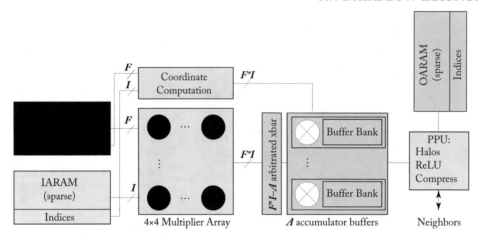

Figure 5.22: IS dataflow as implemented in SCNN [159].

5.7.4   ROW STATIONARY (RS)

A row-stationary dataflow is proposed in [142], which aims to maximize the reuse and accumulation at the RF level for *all* types of data (weights, input activations, and partial sums) for the overall energy efficiency. This differs from weight-stationary, output-stationary, or input-stationary dataflows, which optimize only for reducing the energy of accessing weights, partial sums, or input activations, respectively.

The row-stationary dataflow assigns the processing of a 1-D row convolution into each PE for processing, as shown in Figure 5.23. It keeps the row of filter weights stationary inside the RF of the PE and then streams the input activations into the PE. The PE does the MACs for each sliding window at a time, which uses just one memory space for the accumulation of partial sums. Since there are overlaps of input activations between different sliding windows, the input activations can be kept in the RF and get reused. By going through all the sliding windows in the row, it completes the 1-D convolution and maximizes the data reuse and local accumulation of data in this row.

With each PE processing a 1-D convolution, multiple PEs can be aggregated to complete the 2-D convolution, as shown in Figure 5.24. For example, to generate the first row of output activations with a filter having three rows, three 1-D convolutions are required. Therefore, it can use three PEs in a column, each running one of the three 1-D convolutions. The partial sums are further accumulated vertically across the three PEs to generate the first output row. To generate the second row of output, it uses another column of PEs, where three rows of input activations are shifted down by one row, and use the same rows of filters to perform the three 1-D convolutions. Additional columns of PEs are added until all rows of the output are completed (i.e., the number of PE columns equals the number of output rows).

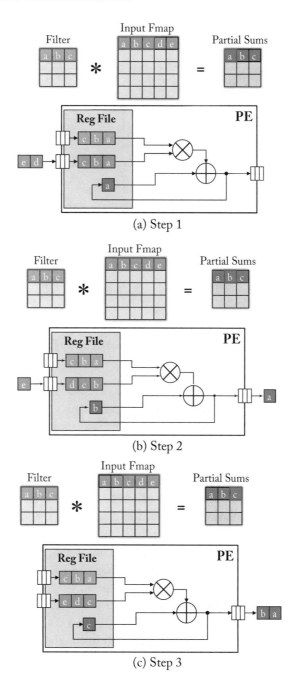

Figure 5.23: 1-D convolutional reuse within PE for row-stationary dataflow [142].

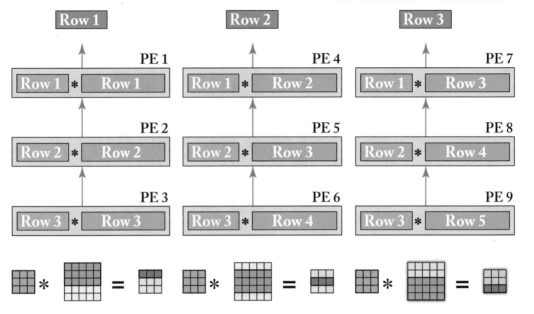

Figure 5.24: 2-D convolutional reuse within spatial array for row-stationary dataflow [142].

This 2-D array of PEs enables other forms of reuse to reduce accesses to the more expensive global buffer. For example, each filter row is reused across multiple PEs horizontally. Each row of input activations is reused across multiple PEs diagonally. And each row of partial sums are further accumulated across the PEs vertically. Therefore, 2-D convolutional data reuse and accumulation are maximized inside the 2-D PE array.

To address the high-dimensional convolution of the CONV layer (i.e., multiple feature maps, filters, and channels), multiple rows can be mapped onto the same PE, as shown in Figure 5.25a. The 2-D convolution is mapped to a set of PEs, and the additional dimensions are handled by interleaving or concatenating the additional data. For filter reuse within the PE, different rows of feature maps are concatenated and run through the same PE as a 1-D convolution (Figure 5.25b). For input feature map reuse within the PE, different filter rows are interleaved and run through the same PE as a 1-D convolution (Figure 5.25c). Finally, to increase local partial sum accumulation within the PE, filter rows and feature map rows from different channels are interleaved, and run through the same PE as a 1-D convolution. The partial sums from different channels then naturally get accumulated inside the PE (Figure 5.25d).

The number of filters, channels, and feature maps that can be processed at the same time is programmable, and there exists an optimal mapping for the best energy efficiency, which depends on the layer shape of the DNN as well as the hardware resources provided, e.g., the number of PEs and the size of the memory in the hierarchy. Since all of the variables are known

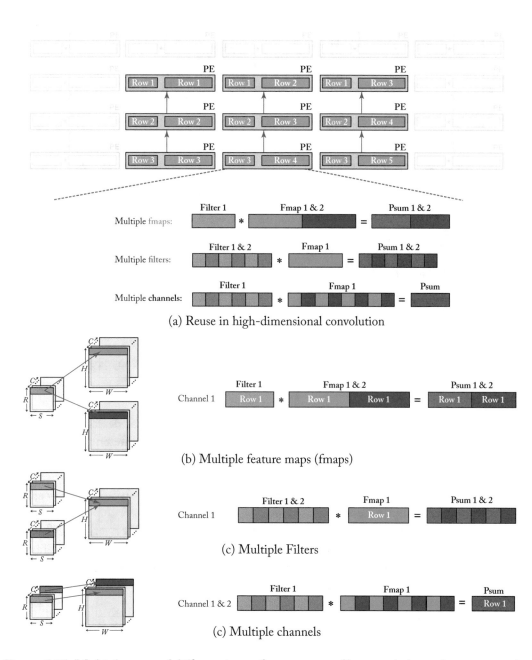

(a) Reuse in high-dimensional convolution

(b) Multiple feature maps (fmaps)

(c) Multiple Filters

(c) Multiple channels

Figure 5.25: Multiple rows of different input feature maps, filters, and channels are mapped to same PE within array for additional reuse in the row-stationary dataflow [142].

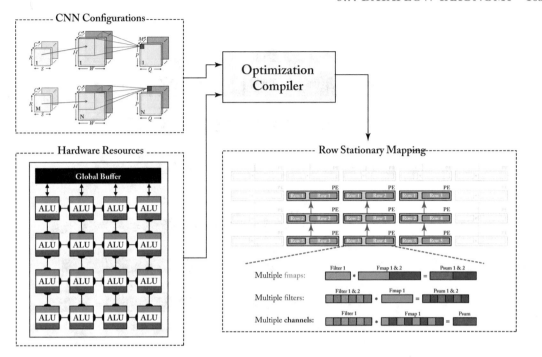

Figure 5.26: Mapping optimization takes in hardware and DNNs shape constraints to determine optimal energy dataflow [142].

before runtime, it is possible to build a compiler (i.e., mapper as described in Chapter 6) to perform this optimization off-line to configure the hardware for different mappings of the row-stationary dataflow for different DNNs, as shown in Figure 5.26. This is analogous to how compilers can optimize the binary for specific CPU or GPU architectures (Section 6.2). In Sections 6.4 and 6.5, we will introduce frameworks to perform analysis on energy efficiency and performance for optimization.

One example that implements the row-stationary dataflow is Eyeriss [160]. It consists of a 14×12 PE array, a 108KB global buffer, ReLU and feature map compression units, as shown in Figure 5.27. The chip communicates with the off-chip DRAM using a 64-bit bidirectional data bus to fetch data into the global buffer. The global buffer then streams the data into the PE array for processing.

In order to support the row-stationary dataflow, two problems need to be solved in the hardware design. First, how can the fixed-size PE array accommodate different layer shapes? Second, although the data will be passed in a very specific pattern, it still changes with different shape configurations. How can the fixed design pass data in different patterns?

Two mapping strategies can be used to solve the first problem, as shown in Figure 5.28. First, replication can be used to map shapes that do not use up the entire PE array. For example,

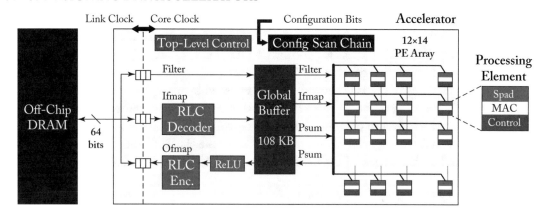

Figure 5.27: Eyeriss DNN accelerator [160].

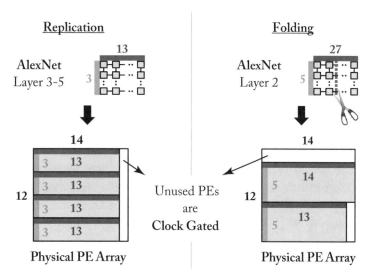

Figure 5.28: Mapping uses replication and folding to maximized utilization of PE array [160].

in the third to fifth layers of AlexNet, each 2-D convolution only uses a 13×3 PE array, where 3 is their filter height (R) while 13 is the output feature map height (P). This structure is then replicated four times, and runs different channels and/or filters in each replication. The second strategy is called folding. For example, in the second layer of AlexNet, it requires a 27×5 PE array to complete the 2-D convolution. In order to fit it into the 14×12 physical PE array, it is folded into two parts, 14×5 and 13×5, and each are vertically mapped into the physical PE array. Since not all PEs are used by the mapping, the unused PEs can be clock gated to save energy consumption.

A custom multicast network is used to solve the second problem related to flexible data delivery. The simplest way to pass data to multiple destinations is to broadcast the data to all PEs and let each PE decide if it has to process the data or not. However, it is not very energy efficient especially when the size of PE array is large. Instead, a multicast network is used to send data to only the places where it is needed. This is achieved by using the multicast controllers on the NoC paths that only pass data when there are destinations that require the data downstream. To determine which data to pass through each controller, each data is sent from the global buffer to the NoC with a tag value. Each multicast controller is also configured with an ID off-line. The controller then checks if the tag matches its local ID to determine the passing of data.

5.7.5   OTHER DATAFLOWS

The dataflows introduced in the taxonomy are often used as building blocks to create new dataflows. Since the stationariness of a dataflow is only relative to a specific level of memory hierarchy, different stationariness can be used at each level of the memory hierarchy.[9] For example, the dataflow can work as weight stationary at the RF storage level but output stationary at the global buffer storage level. From the perspective of a loop nest, it involves tiling the various data types into multiple levels and then reordering the loops at different levels to prioritize different data types. For example, if there are K levels of memory hierarchy, both dimension Q and S in the loop nest of Figure 5.8a can be further divided up into K loops (i.e., for the open ranges $[Q_0, Q_K)$ and $[S_0$ to $S_K)$) and then reordering the loops independently from loop level 0 to level $K - 1$. The same strategy to make reordering decisions can be applied to each storage level; in other words, the design is fractal. This also implies that smaller DNN accelerator designs can be further combined together with a new level of memory to form a larger accelerator.

In addition to curating one dataflow for the accelerator, there are recent works that explore supporting multiple dataflows in a single flexible hardware, including FlexFlow [162], DNA [163], and Maeri [164]. The key to these designs is to propose flexible NoCs and support various memory access patterns in order to execute different dataflows that have different numbers of loop levels and loop ordering in the loop nest.

While there are endless possibilities for creating new dataflows and optimizing the hardware architecture for those dataflows, the combination of dataflow(s) and hardware architecture that results in the best performance and energy efficiency is still an open research question since it heavily depends on the problem size, technology, and amount of hardware resources. Therefore, it is crucial to be able to systematically and efficiently analyze different design decisions. In Chapter 6, we will introduce several tools that can help with such an analysis.

---

[9]In a previous version of the taxonomy introduced in [142], there is an additional dataflow called no local reuse (NLR), which keeps no data stationary locally within the PE. The NLR dataflow, however, can be classified as one of the three stationary dataflows (i.e., WS, OS, or IS) at the next storage level, e.g., the global buffer.

5.7.6   DATAFLOWS FOR CROSS-LAYER PROCESSING

Up until now, we have been looking at dataflows and the associated architectures that focus on processing one layer at a time. The processing of multiple layers has to be scheduled sequentially, which may involve reconfiguration of the hardware to adapt to the varying shapes and sizes of different layers and possibly moving activations to and from DRAM between layers. These designs were based on the assumption that it has been unlikely for the accelerator to fit an entire layer at once for processing; therefore, it does not make much sense to consider processing multiple layers at the same time.

However, as more hardware resources are being devoted for DNN acceleration and the DNN models are becoming more compute and storage efficient to achieve the same accuracy for various applications, dataflows and hardware that can process multiple layers at a time can be beneficial. Specifically, the output activations from one layer are often used directly as the input activations of the next layer. By keeping the activations in the local memory for cross-layer processing, it can save additional memory accesses at higher levels, such as DRAM. However, this is often at the cost of requiring more storage for weights from different layers, and can only process a smaller tile of activations in each layer at a time.

Several previous works have proposed dataflows and architectures for cross-layer processing. For example, Fused-layer [165] focuses on saving memory accesses for activations to process across multiple convolutional layers in a DNN, while Shortcut Mining [166] targets the data reuse from the shortcut connections in residual networks. BRein [167] exploits similar ideas by running 13 layers at a time on the same hardware thanks to using binary/ternary weights for the DNN to keep storage requirements small. Brainwave [168] takes the idea of processing many layers at a time to the extreme by aggregating many FPGAs, each processing one layer at a time, to form a pipelined chain of processing fabric. In this case, each FPGA can be configured with the dataflow that is optimized for a specific layer.

Pipelined computation of multiple layers adds many additional considerations. First, a layer-granularity pipeline can only be exploited if a series of input feature maps need to be processed (i.e., a batch size greater than one). Otherwise, only a single stage of the pipeline would be busy, although tile-level pipelining could still provide some benefit. Second, pipelining layers introduces requirements for inter-stage storage of activations (since a key motivation was avoiding dumping and restoring activations from large expensive storage between layers). This has to be managed by careful tiling and/or selection of dataflows, which may vary between layers. For example, Tangram [169] uses tiling to control inter-stage activation storage and BRein [167] saves intermediate state for some layer pairs by feeding an output-stationary dataflow into an input-stationary dataflow. Finally, the processing and communications in each stage of the pipeline should be balanced to keep throughput high, as was explored in Simba [123]. All these considerations add considerable complexity to the design and mapping optimization process, and we are unaware of any comprehensive evaluation or solution to these issues.

Although dataflows are a key design element of any DNN accelerator, other hardware elements are important as well. In the next two sections, we review important considerations for the design of buffers and NoCs for DNN accelerators.

5.8   DNN ACCELERATOR BUFFER MANAGEMENT STRATEGIES

As was described above, data buffering is one of the key components of creating an efficient DNN accelerator. So, beyond selecting the appropriate dataflow, which control where and when data is buffered, an DNN accelerator needs a buffering scheme that should seek to achieve the following objectives:

- efficiently (e.g., in storage requirements) and in a timely fashion transfer exactly the data that will be needed in the future by the consumer of the data;

- overlap the receipt of data that will be needed in the future with the use of data currently being consumed;

- remove data exactly when it is no longer needed; and

- do all of the above with precise and cheap synchronization.

Generically, striving to achieve these objectives is called achieving good *data orchestration*. A classification of the current approaches for data orchestration from [170] is illustrated in Figure 5.29. In the figure, buffering idioms are split along two axes. At a high level, the implicit/explicit distinction along one axis refers to the degree to which workload knowledge is leveraged to control data buffering decisions, while the coupled/decoupled on the other axis refers to whether memory responses and requests are round-trip (request-response) or flow-forward (data is automatically pushed to consumer).

5.8.1   IMPLICIT VERSUS EXPLICIT ORCHESTRATION

In the general-purpose computing community, caches are the predominant buffering mechanism and are based on load/store (i.e., round-trip) operations. Caches have several desirable properties, such as composing invisibly into hierarchies. Memory-level parallelism—both multiple outstanding fills, as well as concurrency between fills and accesses to current contents—can be achieved using well-studied additional hardware (often called *lockup-free* cache structures).

Caches can be characterized as performing *implicit* data orchestration as the load request initiator does not directly control the cache hierarchy's decisions about whether the response data is retained at any given level of the storage hierarchy, nor when it is removed. Heuristic replacement policies are advantageous in general-purpose scenarios because they are workload agnostic.[10] On the other hand, for DNN accelerators, the area and energy overheads for features

---

[10]As many programmers care more about optimization than portability, they often reverse engineer the details of the cache hierarchy and replacement policy to try to explicitly manipulate them.

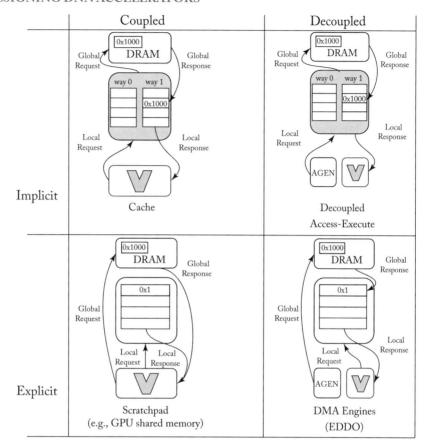

Figure 5.29: Taxonomy of data orchestration approaches. Communication (lines with arrowheads) is assumed to travel on a hardware channel (usually a NoC link). (Figure from [170].)

like tag matches and associative sets are high, and so far as we are aware no contemporary DNN accelerator incorporate caches.

An alternative to caches is to use *scratchpads*, which expose an address range of a particular staging buffer for loads and stores, thereby enabling *explicit* and precise control over the data orchestration. (In Figure 5.29 this is represented by the datapath managing both local and global requests/responses.) A GPU's *shared memory* scratchpad [171] is a widespread contemporary example of this idiom for explicit data orchestration. The size and address range of the scratchpad is exposed architecturally, and the transfer of data into and out of the scratchpad is managed via explicit instructions. While scratchpads avoid the hardware overheads of caches, extracting memory parallelism—both across fills and overlapping fills and accesses—is tedious and error-prone,[11] and as a result they are difficult to compose into hierarchies.

5.8.2 COUPLED VERSUS DECOUPLED ORCHESTRATION

Caches and scratchpads both use a load/store paradigm where the initiator of the request also receives the response. This is referred to as *coupled* staging of data, reflected in the left column of Figure 5.29. With this setup, synchronization between data demand and data availability is efficient and intuitive—the requester is notified when corresponding response returns (load-to-use). The disadvantage to this approach is that it complicates overlapping the fill and access of data tiles (e.g., via double-buffering) as the single requester/consumer must alternate between requesting and consuming responses. Additionally, a "landing zone" for the incoming data tile must be held reserved (and are therefore idle) for the entire round-trip load latency, which increases pressure on memory resources that could otherwise be used for larger tile sizes.[12]

The alternative is to *decouple* the load request initiator from the response receiver. (In Figure 5.29 this is represented by the request/response arrows going to different modules.) In this setup, a separate hardware module (e.g., a DMA engine, or *address generator* (AGEN)) is responsible for *pushing* data into one or more functional units' buffers.[13] To tolerate latency, these are often double-buffered and hence sometimes referred to as ping-pong buffers [172, 173]. The main advantage to this approach is that the requester can run at its own rate, and can multicast data to multiple simultaneous consumers. Additionally, the feed-forward nature of the pipeline means that the tile landing zone only needs to be reserved proportional to the latency between adjacent levels of the hierarchy, rather than the entire hierarchy traversal round-trip, allowing for increased utilization of equivalent sized memory. Finally, this approach often can transmit large blocks of data (i.e., bulk transfers, which are more efficient than small requests) which must dynamically re-coalesce accesses to the same memory line.

This separate producer/consumer approach is similar to Smith's [174] *decoupled access-execute* (DAE) style of general-purpose computing architecture. In a DAE organization two processors are connected by a hardware queue. The access processor is responsible for performing all address calculations and generating loads—analogous to the DMA engine. Load responses are passed to the execute processor—analogous to an accelerator's functional units and their local staging buffers. DAE improves parallelism and reduces the critical paths of instructions while allowing both processors to compute at their natural rate. However, classical DAE does not explicitly control data orchestration buffers—decisions about staging data are still managed by the cache hierarchy, thus Figure 5.29 categorizes DAE as implicit decoupled.

---

[11]GPU shared memory is paired with high multi-threading and loop unrolling to offset these problems, but this complexity is almost certainly unacceptable for a more specialized accelerator.

[12]The magnitude of this effect can be evaluated using Little's Law [118].

[13]Cache pre-fetching can be considered an example of decoupling. Consideration of this large body of work is beyond the scope of this book.

5.8.3   EXPLICIT DECOUPLED DATA ORCHESTRATION (EDDO)

The most common buffering approach in DNN accelerators is *explicit decoupled* data orchestration (EDDO). Hardware FIFOs [175, 176] are one traditional reusable EDDO staging buffer organization. The advantages are that FIFOs cleanly encapsulate synchronization via head and tail pointers, and are easily hierarchically composable. However, in practice FIFOs are not flexible enough to meet the needs of DNN accelerators, which often repeats accesses within a window of a tile (e.g., when performing a convolution). Additionally, for data types such as the partial sums, staged data must be modified several times in place before being drained. This is not possible in single write-port FIFOs without costly re-circulation.

Explicit decoupled data orchestration (EDDO) schemes have been incorporated as a customized buffering mechanism in some DNN accelerators [142, 152, 159, 177, 178] and other specific EDDO buffering schemes, such as DESC [179], have been proposed. However, to illustrate a typical EDDO scheme we will describe *buffets* [170], which are a generalization of the data orchestration scheme in Eyeriss [101].

At its heart, the operation of a buffet is FIFO like in that values are filled from an input NoC link (i.e., a hardware communication channel) into a circular buffer controlled by *head* and *tail* pointers. Values will only be removed from the fill NoC link if the fill occurs. Access to data in the buffer is provided by a *read* command, however unlike a FIFO which can only read at its head, a buffet read is augmented with an address, which is interpreted as an offset from the head. Keeping a set of values in the buffer and reading them multiple times allows for reuse of a tile of data. Analogous to the fill, the read command will only execute if the read value can be sent on the read value NoC link (i.e., the NoC link is not blocked).

Buffets also support updates of values in its buffer. Updates only are allowed at a previously read value at a location read with a *read+update* command. This allows a buffet to support storing and updating partial sums.

Finally, a buffet provides a *shrink* operation that removes a specified number of entries from the head of the buffer. The shrink allows one to easily free the space occupied by the tile. To avoid costly delays when switching tiles, one can define a tile size that is smaller than the size of the buffet. Therefore, fills of the next tile can begin transparently while the previous tile is being processed. However, the extra space only needs to be large enough to avoid the startup transient prior to work starting on the next tile. That is often much less than the space required for double buffering.

Shrinks need not remove an entire tile. Removing only a portion of a tile (e.g., just one value) and then reading sequentially again starting at offset zero allows buffets to support sliding windows.

Figure 5.30 shows a block diagram of a buffet. Actions occur when there are values on all the input NoC links needed by the action (command or fill) and there is room in the output NoC link (only needed for reads). The activity illustrated in the figure is the following.

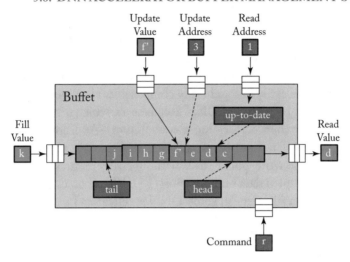

Figure 5.30: Buffet Block Diagram—a block diagram of the major components of a buffet. The principle inputs are a *fill value*; a *read address* and *read value*; an *update address*, and *update value*; and a *command*, which can specify whether to perform a *read*, *read+update*, or *shrink*. The only output is a *read value*. The *head*, *tail*, and *up-to-date* units internally provide synchronization, stalling operations to preserve proper ordering.

- A *read* command (r) is being invoked that takes *read address* (1) as an offset from *head* to produce the *read value* (d).

- An update at *update address* (3) is writing an *update value* (f') into the buffet. Note, this is allowed because an earlier command must have been a *read+update* at offset 3.

- A *fill value* (k) is about to be written into the *tail* of the buffet.

Note that all of the above activity is mediated locally within the buffet by the *head*, *tail*, and *up-to-date* state, which guarantees proper ordering. For example, a *read* must wait until its data has been filled and updated (if there was a prior *read+update*). And a *fill* must wait until there is room in the buffer.

Not illustrated is a *shrink* command, which simply removes a given number of values from the head of the buffet by adjusting the *head* pointer after waiting for outstanding updates.

Figure 5.31 shows a toy example that demonstrates how buffets naturally compose and can be used to process sliding windows and updates. The natural composition of the L1 Input Buffet by the L0 Input Buffet allows external synchronization-free filling, because the filling is controlled by internal ordering controls in each buffet. The shrinks by one in the L0 Input Buffet create a sliding window of inputs, so a relative sequence of inputs would be 0, 1, 2, 1, 2, 3, 2, 3, 4....... Internal synchronization also controls the updates of partials sums.

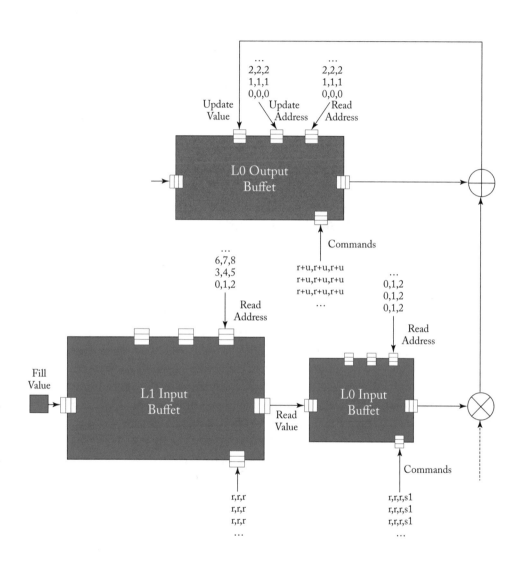

Figure 5.31: Buffet Example—an artificial example of an Eyeriss-like global buffer and PE built with buffets. Reads to the L1 Input Buffet fill the L0 Buffet, which performs reads that pass a sliding window of inputs to the multiplier. The L0 Output Buffer performs a series of read+update commands to generate the partial sums. The weight buffet is not shown.

In summary, efficient data orchestration is needed in DNN accelerator designs and will usually be provided by mechanisms, like buffets, that manage data movement. This generally manifests in the design as explicit control where data is pushed deterministically through the storage hierarchy avoiding costly round-trip communication and minimizing "landing zone" storage requirements. Efficiency is also enhanced by decoupled activity where the hardware provides local determination of the values needed and local synchronization controls. Obviously, the full semantics of a buffet is not needed for every storage buffer, so an optimized implementation of a subset of those semantics or other custom design that provides comparable benefits can be employed.

5.9   FLEXIBLE NOC DESIGN FOR DNN ACCELERATORS

The NoC is an indispensable part of modern DNN accelerators in order to support the data delivery patterns of various dataflows, as described in Section 5.7, and its design has to take the following factors into consideration: (1) support processing with high parallelism by efficiently delivering data between storage and datapaths; (2) exploit data reuse to reduce the bandwidth requirement and improve energy efficiency; and (3) can be scaled at a reasonable implementation cost.

Figure 5.32 shows several NoC designs commonly used in DNN accelerators. Due to the property of DNN that data reuse for all data types cannot be maximally exploited simultaneously, a mixture of these NoCs is usually adopted for different data types. For example, a DNN accelerator can use a 1-D horizontal multicast network to reuse the same weight across PEs in the same row and a 1-D vertical multicast network to reuse the same input activation across PEs in the same column. This setup will then require a unicast network that gathers the unique output activations from each PE. This combination, however, implies that each weight needs to have the amount of reuse with different input activations at least equal to the width of the PE array, and the number of input activation reuse with different weights at least equal to the height of the PE array. If these conditions are not fulfilled, the PE array will not be fully utilized, which will impact both throughput and energy efficiency.

Figure 5.33 shows two designs with the example NoCs that are commonly used in many existing DNN accelerators [135, 145, 156, 157, 180–182]. A spatial accumulation array architecture (Figure 5.33a), which is often used for a weight-stationary dataflow, relies on both output and input channels to map the operations spatially onto the PE array to exploit parallelism. At the same time, each input activation can be reused across the PE array vertically with weights from different output channels, while partial sums from the PEs in the same row can be further accumulated spatially together before written back to the global buffer. Similarly, a temporal accumulation array architecture (Figure 5.33b), which is often used for an output-stationary dataflow, relies on another set of data dimensions to achieve high compute parallelism. In this case, each input activation is still reused vertically across different PEs in the same column, while each weight is reused horizontally across PEs in the same row.

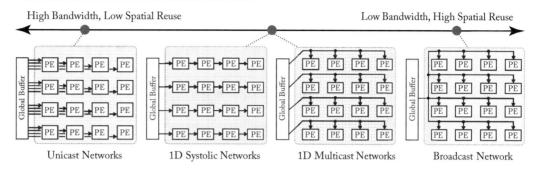

Figure 5.32: Common NoC designs.

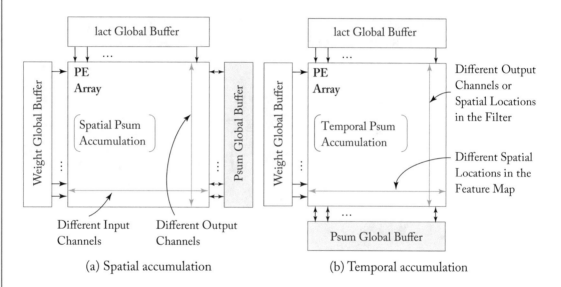

Figure 5.33: Two common DNN accelerator designs: (a) spatial accumulation array [135, 145, 156, 157]: input activations (iacts) are reused vertically and partial sums (psums) are accumulated horizontally; and (b) temporal accumulation array [180–182]: input activations (iacts) are reused vertically and weights are reused horizontally.

When the set of pre-selected data dimensions diminish due to the change in DNN shapes and sizes, e.g., the number of output channels in a layer (M) is less than the height of the PE array, efficiency decreases. Specifically, these spatial mapping constraints result in both reduced array utilization (i.e., fewer PEs are used) as well as lower energy efficiency. Furthermore, these inefficiencies are magnified as the size of the PE array is scaled up, because the diminished dimension is even more likely to be unable to fill the array. For example, as shown in Figure 5.34, the aforementioned spatial and temporal accumulation arrays will find it difficult to fully uti-

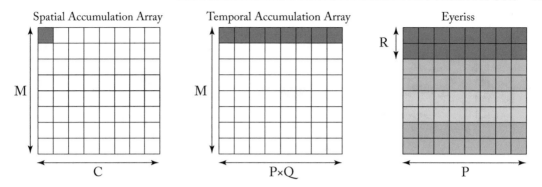

Figure 5.34: Array utilization of different architectures for depth-wise (DW) convolutions in MobileNet. The colored blocks are the utilized part of the PE array. For Eyeriss [101], the different colors denote the parts that run different channel groups (G). Please refer to Table 2.1 for the meaning of the variables.

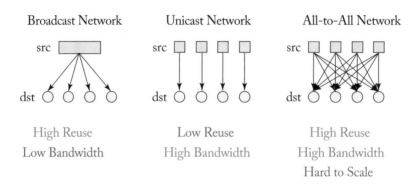

Figure 5.35: The pros and cons of different NoC implementations.

lize the array due to the lack of input and output channels when performing depth-wise (DW) convolutions in MobileNet [183] (see Section 9.1.1 and Figure 9.6 for more on depth-wise convolution). In contrast, Eyeriss [101] can still achieve high array utilization under such circumstances by mapping the independent channel groups onto different part of the PE array due to the flexibility of its row-stationary dataflow.

The varying amount of data reuse for each DNN data type across different layers or models pose a great challenge to the NoC design. As shown in Figure 5.35, the broadcast network can exploit the most data reuse, but its low source bandwidth can limit the throughput when data reuse is low. The unicast network can provide the most source bandwidth but misses out on the data reuse opportunity when available. Taking the best from both worlds, an all-to-all network that connects any data sources to any destinations can adapt to the varying amount of data reuse

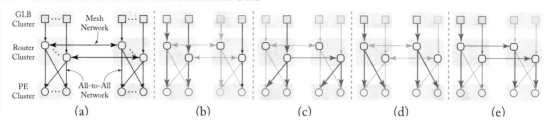

Figure 5.36: (a) High-level structure of the hierarchical mesh network (HM-NoC), and its different operating modes; (b) high bandwidth mode; (c) high reuse mode; (d) grouped-multicast mode; and (e) interleaved-multicast mode. In each mode, the colored arrows show the routing path; different colors denote the path for unique data.

and bandwidth requirements. However, the cost of its design increases quadratically with the number of nodes, e.g., PEs, and therefore is difficult to scale up to the amount of parallelism required for DNN accelerators.

5.9.1   FLEXIBLE HIERARCHICAL MESH NETWORK

To deal with this problem, Eyeriss v2 [161] proposed a new NoC architecture for DNN accelerators, called hierarchical mesh network (HM-NoC), as shown in Figure 5.36a. HM-NoC takes advantage of the all-to-all network, but solves the scaling problem by creating a two-level hierarchy. The all-to-all network is limited within the scope of a cluster at the lower level. There are usually only dozens of PEs within each cluster, which effectively reduce the cost of the all-to-all network. At the top level, the clusters are further connected with a mesh network. While this example shows a 2×1 mesh, an actual design can have a much larger mesh size. Scaling up the architecture at the cluster level with the mesh network is much easier than with the all-to-all network since the implementation cost increases linearly instead of quadratically.

Figure 5.37 shows several example use cases on how HM-NoC adapts different modes for different types of layers. For simplicity, we are only showing a simplified case with 2 PE clusters with 2 PEs in each cluster, and it omits the NoC for partial sums. However, the same principles apply to NoC for all data types and at larger scales.

- Conventional CONV layers (Figure 5.37a): in normal CONV layers, there is plenty of data reuse for both input activations and weights. To keep all four PEs busy at the lowest bandwidth requirement, we need two input activations and two weights from the data source (ignoring the reuse from RF). In this case, either the HM-NoC for input activation or weight has to be configured into the grouped-multicast mode, while the other one configured into the interleaved-multicast mode.

- Depth-wise (DW) CONV layers (Figure 5.37b): for DW CONV layers, there can be nearly no reuse for input activations due to the lack of output channels. Therefore, we can

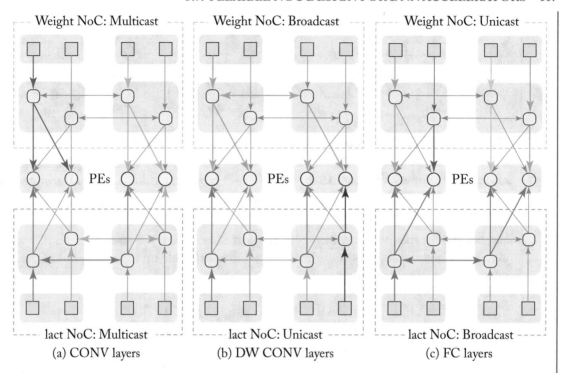

Figure 5.37: Examples of weight and input activation hierarchical mesh networks configured in different modes for different types of DNN layers: (a) CONV layers; (b) depth-wise (DW) CONV layers; and (c) FC layers. Green arrows and blue arrows show the routing paths in the weight and input activation NoC, respectively.

only exploit the reuse of weights by broadcasting the weights to all PEs while fetching unique input activation for each PE.

- FC layers (Figure 5.37c): contrary to the DW CONV layers, FC layers usually see little reuse for weights, especially when the batch size is limited. In this case, the modes of input activation and weight NoCs are swapped from the previous one: the weights are now unicast to the PEs while the input activations are broadcast to all PEs.

While conventional mesh NoC implementations often come with high area and power overhead due to the need to route the data dynamically, it is not the case for HM-NoC since it does not require routing at runtime. Instead, all active routes are determined at configuration time based on the specific data delivery pattern in use. As a result, no flow control is required, and the routers are simply multiplexers for circuit-switched routing that has minimum implementation cost.

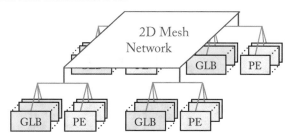

Figure 5.38: A DNN accelerator architecture built based on the hierarchical mesh network.

Figure 5.38 shows an example DNN accelerator built based on the hierarchical mesh network. The router clusters are now connected in a 2-D mesh. The global buffer (GLB) is banked and distributed into each source cluster, and the PEs are grouped into the destination clusters instead of one single array.

5.10   SUMMARY

In this chapter, we presented the motivation, principal objectives and key design alternatives for specialized hardware for DNN accelerators. The tremendous degrees of freedom in the ordering of the MAC operations in DNN computations, the need for efficient data movement and the desire to flexibly handle many DNN workload shapes, leads to a large design space. This chapter concentrated on understanding that design space from the perspective of performance and energy efficiency with a particular emphasis on reuse and how to exploit it spatially and temporally in a multi-level storage hierarchy. Central to that topic was the notion of *dataflows*, how they manifest in various existing designs and how they can be expressed precisely with the use of *loop nests*. Also presented were ideas related to efficient buffering using *explicitly decoupled data orchestration* (EDDO) and flexible NoC design. Left to be explored in the next chapter is the process of finding optimal mapping of specific workload shapes onto a given DNN acclerator.

CHAPTER 6

Operation Mapping on Specialized Hardware

In Chapter 5, we discussed various key design considerations and techniques for the implementation of specialized DNN hardware. Also introduced was the notion of the *mapping* of the computation for a particular workload layer shape onto a specific DNN accelerator design, and the fact that the compiler-like process of picking the right mapping is important to optimize behavior with respect to energy efficiency and/or performance.[1]

Mapping involves the placement and scheduling in space and time of every operation (including delivering the appropriate operands) required for a DNN computation onto the hardware function units of the accelerator.[2] Mapping has the following key steps.

1. *Determine the dataflow:* If the targeted DNN accelerator supports more than one dataflow, then the mapping must select a dataflow. In the loop nest representation of a design introduced in Chapter 5, a dataflow is manifest in the number and order of `for` loops. For multiple levels of storage, the loops can have an independently selectable order at each level. The choice among all these myriad orders can have a major impact on the behavior of an accelerator, and so is an important component of mapping. However, just choosing a dataflow is not sufficient, because it does not define the loop bounds.

2. *Determine the data tile sizes:* After selecting the dataflow, we need to determine the tile of data for each data type that each instance of storage at each level works on within a certain duration of time. Thus, this *tiling* is in both space and time and can have a major impact on the amount of data that needs to be moved. For example, with the weight-stationary (WS) dataflow at storage level $L1$, the sizes of the input activation tile and partial sum tile determine how many times the weight tile will be reused in $L1$, and the storage capacity of $L1$ constrains the sizes of these three data tiles. From the point of view of the loop nest representation of a design, this step determines the `for` loop bounds.

3. *Bind the operations to the hardware:* Given the loop nest order and fixed loop bounds, the final step is to determine where the operations (arithmetic computation or data access) at

---

[1]For DNN accelerators capable of processing multiple layers at once the notion of mapping must be expanded to consider optimizing more globally. Such considerations are beyond the scope of this chapter.

[2]Some papers refer to this activity just as *scheduling*, but we use the word mapping to emphasize its relevance to both space and time.

each iteration of the loop *binds* to the hardware (either PE or storage buffer). For example, one possibility is to simply define that the arithmetic operation at iteration i of the `parallel-for` loop goes to the PE with ID i. Without losing generality, we will assume this simple binding is employed in the rest of this chapter. However, more complicated binding schemes can be beneficial in certain cases, for instance, when the range of the `parallel-for` loop is not equal to the number of PEs, or when some physical attribute of the hardware can be taken into account (e.g., the physical proximity of specific PEs or buffers to one another).

The set of all possible mappings for a specific workload shape on to a specific DNN accelerator is called the *map space* for that workload. Given the number of degrees of freedom in mapping, map spaces can be very large. Furthermore, the fact that many attributes of an accelerator's design are fixed, such as the number of PEs, the number and sizes of buffers and the connectivity of the networks-on-chip (NoCs), lead to complex constraints on which mappings are legal. Therefore, map spaces tend to be irregular (i.e., there are many illegal points in the unconstrained map space).

The large size and irregularity of maps spaces and the fact that different mappings can result in drastically different performance and/or energy efficiency on the hardware, leads to a strong desire to be able to find the optimal mapping in the map space. The tool that is used to find such a mapping is called a *mapper*. To do its job, a mapper needs to both be able to express and search the map space and evaluate and quantify the energy efficiency and performance of different mappings. In this chapter, we will present the mechanics of performing such searches, including some analysis frameworks that can be used to systematically characterize a mapping. But first, we will explore a bit more deeply what constitutes a mapping.

6.1 MAPPING AND LOOP NESTS

Mapping a workload shape to a DNN accelerator involves a number of choices including: picking a dataflow from among those supported by the accelerator and picking both the spatial and temporal tiling of all the operands to the storage and computational components of the accelerator. For both CONV and FC[3] layers, these factors can be represented in terms of the loop nests described in Chapter 5. Therefore, one can view the process of creating a mapping from the perspective of selecting and parameterizing loop nests.

To provide an instructive example of mapping onto a DNN accelerator design, consider the simple DNN accelerator design depicted in Figure 6.1. The architecture is assumed to have two PEs each of which handles processing of a single-input-channel, 1-D convolution for multiple output channels. Each PE's buffers can hold a copy of the input feature map for the one and only input channel, as well as the output feature map and filter weights for the output channel they are currently working on. Since the PEs can only hold the weights for a single output

---

[3]Recall from Section 4.1, the matrix multiply of an FC layer can be represented as a convolution with $R = H$ and $S = W$.

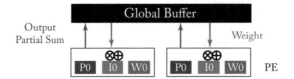

Figure 6.1: Mapping target architecture for single input channel 1-D convolution with multiple output channels. This simple design has a global buffer and two PEs each with buffers for a single channel's worth of input and output feature maps and filter weights. Each PE is configured with the input feature map preloaded and processes each output channel by loading the filter weights for that channel from the global buffer, computing the output feature map for that channel and sending that output feature map back to the global buffer.

channel at a time, processing a new output channel requires that its weights be loaded from the global buffer, and the output feature map from the prior output channel is sent to the global buffer. Finally, we assume the PE can be configured to run either an output-stationary or a weight-stationary dataflow.

A loop nest corresponding to that design is shown in Design 6.1 below. The design performs a single-input-channel, multiple-output-channel 1-D convolution using an input feature map (i[]) and filter weights (f[]) generating a output feature map with multiple output channels (o[]). Those variables are defined (along with their sizes) in lines 1–3. Skipping lines 7–9 for the moment, the outermost for loop (line 13) represents a tiling of the M output channels, and each iteration of that loop results in a tile-sized chunk of weights being read from the global buffer. Those tiles are fed to the parallel PE units that are represented by the **parallel-for** in line 17, where m1 corresponds to the ID of the PE doing the processing. The index of the active output channel is calculated in line 18 from the output channel tile number (m2) and the ID of the PE doing the processing (m1). Finally, the actual convolutions are performed in a PE processing unit using either a weight-stationary dataflow (lines 23–26) or an output-stationary dataflow (lines 28–31).

In the loop nest for Design 6.1, there are a set of variables that control its behavior. First are variables W, M, S, and Q used in lines 1–3 that define the shape of workload. Second, are the variables M2, M1, and PE_dataflow that control the behavior of the loop nest. For this design, those variables constitute the *mapping* of the workload onto the design, because they control the placement and scheduling of activity in both space and time. In the code, those variables are assigned from the associative array with the unsurprising name mapping. The contents of mapping are assumed to have been set by a *mapper* that picked them based on some optimization criteria.

In this design, there are only two components of the mapping. The first component of the mapping is a selection of one of two the dataflows supported. This is controlled by the PE_dataflow mapping variable, which specifies the iteration order of the index variables ("s"

Design 6.1 Example Mapping Target - parallel output channels

```
1    i = Array(W)         # Input feature map
2    f = Array(M, S)      # Filter weights
3    o = Array(M, Q)      # Output feature map
4
5    # Mapping
6
7    M2 = mapping["M2"]
8    M1 = mapping["M1"]
9    PE_dataflow = mapping["PE_dataflow"]
10
11   # Level 2 - Global Buffer
12
13   for m2 in [0, M2):
14
15   # Level 1 - PE array
16
17       parallel -for m1 in [0, M1):
18           m = m2 * M1 + m1
19
20   # Level 0 - PE
21
22           if PE_dataflow == "sq"):
23               for s in [0, S):
24                   for q in [0, Q):
25                       w = q + s
26                       o[m, q] += i[w] * f[m, s]
27           elif PE_dataflow == "qs":
28               for q in [0, Q):
29                   for s in [0, S):
30                       w = q + s
31                       o[m, q] += i[w] * f[m, s]
```

---

and "q") used in the `for` loops. Thus, `PE_dataflow` essentially sets a permutation order the for the `for` loops representing the PE's dataflow as either weights-outputs ("sq") or outputs-weights ("qs"), which correspond to weight-stationary and output-stationary dataflows, respectively.

The second component is the spatio-temporal tiling of output channel-related data controlled by the mapping variables M2 and M1. The product of these variables must be at least equal to the number of output channels, so M <= M2 × M1. In addition, the spatial tiling size (M1) is limited by the actual number of PEs in the design, so M1 <= 2.

The set of conditions that any mapping must satisfy are referred to as *mapping constraints*, and the objective of the *mapper* is to optimize system behavior within those constraints. In this simple example, it is clear that to optimize throughput one would like to set M1 = 2 to achieve full utilization of the PEs, but in more complex scenarios one cannot always achieve perfect utilization. Note that even with M1 = 2 an odd number of output channels (M) would result in under-utilization of the PEs in the last iteration of the m2 loop (line 13).

The above example was very simple as there were few index variables at each level of the storage hierarchy (and none for additional dimensions of the input feature maps or filter weights, batch size or input channels) and only the output channels were tiled. Therefore, the mapping choices were both limited and generally obvious. In more realistic designs, many (and sometimes all) of the index variables will appear at each level of the hierarchy and a large number of loop order permutations might be available. This would make mapping space much larger.

Another characteristic of more realistic designs would be the number and complexity of the constraints on mappings. Sources of such constraints include buffer capacity limits at each storage unit (either partitioned per operand type or allocated from a shared pool) and incomplete NoC connectivity between units (including options for bypassing of storage levels). This results in a large and irregular mapping space which makes for a correspondingly more complex search. This process, however, has some similarities to the compilation process for conventional processors, which will be discussed next.

6.2 MAPPERS AND COMPILERS

The determination of a good mapping for a DNN accelerator can be viewed as being analogous to that of compiling for a general-purpose processor, as illustrated in Figure 6.2 [184]. In conventional computer systems, the *compiler* translates the program into machine-readable binary codes for execution; in the processing of DNNs, the *mapper* translates the desired DNN layer computation (i.e., problem specification) along with its shape and size[4] into a hardware-compatible mapping for execution. While the compiler usually optimizes just for performance, the mapper will typically optimize for performance and/or energy efficiency.

As described in Chapter 5, the dataflow(s) that a DNN accelerator supports is a key attribute of the design. Therefore, the supported dataflow(s) can be thought of as analogous to one of the most salient attributes of the *architecture* of a general-purpose processor in the sense that it prescribes what constitutes a properly formed program for the system. Similar to the role of

---

[4]While this chapter largely considers CONV layer computations so the problem specification consists of just a shape and size, it is natural to imagine this specification being extended to something more general, like an expression in the tensor index notation described in Chapter 2.

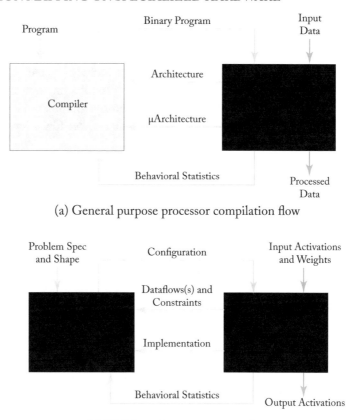

(a) General purpose processor compilation flow

(b) DNN accelerator mapping flow

Figure 6.2: An analogy between (a) the compilation process for general-purpose processors and (b) the mapping process for DNN accelerators. (Figure adapted from [184].)

an instruction set architecture (ISA) or memory consistency model, dataflow characterizes the hardware implementation and defines the many of the rules that the mapper has to follow in order to generate hardware-compatible mappings.

In many cases, a key characteristics of an architecture is that it remains stable (or only is evolved in an "upward compatible" fashion) across implementation generations. In this way, we also consider the dataflow as analogous to the architecture, since we believe that within a DNN accelerator family the set of available dataflows is going to largely remain invariant across implementations. However, like GPUs, the stability of some aspects of the architecture, including supported dataflow(s), could diminish for DNN accelerators due to their rapid evolution and increased reliance on the compiler/mapper to mask differences between designs.

In addition to the dataflows, other constraints on the mapping space would be included in the characteristics analogous to architecture. These reflect features such as buffer sizes and NoC connectivity that the mapper must consider if the resulting mapping is going to function correctly. The fact that attributes like buffer sizes are included as part of the architecture for the accelerator make it quite different from a processor's architecture because storage sizes in a processor (like cache sizes) are generally not included in the architecture.[5] Thus, DNN accelerator buffer sizes are more akin to the size of a processor's register file, which manifests as inviolable constraints for creating legal binaries.

Detailed information about the hardware implementation, including latency, throughput and energy cost for storage accesses and NoC traffic at each level of the storage hierarchy, is analogous to the micro-architecture of processors for the following reasons: (1) they can vary considerably across implementations; and (2) although they play a vital part in performance and energy efficiency optimization, considering their characteristics is not essential, since even if ignored the mapper will generate functional, but sub-optimal mappings.

The final input into either the compiler or mapper is behavioral statistics from the hardware. By using such statistics, the compiler/mapper can better decide on good optimizations via iterative refinement. However, in many cases this information is not available and the compiler/mapper must make its own projections of the impact of a choice it makes. For DNN accelerators, this involves modeling the design to project metrics of interest for a given mapping. An example of how to generate such projections will be described in Section 6.4.

Given the above information, the output of the compiler is an optimized binary program that can run on the processor. By analogy, the goal of the mapper is to search in the mapping space for the mapping that optimizes the metric(s) of interest and generate a *configuration* for the DNN accelerator that embodies that optimal mapping. Generally, a configuration will consist of a set of values that will be loaded into configuration registers of the DNN accelerator. Those configuration registers will then control the operation of the accelerator including the read/write access patterns to all the buffers, the NoC data transfer patterns and the sequence of computations at the PEs.

6.3   MAPPER ORGANIZATION

Given all the above inputs and outputs of the mapper, an abstract internal organization of a mapper can be envisioned using Figure 6.3. The flow involves creating a representation of the map space for the given workload and DNN accelerator, searching that space for optimal mappings based on the metrics of interest (e.g., performance, energy efficiency, or a combination), evaluating the metrics for mappings proposed by the search, and after selecting a desired mapping creating a configuration for the accelerator that will result in an execution with that mapping.

---

[5]The inclusion of buffer sizes in the architecture result from DNN accelerators typically using explicit (as opposed to implicit) data orchestration as will be discussed in Section 5.8.

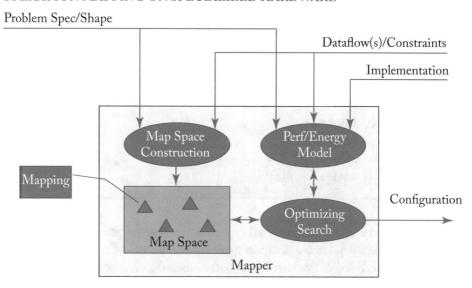

Figure 6.3: Block diagram of a mapper - Given a DNN *problem specification/shape* and the *dataflow(s)/constraints* (architecture) of a DNN accelerator, the *map space construction* step creates a map space. The *optimizing search* step iterates over mappings from the map space. It evaluates their characteristics by sending the mapping to a *performance/energy model* that also takes in the problem specification/shape, dataflows(s)/constraints (architecture), and implementation details (micro-architecture) of the DNN accelerator and returns the performance and/or energy characteristics of that mapping. Ultimately the optimizing search generates a "best" mapping that the *mapping to configuration* step converts into a *configuration*.

6.3.1   MAP SPACES AND ITERATION SPACES

In Figure 6.3, two of the inputs to the mapper—DNN problem specification/shape and dataflow/constraints (DNN accelerator architecture[6])—are inputs to the first step, which is labeled *map space construction*. The responsibility of this step is to create an enumeration of all the possible legal mappings for the given problem shape and architecture. The gray box in the figure indicates the map space containing all the legal mappings in the map space.

For some DNN accelerators (e.g., like the simple design in Section 6.1) it is possible to exhaustively enumerate the mappings in the map space. However, that is not generally feasible, so a more abstract representation of the map space can be useful. A common abstraction used to characterize a map space is to employ a concept commonly used in the compiler community called an *iteration space*, whose origin can be traced back to [185, 186]. For a problem specification consisting of a single loop nest, such as used for 1-D convolution (see Design 6.2), an

---

[6]To simplify the descriptions in the rest of the chapter, the term *architecture* will be used to refer to the DNN dataflow/-constraints, and the term DNN *micro-architecture* will be used to refer to the DNN implementation details.

Design 6.2 1-D Convolution

```
1       i = Array(W)        # Input  feature  map
2       f = Array(S)        # Filter  weights
3       o = Array(Q)        # Output  feature  map
4
5       for q in [0, Q):
6           for s in [0, S):
7               w = q + s
8               o[q] += i[w] * f[s]
```

iteration space is a multi-dimensional space with one point for each execution of the body of the loop nest. Figure 6.4 is the iteration space for this simple 1-D convolution. Each yellow oval represents an execution of the loop body (MAC operation in line 8) at the given indices (w, s, q) of the convolution.[7] Each point in the iteration space is also associated with points in the *data spaces* of the problem. In specific, each point in the data space is connected to the operands of the computation at that point. In this example, the data spaces are the inputs (i[]), the filter weights (f[]) and the outputs (o[]. Thus, the point $(w, s, q) = (4, 1, 3)$ has data space operands of i[4], f[1], and o[3].

Given an iteration space, an execution of the problem requires a visit to each point in the iteration space. For convolution, where the operations can be performed in any order (because there are no dependencies and the addition in the sum of products is commutative) one could visit the points in the iteration space in any order. Note, it is important to observe that the visit order is not determined by the specification that defined the layer's computation, but by intrinsic characteristics of the desired computation. This can be confusing because the language used to specify a layer's computation (e.g., using loop nests as we did here) might have the same syntax as the language used to described a mapped version of the computation, and thus the specification appears to define a visit order. Using a more general notation like the tensor index notation (see Section 2.3.1), which doesn't define a visit order, can help. In other cases, the presence of tiling can be a hint that one is looking at a mapping, but sometimes context is the only way to distinguish between a problem specification and a mapped computation.

A specific visit order through the iteration space does, however, correspond to a specific mapping, including its dataflow. For example, a mapping of a weight-stationary dataflow would be characterized by the iteration space visit order in Figure 6.5. In this case, the sequence of points to visit in the iteration space can be determined by following the values of the variables in a weight-stationary loop nest, such as Design 6.3.

---

[7]To make things clearer in this example, all the indices used to access the data are included at each point in the iteration space, even though any two could be used to compute the third.

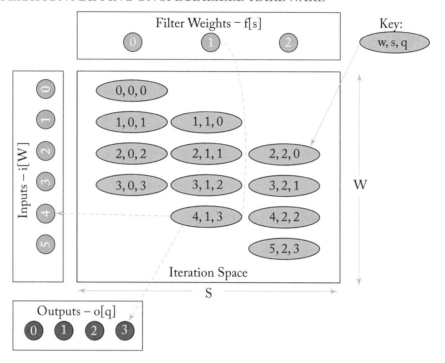

Figure 6.4: Iteration space for a 1-D convolution – each yellow oval represents a point in the iteration space corresponding to a calculation at the (w, s, q) point in the space. The values of the input and output operands from the i[], f[], and o[] data arrays are determined by those coordinates. For example, the iteration point $(w, s, q) = (4, 1, 3)$ accesses values at i[4], f[1], and o[3].

As is illustrated by this example, it should be evident that many useful traversal orders can be specified by the loop nests described in Chapter 5. In those loop nests, the iteration variables are used directly (or through very simple functions) as the indices of the data tensors. There are, however, traversal sequences that do not precisely follow the loop nest pattern and these more complex patterns might be created through the use of more complex computations on the **for** loop variables to create references to the data tensors. The result of such traversals would be a more diverse set of reference patterns to the data tensors. The benefits and hardware costs of such traversal orders are still largely an open question.

Parallelism in the hardware can be represented in the iteration space as multiple simultaneous traversal paths through the nodes in the iteration space. The **parallel-for** loops introduced in Chapter 5, would naturally result in such multiple traversals.

One interesting phenomenon associated with multiple simultaneous traversals through the iteration space is that the number of parallel traversals might not equal the number of hard-

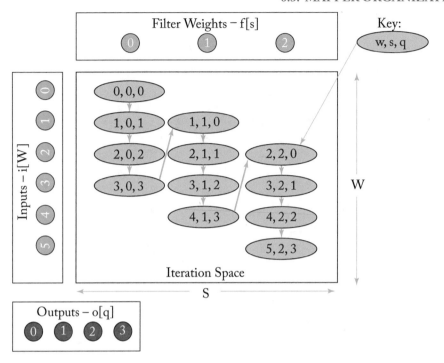

Figure 6.5: Traversal (green arrows) of the iteration space of a 1-D convolution for a weight-stationary dataflow.

Design 6.3 1-D Convolution with weight-stationary visit order

```
1    i = Array(W)         # Input  feature  map
2    f = Array(S)         # Filter  weights
3    o = Array(Q)         # Output  feature  map
4
5    for s in [0, S):
6        for q in [0, Q):
7            w = q + s
8            o[q] += i[w] * f[s]
```

---

ware units (e.g., PEs) or the lengths of the traversals might not be all be equal. These phenomena can be the consequence of hardware constraints that result in the folding and repeating mapping patterns described in the previous chapter in Section 5.7.4. In terms of loop nests, such situations arise when a problem shape parameter (e.g., partial sum width Q) is not factored perfectly (i.e., *imperfect factorization*) into the per layer loop limits (e.g., Q2, Q1, and Q0). In any case, imperfect factorization may result in underutilization of hardware resources, and hence lower performance. It is the job of the mapper to decide when an imperfect factorization, despite lower utilization, is the right choice to optimize the mapping.

In a DNN accelerator with a hierarchy of storage levels and interconnect networks, the iteration space will also have a hierarchical structure with iteration space points corresponding to the data movement actions and computation throughout the hierarchy. This hierarchy also needs to be considered in combination with the existence of simultaneous traversal paths introducing opportunities for inter-layer and intra-layer data transfers. And if that is not complicated enough, there is the further opportunity for the execution of those distinct paths to be *skewed* in time. This time skew can result in changes to the data reference and communication patterns at all levels of the storage hierarchy. The consideration of these issues is a area of current research and is, unfortunately, well beyond the scope of this book.

In conclusion, given all the above considerations, rules for specifying the allowable visit orders in the iteration space, accounting for allowable dataflows and other constraints, can be used to specify all the legal mappings (i.e., the mapping space).

6.3.2   MAPPER SEARCH

After creating a map space, the responsibility of the mapper turns to the *optimizing search* step in Figure 6.3. The aim of the optimizing search step is to find a mapping that optimizes for some objective such as performance, energy, or energy/delay product.

The search can be conducted by picking a mapping from the map space, and then determining how well that mapping meets the optimization objectives. That determination is typically conducted by a performance and/or energy model. In addition to a mapping, those models require the DNN problem specification/shape, the dataflow(s)/constraints of the DNN accelerator, and the micro-architectural implementation details in order to project the performance and/or energy consumption for that mapping.

6.3.3   MAPPER MODELS AND CONFIGURATION GENERATION

Performance modeling approaches for DNN accelerators can range from detailed cycle-level simulation to simple analytic models. In addition, in a recursive application of DNN processing, the model can be implemented as a neural network, which predicts performance.

The energy cost for a DNN accelerator can be estimated using an architecture-level energy estimation methodology such as Accelergy [187] or with an analytic model for energy projections such as is described next in Section 6.4.

Interpreting that result and deciding on the search pattern is search algorithm specific, and beyond the scope of this discussion. But after finding a "best" mapping, that mapping is converted into a configuration for the DNN accelerators by the *mapping-to-configuration* step of the mapper.

6.4   ANALYSIS FRAMEWORK FOR ENERGY EFFICIENCY

Unlike conventional compilers, which typically only focus on optimizing performance, a mapper often needs to optimize for energy efficiency. Conducting such an optimization typically requires a model capable of projecting the energy consumption of a particular mapping on a DNN accelerator. In this section, we will introduce a framework for the evaluation of energy consumption of DNN accelerators based on a spatial architecture. The analysis methodology is lightweight yet general, such that it can be applied to the analysis of many DNN accelerator architectures.

The way each MAC operation fetches inputs (filter weights and input activations) and accumulates partial sums introduces different energy costs due to two factors:

- how the dataflow exploits input data reuse and partial sum accumulation scheduling; and

- fetching data from different storage elements in the DNN accelerator have different energy costs.

The goal of an energy-efficient dataflow is then to perform most data accesses using the data movement paths with lower energy cost. This is an optimization process that takes all data accesses into account, and will be affected by the layer shape and available hardware resources.

In this section, we will describe a framework that can be used by the search step in a mapper to optimize the dataflows for spatial architectures in terms of energy efficiency. Specifically, it defines the energy cost for each level of the storage hierarchy in the DNN accelerator. Then, it provides a simple methodology to incorporate any given dataflow into an analysis using this hierarchy to quantify the overall data movement energy cost. This allows for a search for the optimal mapping for a dataflow that results in the highest energy efficiency for a given DNN layer shape.

Data Movement Hierarchy: We assume a spatial architecture that provides four levels of storage hierarchy. Sorting their energy cost for data accesses from high to low, it includes DRAM, global buffer, NoC, and RF. Fetching data from a higher-cost level to the ALU incurs higher energy consumption. Also, the energy cost of moving data between any of the two levels is dominated by the one with higher cost. Similar to the energy consumption quantification in previous experiments [121, 188, 189], Figure 6.6 shows the normalized energy consumption of accessing data from each storage level relative to the computation of a MAC at the ALU. The numbers are extracted from a commercial 65 nm process.

Analysis Methodology: Given a dataflow, the analysis is formulated in two parts: (1) the input data access energy cost, including filter weights and input activations; and (2) the partial sum accumulation energy cost. The energy costs are quantified through counting the number of accesses at each level of the previously defined hierarchy, and weighting the accesses at each level with a cost from Figure 6.6. The overall data movement energy of a dataflow is obtained through combining the results from the two types of input data and the partial sums.

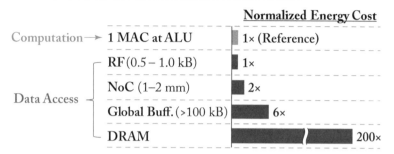

Figure 6.6: Normalized energy cost relative to the computation of one MAC operation at ALU. Numbers are extracted from a commercial 65 nm process.

6.4.1   INPUT DATA ACCESS ENERGY COST

If an input data value is reused for many operations, ideally the value is moved from DRAM to RF once, and the ALU reads it from the RF many times. However, due to limited storage and operation scheduling, the data is often kicked out of the RF before exhausting reuse. The ALU then needs to fetch the same data again from a higher-cost level to the RF. Following this pattern, data reuse can be split across the four levels. Reuse at each level is defined as *the number of times each data value is read from this level to its lower-cost levels during its lifetime.* Suppose the total number of reuses for a data value is $a \times b \times c \times d$, it can be split into reuses at DRAM, global buffer, array and RF for a, b, c, and d times, respectively. An example is shown in Figure 6.7, in which case the total number of reuse, 24, is split into $a = 1$, $b = 2$, $c = 3$, and $d = 4$. The energy cost estimation for this reuse pattern is:

$$
\begin{aligned}
&a \times EC(\text{DRAM}) + ab \times EC(\text{global buffer}) + \\
&abc \times EC(\text{array}) + abcd \times EC(\text{RF}),
\end{aligned}
\tag{6.1}
$$

where $EC(\cdot)$ is the energy cost from Figure 6.6.[8]

6.4.2   PARTIAL SUM ACCUMULATION ENERGY COST

Partial sums travel between ALUs for accumulation through the four-level hierarchy. In the ideal case, each generated partial sum is stored in a local RF for further accumulation. However, this is often not achievable due to the overall operation scheduling, in which case the partial sums have to be stored to a higher-cost level and read back again afterward. Therefore, the total number of accumulations, $a \times b \times c \times d$, can also be split across the four levels. The number of accumulations at each level is defined as *the number of times each data goes in and out of its lower-cost levels during its lifetime.* An example is shown in Figure 6.8, in which case the total number of accumulations, 36, is split into $a = 2$, $b = 3$, $c = 3$, and $d = 2$. The energy cost can then be

---

[8]Optimization can be applied to Equation (6.1) when there is no reuse opportunity. For instance, if $d = 1$, the data is transferred directly from a higher level to the ALU and bypasses the RF, and the last term in Equation (6.1) can be dropped.

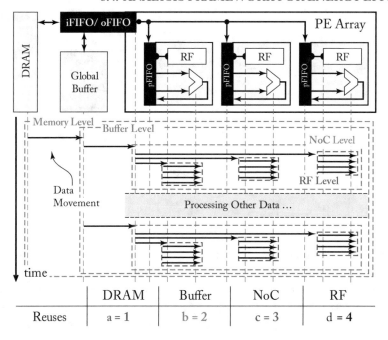

Figure 6.7: An example of the input activation or filter weight being reused across four levels of the memory hierarchy.

estimated as

$$\begin{aligned}(2a - 1) \times EC(\text{DRAM}) + 2a(b - 1) \times EC(\text{global buffer}) + \\ ab(c - 1) \times EC(\text{array}) + 2abc(d - 1) \times EC(\text{RF}).\end{aligned} \tag{6.2}$$

The factor of two accounts for both reads and writes. Note that in this calculation the accumulation of the bias term is ignored, as it has negligible impact on the overall energy.

6.4.3   OBTAINING THE REUSE PARAMETERS

For each mapping, there exists a set of reuse parameters (a, b, c, d) for each of the three data types, i.e., input activations, filter weights, and partial sums. These parameters are a function of the variables in the loop limits of the dataflow. For example, the set of reuse parameters for the simple 1-D convolution dataflow shown in Figure 6.9 is a function of $Q2, Q1, Q0$ and $S2, S1, S0$, and can be summarized in Table 6.1 (ignoring halos on the edges of the convolution).

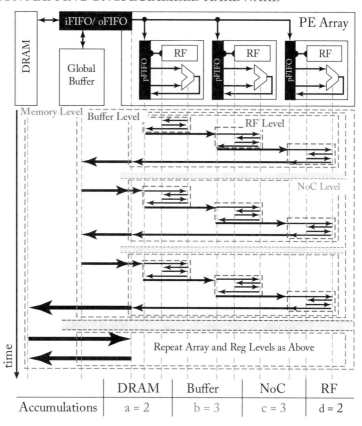

Figure 6.8: An example of the partial sum accumulation going through four levels of the memory hierarchy.

6.5   EYEXAM: FRAMEWORK FOR EVALUATING PERFORMANCE

Thus far in this chapter, we have studied mapping as a relatively heavy-weight process to find a configuration for a particular workload on a completely specified DNN accelerator design. Sometimes, it is important to be able to understand mapping and performance projections earlier in the design process or simply to gain insights into what is limiting performance. The Eyexam framework is a step-by-step process for doing just that.

Eyexam provides a systematic way of understanding the performance limits for DNN processors as a function of specific characteristics of the workload (i.e., DNN model) and accelerator design (i.e., architecture and micro-architecture); it applies these characteristics as sequential steps to increasingly tighten the bound on the performance limits. Specifically, instead of comparing the overall performance of different designs, which can be affected by many non-

```
Input Fmaps:     i[W]
Filter Weights: f[S]
Output Fmaps:    o[Q]

for (q2=0; q2<Q2; q2++) {
    for(s2=0; s2<S2; s2++) {
        parallel-for (q1=0; q1<Q1; q1++) {
            parallel-for (s1=0; s1<S1; s1++) {
                for(q0=0; q0<Q0; q0++) {
                    for(s0=0; s0<S0; s0++) {
                        o[q2*Q1*Q0+q1*Q0+q0] +=
                            i[q2*Q1*Q0+q1*Q0+q0 + s2*S1*S0+s1*S0+s0] ×
                            f[s2*S1*S0+s1*S0+s0];
}}}}}}
```

Figure 6.9: An example dataflow for a 1-D convolution.

Table 6.1: Reuse parameters for the 1-D convolution dataflow in Figure 6.9

Data Type	Reuse Parameters			
	a	b	c	d
Input activation	$S/(S0 \times S1 \times S2)$	$S2$	$S1$	$S0$
Weight	$Q/(Q0 \times Q1 \times Q2)$	$Q2$	$Q1$	$Q0$
Partial sums	$S/(S0 \times S1 \times S2)$	$S2$	$S1$	$S0$

architectural factors such as system setup and technology differences, Eyexam provides a step-by-step process that associates a certain amount of performance loss to each architectural design decision (e.g., dataflow, number of PEs, NoC, etc.) as well as the properties of the workload, which for DNNs is dictated by the layer shape and size (e.g., filter shape, feature map size, batch size, etc.).

Eyexam focuses on two main factors that affect performance: (1) the *number of active PEs* due to the mapping as constrained by the dataflow; and (2) the *utilization of active PEs*, i.e., percentage of active cycles for the PE, based on whether the NoC has sufficient bandwidth to deliver data to PEs to keep them active. The product of these two components can be used to compute the *utilization of the PE array* as follows:

$$\text{utilization of the PE array} = \text{number of active PEs} \times \text{utilization of active PEs.} \qquad (6.3)$$

Later in this section, we will see how this approach can use an adapted form of the well-known roofline model [119] for the analysis of DNN processors.

We will perform this analysis on a generic DNN processor architecture based on a spatial architecture that consists of a global buffer and an array of PEs. Each PE can have its own register file (RF) and control logic, and the PE array communicates with the global buffer through the NoCs. Separate NoCs are used for the three data types.

As described in Chapter 5, the dataflow of a DNN processor is one of the key attributes that define its architecture [184]. We will feature architectures that support the following four popular dataflows [142, 159]: weight stationary (WS), output stationary (OS), input stationary (IS), and row stationary (RS).

To help illustrate the capabilities of Eyexam, we will re-examine the simple 1-D convolution example in Section 6.5.1, and walk through the key steps of Eyexam in Section 6.5.2 with the 1-D convolution. We will then highlight various insights that Eyexam gives on real DNN workloads and architectures in Section 6.5.

6.5.1   SIMPLE 1-D CONVOLUTION EXAMPLE

We will re-examine the simple 1-D convolution example. This example illustrates the two components of the problem. The first is the *workload*, which is represented by the shape of the layer for a 1-D convolution. This comprises the filter size S and the input feature map size W and the output feature map size Q. The second is the *architecture* of the processing unit, for which a key characteristics is the dataflow shown in Figure 6.9. In this example, the two `parallel-fors` represent the distribution of computation across multiple PEs (i.e., spatial processing); the inner two `for loops` represent the temporal processing and RF accesses within a PE, and the outer two `for loops` represent the temporal processing of multiple passes across PE array and global buffer (GLB) accesses. For this example, we assume the input activations and weights fit in the GLB, i.e., the reuse parameter a is 1 for both the input activation and weight.

A mapping assigns specific values to loop limits $Q0$, $Q1$, $Q2$ and $S0$, $S1$, $S2$ to execute a specific workload shape and loop ordering. This assignment of $Q0$, $Q1$, $Q2$ and $S0$, $S1$, $S2$ is constrained by the shape of the workload and the hardware resources. The workload constraints in this example are $Q0 \times Q1 \times Q2 = Q$ and $S0 \times S1 \times S2 = S$.[9] The architectural constraint in this example is that $Q1 \times S1$ must be less than the number of PEs (later we will see that the NoC can impose additional restrictions). The size of the RF allocated to input activations, partial sums and weights will restrict $Q0$ and $S0$, and the space in the GLB allocated to partial sums restricts $Q1$ and $S1$.

While this is a simple 1-D example, it can be extended to additional levels of buffering by adding additional levels of loop nest (Section 5.6). Furthermore, extending it to support additional dimensionality (e.g., 2-D and channels) will also results in additional loops.

---

[9]We assume perfect factorization in this example. Imperfect factorization will lead to cycles where no work is done, as discussed in Section 5.5.

6.5.2   APPLY PERFORMANCE ANALYSIS FRAMEWORK TO 1-D EXAMPLE

The goal of Eyexam is to provide a fine-grain performance profile for an architecture. It is a sequential analysis process that involves seven major steps. The process starts with the assumption that the architecture has infinite processing parallelism, storage capacity and data bandwidth. Therefore, it has infinite performance (as measured in MACs/cycle).

For each of the following steps, certain constraints will be added to reflect changes in the assumptions on the architecture or workload. The associated performance loss can therefore be attributed to that change, and the final performance at one step becomes the upper-bound for the next step.

Step 1 (Layer Shape and Size): In this first step, we look at the impact of the workload constraint, specifically the layer shape (S, W, and Q), assuming unbounded values for $S1$ and $Q1$ since there is no architectural constraints. This allows us to set $S1 = S$, $Q1 = Q$, and $Q2 = Q0 = 1$, $S2 = S0 = 1$, so that there is all spatial (i.e., parallel) processing, and no temporal (i.e., serial) processing. Therefore, the performance upper bound is determined by the finite size of the workload (i.e., the number of MACs in the layer, which is $Q \times S$).

Step 2 (Dataflow): In this step, we define the dataflow and examine the impact of this architectural constraint. For example, to configure the example loop nest into a weight-stationary dataflow, we would set $Q1 = 1$, $Q0 = Q$ and $S1 = S$, $S0 = 1$. This means that each PE stores one weight, that weight is reused $Q0$ times within that PE, and the number of PE equals the number of weights. This forces the absolute maximum amount of reuse for weights at the PE. The forced serialization of $Q0 = Q$ reduces the performance upper bound from $Q \times S$ to S, which is the maximum parallelism of the dataflow.

Step 3 (Number of PEs): In this step, we define a finite number of PEs, and look at the impact of this architectural constraint. For example, in the 1-D WS example, where $Q1 = 1$ and $Q0 = Q$, $S1$ is constrained to be less than or equal to the number of PEs, which dictates the theoretical peak performance. As hinted in Section 5.5, there are two scenarios when the actual performance is less than the peak performance. The first scenario is called spatial mapping fragmentation, in which case S, and therefore $S1$, is smaller than the number of PEs. In this case, some PEs are completely idle throughout the entire period of processing. The second scenario is called temporal mapping fragmentation, in which case S is larger than the number of PEs but not an integer multiple of it. For example, when the number of PEs is 4, $S = 7$, and $S1 = 4$, it takes two cycles to complete the processing, and none of the PEs are completely idle. However, one of the 4 PEs will only be 50% active. Therefore, it still does not achieve the theoretical peak performance. In general, however, if the workload does not map into all of the PEs in all cycles, then some PEs will *not* be used at 100%, which should be taken into account in performance evaluation.

Step 4 (Physical dimensions of the PE array): In this step, we consider the physical dimensions of the PE array (e.g., arranging 12 PEs as 3×4, 2×6, or 4×3, etc.). The spatial partitioning is constrained per dimension which can cause additional performance loss. To explain this step with the simple example, we need to relax the WS restriction. Let us assume $Q1$ is mapped to the width of the 2-D array and $S1$ is mapped to the height of the 2-D array. If $Q1$ is less than the width of the array or $S1$ is less than the height of the array (spatial mapping fragmentation), not all PEs will be utilized even if without the constraint that $Q1 \times S1$ is smaller or equal to the number of PE. A similar case can be constructed for the temporal mapping fragmentation as well. This architectural constraint further reduces the number of active PEs.

Step 5 (Storage Capacity): In this step, we consider the impact of making the buffer storage finite. For example, for the WS dataflow example, if the allocated storage for partial sums in the GLB is limited, it limits the number of weights that can be processed in parallel, which limits the number of PEs that can operate in parallel. Thus, an architectural constraint on how many partial sums can be stored in the GLB restricts $Q1$ and $S1$, which again can reduce the number of active PEs.

Step 6 (Data Bandwidth): In this step, we consider the impact of a finite bandwidth for delivering data across the different levels of the loop nest (i.e., memory hierarchy). The amount of data that needs to be transferred between each level of the loop nest and the bandwidth at which we can transmit the data dictate the speed at which the index of the loop can increment (i.e., number of cycles per MAC). For instance, the bandwidth of the RF in the PE dictates the increment speed of $s0$ and $q0$, the bandwidth of the NoC and GLB dictates the rate of change of $s1$ and $q1$, and the off-chip bandwidth dictates the rate of change of $s2$ and $q2$.

To quantify the impact on performance from insufficient bandwidth, we can adapt the well-known roofline model [119] for the analysis of DNN processors. The roofline model, as shown in Figure 6.10, is a tool that visualizes the performance of an architecture under various degrees of operational intensity. It assumes a processing core, e.g., PE array, that has insufficient local memory to fit the entire workload, and therefore its performance can be limited by insufficient bandwidth between the core and the memory, e.g., GLB. When the operational intensity is lower than that at the inflection point, the performance will be bandwidth-limited; otherwise, it is computation-limited. The roofline indicates the performance upper-bound, and the performance of actual workloads sit in the area under the roofline.

For this analysis, we adapt the roofline model as follows:

- We use three separate rooflines for the three data types instead of one with the aggregated bandwidth and operational intensity.[10] This helps to identify the performance bottleneck and is also a necessary setup since independent NoCs are used for each data type. However, the performance upper-bound will be the worst case of the three rooflines.

- The roofline is typically drawn with the peak performance of the core and the total bandwidth between the core and memory. However, since we have gone through the first 5 steps in Eyexam, it is possible to get a tighter bound (Figure 6.11). The leveled part of the roofline

---

[10]Ideally, we should draw a *roof-manifold* with the operational intensity of each data type on a separate axis; unfortunately, it will be a 4-D plot that cannot be visualized.

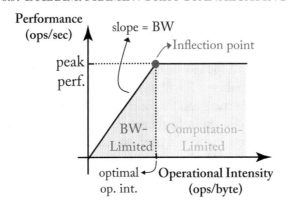

Figure 6.10: The roofline model.

is now at the performance bound from step 5; the slanted part of the roofline should only consider the bandwidth to the active PEs for each data type. Since performance is measured in MACs/cycle, the bandwidth should factor in the clock rate differences between processing and data delivery.

- For a workload layer, the operational intensity of a data type is the same as its amount of data reuse in the PE array, including both temporal reuse with the RF and the spatial reuse across PEs. It is measured in MACs per data value (MAC/data) to normalize the differences in bitwidth.

Step 7 (Varying Data Access Patterns): In this step, we consider the impact of bandwidth varying across time due to the dynamically changing data access patterns (Step 6 only addresses average bandwidth). For the WS example, during ramp up, the weight NoC will require high bandwidth to load the weights into the RF of the PEs, but in steady state, the bandwidth requirements of the weight NoC will be low since the weights are reused within the PE. The performance upper bound will be affected by ratio of time spent in ramp up versus steady state, and the ratio of the bandwidth demand versus available bandwidth. This step causes the performance point to fall off the roofline, as shown in Figure 6.11. There exist many common solutions to address this issue, including using double buffering or increased bus width for the NoC.

Table 6.2 summarizes the constraints applied at each step. While Eyexam is useful for examining the impact of each step on performance, it can also be used in the architecture design process to iterate through a design. For instance, if one selects a dataflow in Step 2 and discovers that the storage capacity in Step 5 is not a good match causing a large performance loss, one could return to Step 2 to make a different dataflow design choice and then go through the steps again. Another example is that double buffering could be used in Step 7 to hide the high bandwidth during ramp up, however, this would require returning to Step 5 to change the effective storage capacity constraints. Eyexam can also be applied to consider the trade-off between performance

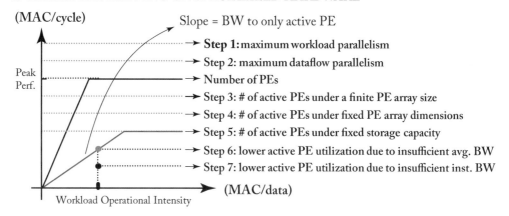

Figure 6.11: Impact of steps on the roofline model.

and energy efficiency in combination with the framework for evaluating energy efficiency as discussed in Section 6.4, as well as consider the impact of sparsity and workload imbalance on performance.

6.6   TOOLS FOR MAP SPACE EXPLORATION

There are a number of efforts that have developed tools to address the DNN accelerator mapping problem. Some well-known mappers are: the Eyeriss mapper [184], MAESTRO [190], R-Stream [191], Timeloop [192], and TVM [130]. Other notable tools that address a similar problem, but were not originally intended to target DNN accelerators are: Halide [193] and Tiramisu [194]. Halide introduced the separation of algorithm from schedule (or mapping) and Tiramisu, although targeted only at CPUs and GPUs, is particularly interesting because it optimizes sparse computations, which is discussed in Chapter 8.

Mappers and mapper-like tools can typically be characterized by how they approach the following issues: (1) The range of architectures they target. (2) The range of computations they can map (e.g., just DNN layers or more general computations). This is generally dictated by the framework they use to describe the map space (e.g., distinguish legal and illegal mappings). These frameworks also usually facilitate the analysis of the behavior of the computation on the specified architecture. (3) The heuristic they use to search the map space. (4) The technique they use to predict the performance and/or energy for a given mapping. Below, we will characterize the approaches used by some popular tools. Note, however, that many of these tools are still evolving rapidly, and therefore this characterization is just a snapshot as of the time this is being written.

There is a considerable range of architectures that existing mappers target. Some target only a specific DNN accelerator. An example is the Eyeriss mapper, which only targets Eye-

Table 6.2: Summary of steps in Eyexam

Step	Constraint	Type	New Performance Bound	Reason for Performance Loss
1	Layer size and shape	Workload	Max workload parallelism	Finite workload size
2	Dataflow loop nest	Architectural	Max dataflow parallelism	Restricted dataflow mapping space by defined by loop nest
3	Number of PEs	Architectural	Max PE parallelism	Additional restriction to mapping space due to shape fragmentation
4	Physical dimensions of PEs array	Architectural	Number of active PEs	Additional restriction to mapping space due to shape fragmentation for each dimension
5	Fixed storage capacity	Architectural	Number of active PEs	Additional restriction to mapping space due to storage of intermediate data (depends on dataflow)
6	Fixed data bandwidth	Microarchitectural	Max data bandwidth to active PEs	Insufficient average bandwidth to active PEs
7	Varying data access patterns	Microarchitectural	Actual measure performance	Insufficient instant bandwidth to active PEs

riss' row-stationary dataflow. In order to support a wider range of target designs others accept a *template* that describes a DNN accelerator architecture.[11] These templates include factors such as the dataflows supported, the number of PEs and levels of storage hierarchy and the storage sizes and network connectivity. The MAESTRO, Timeloop, and TVM mappers all accept such templates. Some tools, (e.g., TVM) extend to target general purpose processors by characterizing certain hardware functionality (e.g., vectors or Nvidia's tensor cores) as highly constrained templates.

The range of problem specifications accepted by these tools also varies. Some, like the Eyeriss mapper, assume the standard expression for CONV/FC layers (see Chapter 2). Others, accept expressions equivalent to the tensor index expressions described in Chapter 2. These in-

---

[11]Architecture templates will typically be expressed in a human-readable configuration language, such as YAML.

clude Tiramisu and Timeloop. TVM accepts problem specifications generated by a variety of standard machine learning libraries, such as Tensorflow [98] and PyTorch [99]. Finally, Halide operates as an embedded domain specific language (EDSL) in C++.

Another requirement of a mapper is having a way to represent the map space. Some use a well-known compiler framework called the polyhedral model [128]. This model allows one to represent sequences of loop nests that form what are called *static control parts* or SCoPs. Roughly, a SCoP is a sequence of loop nests where the loops can have dynamic loop bounds and where the statements in the body of the loops can have conditions that control whether the statement executes or not. However, the loop bounds and conditions must be linear functions of the loop index variables and constant parameters (i.e., they are not data dependent). Polyhedral models are often paired with an analysis framework that computes behavioral characteristics like reuse for specific architectures. Tools that use a polyhedral model include R-Stream and Tiramisu.

So far as we know, there is no polyhedral-based cost model for some of the more complex characteristics of DNN accelerators. These more complex characteristics include multi-level explicit decoupled data orchestration-style (EDDO) buffering (see Section 5.8) or complex NoC topologies, such as those with support for multi-level multicast or spatial reduction. Furthermore, existing frameworks that implement the polyhedral model are quite heavyweight and can be overkill for some situations. Therefore, some mappers use custom representations for the map space. These often support a more restricted set of loop nests (e.g., no conditionals or fixed loop bounds). Examples of such mappers are: the Eyeriss mapper, Halide, Timeloop, and TVM. These systems tend to determine data reuse using a custom analysis that tracks the movement of data tiles, which is described in Section 6.4.

Search heuristics vary greatly. Some current approaches include: genetic algorithm-based search (Eyeriss mapper), beam search (Halide, Tiramisu), parallel random search of legal mappings (Timeloop), and parallel simulated annealing (TVM).

There are two major approaches to projecting the performance and/or energy consumption for a mapping: deep-learning based and analytical.[12] To use a deep learning-based approach, a DNN model is trained to project operational metrics for a mapping on a given architecture. Halide, Tiramisu and TVM use this approach. Such models are relatively fast and accurate. However, deep learning-based models tend to only be useful for the architectures they were trained on and are not amenable to design space exploration because the training does not reflect the characteristics of designs that were not in the training set (i.e., those one would like to discover). Also, most works focus on projecting performance rather than energy.

To allow more flexibility in the range of architectures supported and to project energy (as well as performance) some tools use custom analytical models. Such tools include the Eyeriss mapper (which uses the model described in Section 6.4), MAESTRO, and Timeloop. Timeloop is paired with a discrete energy evaluation framework, Accelergy [187], which al-

---

[12]A cycle-level simulation approach would almost certainly be too slow for projecting metrics for a mapping during a search.

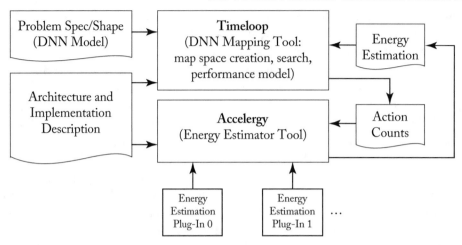

Figure 6.12: Timeloop [192] with integration of Accelergy [187] as energy estimation model. Timeloop sends projected action counts for a mapping to Accelergy and receives an energy estimation to guide its search. Accelergy plug-ins allow for customization of component energy estimation. These tools are available at http://accelergy.mit.edu/tutorial.html.

lows for template-based descriptions of the architecture and component energy costs, as shown in Figure 6.12. This later capability is especially useful for understanding the impact of new technologies like those discussed in Chapter 10.

In summary, there are a variety of mapper (or mapper-like) tools that target a variety of DNN accelerator architectures to find an optimal mapping. They typically use some variant of a loop nest (or sequence of loop nests) to describe the mappings in the map space and use analysis tools to project performance and/or energy to guide a search of the map space to find an optimal mapping.

PART III

Co-Design of DNN Hardware and Algorithms

CHAPTER 7

Reducing Precision

As highlighted in the previous chapters, data movement dominates energy consumption and can affect the throughput for memory-bound systems. One way to address this issue to reduce the number of bits (bit width) required to represent the weights and activations of the DNN model. Using fewer bits per weight and/or activation effectively reduces the number of unique values that they can take on and thus is often referred to as *reducing precision*. Benefits of reduced precision can include reduced data movement (i.e., reduce memory bandwidth), reduced storage cost (i.e., reduce chip area), reduced energy per memory access (due to smaller memories), and reduced energy and time per MAC operation. However, reducing precision can also affect accuracy and thus its impact on accuracy must be carefully evaluated.

Therefore, the overarching goal in reducing precision is to minimize number of bits while also maintaining accuracy and minimizing any hardware overhead. In this chapter, we will discuss various forms of quantization, which involves mapping a larger set of values to a smaller set of values. Thus, quantization can be applied to reduce precision. We will discuss various design considerations when deciding the number of unique values (and number of bits) to allow per weight and/or activation (which affects the accuracy), the relationship between these values (which affects how they are computed upon and stored—i.e., hardware overhead), and whether these properties (i.e., number of values and relationship between them) are allowed to vary across different parts of the DNN model (e.g., layers and/or filters), which can further reduce the number of bits at the cost of hardware overhead to support this flexibility.

7.1 BENEFITS OF REDUCE PRECISION

Reducing the number of bits per weight and/or activation (and consequently partial sum) has several benefits.

First, it reduces the amount of data movement resulting in lower energy consumption. Reducing data movement can also increase throughput since it reduces memory bandwidth requirements, which reduces the likelihood that processing elements (PEs) become idle while waiting for data.

Second, it reduces the amount of storage required for a given number of weights, activations, and/or partial sums. This can be exploited in two ways.

- It can reduce the amount of on-chip memory, which reduces the area cost of the chip; alternatively, for a fixed chip area, more area can be allocated to the PEs to allow for more

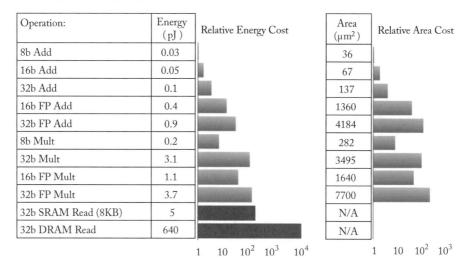

Operation:	Energy (pJ)	Relative Energy Cost	Area (μm²)	Relative Area Cost
8b Add	0.03		36	
16b Add	0.05		67	
32b Add	0.1		137	
16b FP Add	0.4		1360	
32b FP Add	0.9		4184	
8b Mult	0.2		282	
32b Mult	3.1		3495	
16b FP Mult	1.1		1640	
32b FP Mult	3.7		7700	
32b SRAM Read (8KB)	5		N/A	
32b DRAM Read	640		N/A	

Figure 7.1: The area and energy cost for additions and multiplications at different precision, and memory accesses in a 45 nm process. The area and energy scale different for multiplication and addition. The energy consumption of data movement (red) is significantly higher than arithmetic operations (blue). (Figure adapted from [121].)

parallel compute, which can help increase throughput. In addition, smaller memories tend to consume less energy.

- It can keep the same amount of on-chip memory, but now each memory can store more weights, activations and/or partial sums, which can potentially allow for more data reuse and thus reduce the amount of off-chip data movement (and on-chip data movement between the different levels of the memory hierarchy).

Finally, it reduces the cost of the MAC hardware. The cost of multiplication and accumulation scales differently with bit width. The area and energy cost of a multiplier scales quadratically $O(n^2)$ with bit width (n). The delay of the critical path of a multiply typically scales linearly $O(n)$ with bit width; therefore, reducing the number of bits per input operand (i.e., weights and activations) can also help increase throughput. The energy and area cost of an accumulator (adder) scales linearly $O(n)$ with bit width, while the delay of the critical path can scale either linearly $O(n)$ or square-root $O(\sqrt{n})$ depending on the implementation [195, 196]. Figure 7.1 shows the energy and area cost for multipliers and adders at different precisions. From this figure, we observe that multiplications are more expensive than additions, and reading from memory (data movement) is more expensive than both; therefore, reducing data movement is a very important benefit of reducing precision.

Note that the bit widths of integer multiplication and accumulation are different within a MAC than in storage, as shown in Figure 7.2. The bit width of the multiplier is dictated

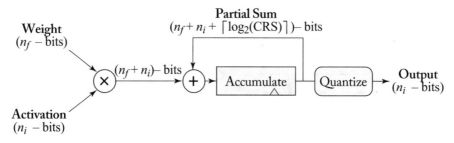

Figure 7.2: Various bit widths in a multiply and accumulate (MAC).

by the bit width of the filter weights (n_f) and input activations (n_i). To be correct to the last bit, the bit width of the output (product) of the multiplier must be $n_f + n_i$. These products are then accumulated across the size of the filters in the DNN model (i.e., $C \times R \times S$). To be correct to the last bit, the bit width of the accumulator that generates the partial sum must be $n_f + n_i + \log_2(C \times R \times S)$, where $C \times R \times S$ is the maximum number of weights in a filter for the DNN model.[1] After the partial sums are fully accumulated, they are typically reduced to the bit width of the activations (n_i) as the accumulated output will eventually become an input activation to the next layer. Reducing the bit width of the fully accumulated partial sum will not have a significant impact on accuracy if the distribution of the weights and activations are centered near zero with a limited standard deviation; batch normalization may help achieve this effect. Since the bit width of the final output activation will ultimately be less than the maximum possible partial sum, some implementations (e.g., processing-in-memory accelerators discussed in Chapter 10) will use reduced bit width computation (i.e., less than $n_f + n_i + \log_2(C \times R \times S)$) during the accumulation to reduce cost.

7.2   DETERMINING THE BIT WIDTH

The previous section highlighted the potential benefits of reducing precision and how the cost of a MAC operation scales with the bit width of the weights, activations, and partial sum. In this section, we will discuss the factors that affect how we determine the bit width for these various data types.

7.2.1   QUANTIZATION

The number of unique values determine the bit width of a given data type. The process of mapping data to from a large number of possible values (full precision) to a reduced set of values (reduced precision) is referred to as *quantization*.

Design decisions for quantization include (1) the number of values (often referred to as the number of *quantization levels*) that should be represented at reduced precision (which affects

[1]For the popular CNN models described in Chapter 2, $\log_2(C \times R \times S)$ is typically on the order of 10 to 16 bits.

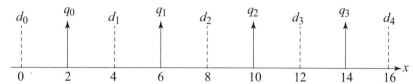

(a) Quantization function $Q(\cdot)$ defined by its quantized values q_i and its decision boundaries d_i

$$x = 1, 3, 7, 8, 9, 15, 6, 2 \longrightarrow \boxed{\begin{array}{c} \text{Quantization} \\ Q(\cdot) \end{array}} \longrightarrow \hat{x} = 2, 2, 6, 6, 10, 14, 6, 2$$

(b) Examples of values quantized using $Q(\cdot)$

Figure 7.3: Example of quantization (adapted from [137]): The input data x lies between 0 and 16. Using the quantization function $Q(\cdot)$ shown in (a), x is quantized to $\hat{x}$ by mapping it to one of $L = 4$ possible quantized values (q_i). In this example, $Q(\cdot)$ performs *uniform* quantization, which means that the quantized values are equally spaced out such that the four possible values that $\hat{x}$ can take on are {2, 6, 10, 14}. Decision boundaries d_i are used to decide the quantization value that x should be mapped to. The quantization error is computed as the mean squared error between x and $\hat{x}$, i.e., $E[(x - \hat{x})^2]$. For the example sequence in (b), the quantization error is 0.625.

the bit width); and (2) the actual values (often referred to as the *quantized values*) that should be represented at the quantization levels (which is affected by the distribution of the data being quantized), and the relationship between these values (which affects how computation is performed on the quantized values).

The typical design goal for quantization is to reduce the number of values while at the same time reduce the *quantization error*, which is a measure of the *average* difference between the original full precision and the quantized reduced precision representation. In the context of DNNs, the quantization error can be used to estimate the impact on accuracy, although the ultimate goal is to reduce the impact of quantization on the outputs (i.e., output activations via the partial sums), rather than the inputs (i.e., input activations or weights).

The quantization method and quantization error can be more precisely defined as follows: Let x denote the data at the original full precision. Quantization involves mapping data, x, to a smaller set of quantization levels, q_i, where $0 \leq i \leq L - 1$ and L is the number of levels; $\hat{x}$ denotes the quantized value of x as defined by $\hat{x} = Q(x)$, where $Q(\cdot)$ is a function that defines the quantization method as $Q(x) = q_i$, for $d_i < x \leq d_{i+1}$, where d_i are the decision boundaries in the function. Thus, $Q(\cdot)$ defines how x is mapped to q_i. An example of quantization is shown in Figure 7.3.

The quantization error is measured between the quantized data and the original data. This error depends on the distribution of the data, the quantized values (q_i), and the decision bound-

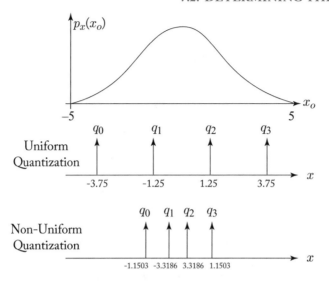

Figure 7.4: Quantization if distribution of x is not uniform. In this example, x has a Gaussian distribution $(p_x(x_o))$. Accordingly, uniform quantization with $L = 4$ levels results in a quantization error of 0.5477, whereas with non-uniform quantization with $L = 4$ levels results in a quantization error of 0.1467.

aries (d_i). More formally, given that $x_{\min} \leq x \leq x_{\max}$, and the probability density function for x is $p_x(x_o)$, the optimal q_i and d_i that minimize the quantization error can be determined by solving

$$\min_{r_i, d_i} \quad E\left[(x - \hat{x})^2\right] = \int_{x_o = x_{\min}}^{x_{\max}} (\hat{x} - x_o)^2 \cdot p_x(x_o) \cdot dx_o. \tag{7.1}$$

Therefore, the optimal quantization function $Q(\cdot)$ that minimizes the quantization error depends on the distribution of x. While *uniform quantization*, which has equal spacing between the quantized values (q_i), can be operated on directly by standard ALU hardware, it is only optimal if x has a uniform distribution (i.e., all possible values of x have equal probability). Designing a $Q(\cdot)$ that better matches the distribution of the original data x can either reduce the quantization error (see Figure 7.4), or maintain the same quantization error with fewer quantization levels, which reduces the bit width of $\hat{x}$.

As it turns out, the distributions of the weights and activations in DNN models are not uniform [197, 198]. Accordingly, a popular approach to reduce the number of quantization levels is to use *non-uniform* quantization, where the spacing between levels varies.

The relationship between the quantized values can be "computable," where the relationship between the index (i) to the quantization value (q_i) can be implemented using simple computation or logic. For instance, the quantized values can be assigned to be powers of two [199],

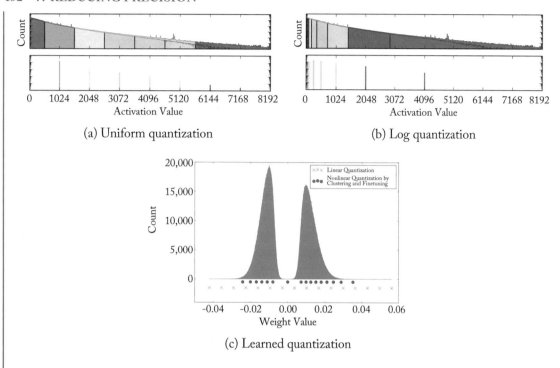

(a) Uniform quantization   (b) Log quantization

(c) Learned quantization

Figure 7.5: Various methods of quantization. (Figures adapted from [197, 199].)

as shown in Figure 7.5b, where the mapping between the index and the quantized values can be performed with a simple shift. For this mapping, an additional benefit is that any multiplication with the quantized value can be replaced with a bit-shift [198, 200]. This is often referred to as computing in the log domain or log quantization, and the resulting DNN model is often referred to as "multiplier-free."

Alternatively, the relationship between the quantized values can be "non-computable," where the relationship between the index (i) to the quantization value (q_i) is unconstrained and thus requires the use of a hash or look up table. For instance, the quantized values (q_i) can be learned from the data (Figure 7.5c), e.g., using k-means clustering to minimize the quantization error, and thus can take on any value. However, any multiplication with the quantization value becomes a three step process (see Figure 7.6): (1) determine the index of the quantized value; (2) perform a table look up [197] or compute a hash function [201] using the index to obtain the quantized value; and (3) perform the multiplication on the quantization value.

Compared to the uniform quantization or non-uniform quantization that use computable quantized values (e.g., log quantization), non-uniform quantization that uses non-computable quantized values (e.g., learned quantization) can achieve a lower quantization error for a given number of quantization levels, or require fewer quantization levels for a given quantization er-

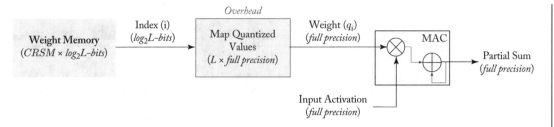

Figure 7.6: An example of performing a multiplication, where the weights are quantized using learned quantization.

ror. However, this comes at a cost of increased complexity when performing computations on the quantized values. Therefore, selecting a method of quantization often involves a trade-off between implementation complexity, the number of quantization levels, and the quantization error (which can impact the accuracy of the DNNs).

Regardless of the form of quantization, an additional benefit of reducing the number of unique values is that it effectively increases redundancy in the data, which can make approaches that exploit sparsity (as described in Chapter 8) more effective. In fact, quantization of the weights can also be thought of as a form of *weight sharing* as it reduces the number of unique weights. This can reduce the memory required to store the weights from $CRSM \times$ full precision to $CRSM \times \log_2 L$, as shown in Figure 7.6. In this example, if non-uniform quantization with non-computable quantized values is applied, the hardware overhead is an additional look up table of size $U \times$ full precision.

The previous example illustrates a cost of doing arithmetic operations on quantized values. These costs vary depending upon the type of quantization. Values that undergo uniform quantization can be directly computed upon using conventional arithmetic hardware. Values that undergo non-uniform quantization typically need additional hardware to convert back to the uniform domain for computation. For non-uniform quantization that uses non-computable quantized values, a look up as in Figure 7.6 is typically used for the conversion. However, for non-uniform quantization that uses computable quantized values, arithmetic operations can sometimes be performed directly in the quantized domain such as the log domain; other times, a computation is required to convert back to the uniform domain.

Another important characteristic of data distribution is the *range* of the data, which is the ratio of the largest and smallest non-zero value (magnitude) (i.e., $x_{\max}/x_{\min}$). A wide range may result in higher quantization error since the quantized values either need to be spread out to cover the wide range or overflow/underflow will occur (i.e., quantization will clip at the maximum and minimum values). While increasing the number of quantization levels can help reduce the error, it may be more advantageous to address a wide range by introducing the notion of a scale factor.

Table 7.1 shows an example of both uniform and scale factor quantization for $L = 16$ levels. Using a scale factor can be thought of as a form on non-uniform quantization, where the

Table 7.1: Example values for uniform and scale factor-based quantization with $L = 16$ quantization levels. For uniform, the quantized values are $q_i = 44{,}000 \times i/16$. For scale factor quantization, the quantized values uniform within each scale as computed with $q_i = (4^{i/2+1} - 4^{i/2}) \times (i\%2)/8 + 4^{i/2} - 1$.

i	q_i - Uniform	q_i - Scale Factor
0	0	0
1	2750	1.5
2	5500	3
3	8250	9
4	11000	15
5	13750	39
6	16500	63
7	19250	159
8	22000	255
9	24749	639
10	27499	1023
11	30249	2559
12	32999	4095
13	35749	10239
14	38499	16383
15	41249	40959

quantization is a function of the magnitude of the original full precision value. The benefit of using such quantization is illustrated in Figure 7.7, where the quantization error for multiplication of values is much lower for the scale factor quantization than uniform quantization.

7.2.2   STANDARD COMPONENTS OF THE BIT WIDTH

Once we have determined the quantized values that we want to represent at reduced precision, we need to map them to a numerical representation in the form of bits. This can be done using the standard format for numerical representations, where the bit width is determined based on several factors:

- **The range of the values that are represented.** As discussed in Section 7.2.1, adding a scale factor can be used to better represent data with a large range. As a result, representing values with a large range (e.g., 10^{-38} to 10^{38}) often require more bits to support the scale

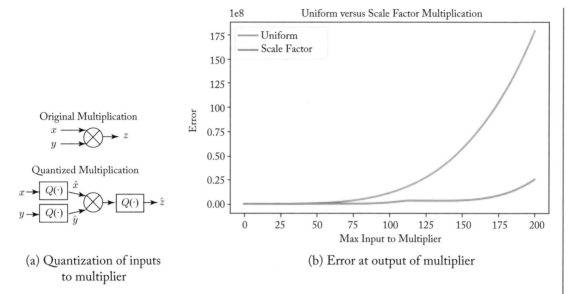

(a) Quantization of inputs
to multiplier

(b) Error at output of multiplier

Figure 7.7: Comparison of the quantization error of scale factor quantization versus uniform quantization. (a) Quantization applied to inputs and output of multiplier. (b) Average quantization error ($E[(z - \hat{z})^2]$) computed at output of multiplier across all multiplications from 0×0 to max input $\times$ max input.

factor as compared to values with a small range (e.g., 0 to 127). For instance, n_e-bits can be used to scale values by a factor of 2^{e-127}, where e is the value represented by the n_e-bits. Note that e can be either positive or negative, where a negative e allows representation of small values. In standard numerical representations that employ scaling, these bits are often referred to as the *exponent*.

- **The number of unique values that are represented for each scaling factor.** As we increase the number of unique values, we will need more bits to index the values. For instance, n_m-bits can be used to represent 2^{n_m} values.[2] In standard numerical representations (e.g., IEEE Standard for Floating-Point Arithmetic (IEEE 754)), these bits are referred to as the *mantissa*.

- **Whether the values are signed or unsigned.** Supporting signed values requires one extra bit ($n_s = 1$) compared to unsigned values. Note, that negative numbers are often represented as two's complement, where the mantissa also changes (but n_m does not change) when representing a negative number.

[2]Using the example from Section 7.2.1, L levels will require $\lceil \log_2 L \rceil$ bits.

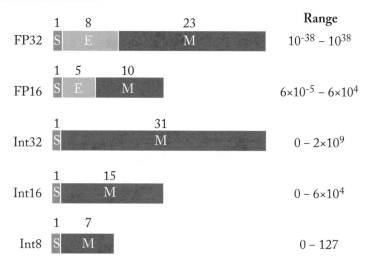

Figure 7.8: Common numerical representations. (Figure adapted from [202].)

Accordingly, the total number of bits required for common numerical representations is $n_s + n_e + n_m$. Figure 7.8 shows examples of common numerical representations and their ranges.

Standard Numerical Representations

To ensure compatibility between different computation systems, there are standards (e.g., IEEE Standard for Floating-Point Arithmetic (IEEE 754)) that specify the arithmetic formats of the numerical representations (including the number bits allocated for the mantissa (n_m), the exponent (n_e), and the sign (n_s)), and methods to perform arithmetic on these formats. These formats are widely used on general computing applications are supported on general-purpose compute processors such as CPUs and GPUs. As such, much of the earlier work in DNN processing used these formats, and they tend to serve as a common baseline for comparison in reduce precision DNN research. We will briefly describe some of these formats.

Fixed point format (also referred to as integer (int) format) has a fixed range and does not required any n_e-bits. Figure 7.9a shows how an 8-bit fixed point (int8) can be represented by $(-1)^s \times m$, where s is the sign bit, m is the $n_m = 7$-bit mantissa, and covers the range of 0 to 127.

Floating point (FP) format, refers to the case where range can be changed for each value, and thus includes n_e-bits.[3] Figure 7.9a shows how 32-bit floating point is represented by $(-1)^s \times m \times 2^{(e-127)}$, where s is the sign bit, e is the value of the $n_e = 8$-bit exponent, and m is the value of the $n_m = 23$-bit mantissa, and covers the range of 10^{-38} to 10^{38}.

---

[3]The name is derived from the fact that changing the scale can be thought of as moving the decimal point for each value.

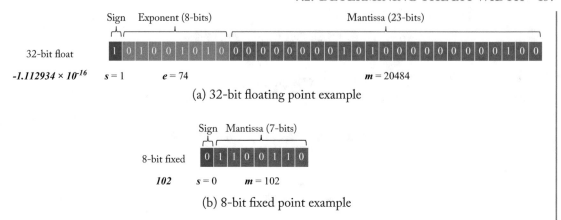

(a) 32-bit floating point example

(b) 8-bit fixed point example

Figure 7.9: Example of different numerical representations.

Figure 7.10 shows how additional hardware can be added on top of a fixed point multiplier to make it a floating point multiplier; this explains the increase in area and energy for floating point multiplication that was observed in Figure 7.1. Various works have explored reducing the overhead of the additional hardware to support floating point (e.g., adders for the exponent update, and adders, shifters, and control logic for the normalizer) by sharing the exponent (i.e., scale factor) across multiple variables; this is often referred to as *block floating point* [203] or *dynamic fixed point* [204] as it can be viewed as a format that lies between floating point and fixed point. For instance, Microsoft's Brainwave Project shares the exponent across a 128-element vector [168], while Intel Nervana Systems' Flexpoint shares the exponent across all elements within a tensor (i.e., all weights in a filter, or all activations in feature map) [205]. The average number of bits for block floating point representation can also be reduced relative to conventional floating point, since n_e-bits can be used across multiple variables.

The default precision used on CPUs and GPUs is 32-bit floating point (fp32) (also referred to as single precision), with $n_s = 1, n_e = 8, n_m = 23$. In general-purpose platforms such as CPUs and GPUs, the main benefit of reducing precision is an increase in throughput; for instance, for the same memory bandwidth and within the same clock cycle, four 8-bit operations can be performed instead of one 32-bit operation. Accordingly, several commercial products that target deep learning have added support for reduced precision. This includes Nvidia's Pascal [206] and Google's TPU [207] which announced support for 8-bit fixed point for inference in 2016. In the following years, several products including Nvidia's Volta, Google's TPUv2, Intel's NNP-L announced support for 16-bit floating point for training. Accordingly, using 8-bit fixed point for inference and 16-bit floating point for training has become common practice. Note however that the above refers to the multiplier precision; in most work, the accumulation remains at 32-bit precision.

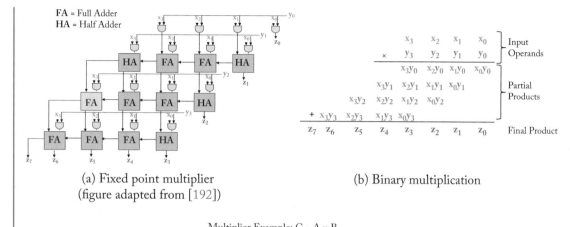

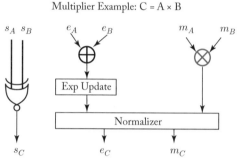

(a) Fixed point multiplier
(figure adapted from [192])

(b) Binary multiplication

(c) Components of a floating point multiplier

Figure 7.10: Fixed point versus Floating point multiplication. The fixed point multiplier in (a) performs the binary multiplication in (b). In the floating point multiplier, the mantissa (m_A, m_B) dictates the number of bits in the fixed point multiplier, contained within the floating point multiplier, highlighted in red. The exponent (e_A, e_B) dictates the number of shifters in the exponent update.

Custom Numerical Representations

In order to further reduce the number of bits required to represent a given value, there has been extensive research on *custom* numerical representations. In particular, many works have explored the trade-off between allocating bits to the manitissa (n_m) versus the exponent (n_e) to reduce the overall number of bits required.

For instance, 16-bit floating point (fp16) as defined by the IEEE floating point standard, also commonly referred to as half precision, uses $n_s = 1, n_e = 5, n_m = 10$, resulting in a range between $\sim 5.9e^{-8}$ to $\sim 6.5e^4$. In contrast, the recently proposed "brain floating-point format" (bfloat16) [98] is also 16-bits, but distributes the bits as $n_s = 1, n_e = 8, n_m = 7$, resulting in a range between $\sim 1e^{-38}$ to $\sim 3e^{38}$.[4] In other words, bfloat16 trades off fewer unique values for a larger range. Supporting the larger range is particularly useful for representing the gradients during training. In comparison, gradients can fall outside the range of fp16, and thus using the $n_e = 5$ in fp16 (rather than $n_e = 8$ in bfloat16 and fp32), would require loss scaling [98]. Training with bfloat16 can achieve the same state-of-the-art accuracy as 32-bit floating point (with the same number of iterations and without changing the hyperparameters) [208].

Another example of custom precision is "ms-fp8" and "ms-fp9" from Microsoft's Brainwave Project [209], which are 8- and 9-bit floating point formats with $n_m = 2$ and $n_m = 3$, respectively; both have $n_s = 1$ and $n_e = 5$. "ms-fp8" and "ms-fp9" provide a larger range than 8-bit integer at a comparable area cost due to their smaller mantissa. This is particularly important for Microsoft's Brainwave Project, which aims to pack as many weights and MACs as possible onto an FPGA, so that it can deliver low-latency high-throughput DNN inference.[5] In fact, the small mantissas result in narrow bitwidth multiplications that can efficiently map to look up tables and DSPs (e.g., 5 narrow bitwidth multiplies can be mapped to the same 18×19 multiplier in a DSP). It should be noted that custom numerical precision are well suited for FPGAs as the specialization can be applied during synthesis.

7.3   MIXED PRECISION: DIFFERENT PRECISION FOR DIFFERENT DATA TYPES

The optimal method of quantization depends on the distribution of the data. Therefore, one way to further reduce the required bit width (while maintaining accuracy) is to tailor the quantization method to the each of the different data types, which have different distributions. Using different precision on different data types is commonly referred to as *mixed precision*.

For inference, the data types that need to be considered include weights, activations, and partial sums. For training, the data types that need to be considered include weights, activations, partial sums, gradients, and weight updates. Training typically requires higher precision than inference since the gradients and weights updates have a larger range then the weights and activations. This is the reason why inference can be performed with 8-bit fixed point while training requires 16-bit floating point.[6]

Supporting mixed precision in custom hardware does not require any additional overhead if the precision for each data type is fixed. However, if the precision *varies* for different regions

---

[4]The conversion between IEEE 32-bit floating point and bfloat16 is simple, as bfloat16 is simply fp32 with the matissa truncated from 23 to 7. With $n_m = 7$, bfloat16 can also represent all 8-bit integers, which means int8 can be converted to bfloat16 without loss.

[5]The low latency requirement means that no batching is used, which reduces the weight reuse and thus increases the off-chip memory bandwidth for weight reads. To address this, the weights are pinned (i.e., "hard coded" during synthesis) onto the on-chip memory of the FPGAs to reduce off-chip bandwidth.

[6]Sometimes for reduced precision training, replicas of the weights and activations will be stored at full precision, and be used during back propagation, while in forward propagation the reduced precision version will be used [157, 205, 210, 211].

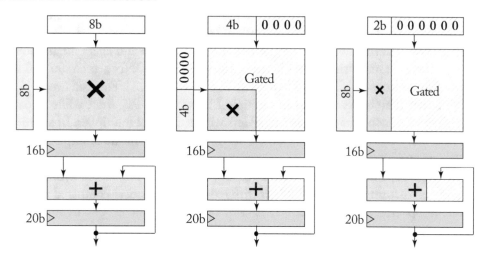

Figure 7.11: Example of a data-gated MAC. For simplicity, in this example only one input is precision scalable (weights). (Figure adapted from [212].)

of the DNN model (e.g., different layers or different weights), then additional hardware support may be required, which will be discussed in the next section.

7.4 VARYING PRECISION: CHANGE PRECISION FOR DIFFERENT PARTS OF THE DNN

Just as different data types have different distributions, their distributions can also vary across different parts of the DNN (e.g., layer, filter, channels). Therefore, to even more aggressively reduce the bit width, the quantization method can adapt to the varying distribution. Allowing the precision to vary across the different parts of the DNN model is commonly referred to as *varying precision*.

Although some systems simply build separate MAC units per precision, varying precision requires the use of a precision-scalable MAC in order to translate the reduced precision into improvement in energy-efficiency or throughput without a significant increase in area. A conventional approach is to use a data-gated MAC, where the unused logic (e.g., full adders) are gated, as shown in Figure 7.11. This reduces unnecessary switching activity, and consequently reduces energy consumption. The data-gated MAC can also be combined with voltage scaling to exploit the shortened critical path for additional energy savings [155].

While a data-gated MAC is a simple approach, it leaves many idle gates without increasing the throughput, making it inefficient in terms of throughput per area. Accordingly, there has been a lot of recent works that look at adding logic gates to increase the utilization of the full adders for higher throughput per area. One of the key challenges is to reduce the overhead of the additional logic gates, while at the same time efficiently mapping the multiplication workload to the adders and maximize their utilization. In most of these works, the focus of the scalability exploration is in the multiplier since it consumes more area and energy than the accumulator.

There are many similarities between the design of a DNN accelerator and the design of a precision-scalable MAC as highlighted in Camus et al. [212]. For instance, DNN accelerators with a spatial architecture contain multiple PEs within a PE array, while a spatial precision-scalable MAC contains multiple full adders within a spatial multiplier. In addition, the PEs in the PE array accumulate partial sums, while the full adders in the multiplier accumulate partial products (see Figure 7.10b).

In the spatial precision-scalable MACs, the array of full adders within the multiplier are regrouped to form *multiple* multipliers with reduced precision, as compared to one reduce precision multiplier in the data-gated MAC case. For instance, an 8b × 8b multiplier can potentially be converted into two 4b × 8b, four 4b × 4b, four 2b × 8b, eight 2b × 4b, or sixteen 2b × 2b multipliers. Regrouping means that the partial products of the full adders belonging to the same multiplier are accumulated together. The accumulation of the partial products can occur *temporally* within a full adder, as shown in Figure 7.12a (e.g., [180, 213]). The temporal accumulation of partial product approach can exploit data reuse at the inputs (low input bandwidth), but generates many partially accumulated values in parallel (e.g., four 14-bit accumulator registers in rightmost sub-figure of Figure 7.12a, and high register update bandwidth). Alternatively, the partial products can be accumulated *spatially* across different full adders, as shown in Figure 7.12b (e.g., [214, 215]). The spatial accumulation of partial product approach requires unique inputs (high input bandwidth), but generates only one partially accumulated value at a time (e.g., only one 14-bit accumulator register in rightmost sub-figure of Figure 7.12b, and low register update bandwidth).

For spatial precision-scalable MACs, the hardware overhead is due to (1) the additional adders and shift logic required to support parallel accumulations of partial products into separate final products and (2) the configuration logic around the adders to support accumulation across different groups of full adders. Additional output registers are also required for the "output stationary" approach.

There are also similarities between DNN accelerators with a temporal architecture and temporal precision-scalable MACs. DNN accelerators with a temporal architecture generate and temporally accumulate multiple partial sums with a single PE, while temporal precision-scalable MACs generate and temporally accumulate multiple partial products for a multiplier are with a single full adder and shifter, as shown in Figure 7.13. Accordingly, the number of cycles required to complete the multiplication is proportional to the number of bits; in other word, the multiplier scales temporally with precision; this is often referred to as a bit-serial processing since the multiplication is computed bit-by-bit. To combat the fact that a bit-serial multiplier takes multiple cycles (rather than a single cycle in conventional multiplier), the bit-serial multiplier can be clocked at a higher clock frequency, since a single full adder and shifter has a shorter

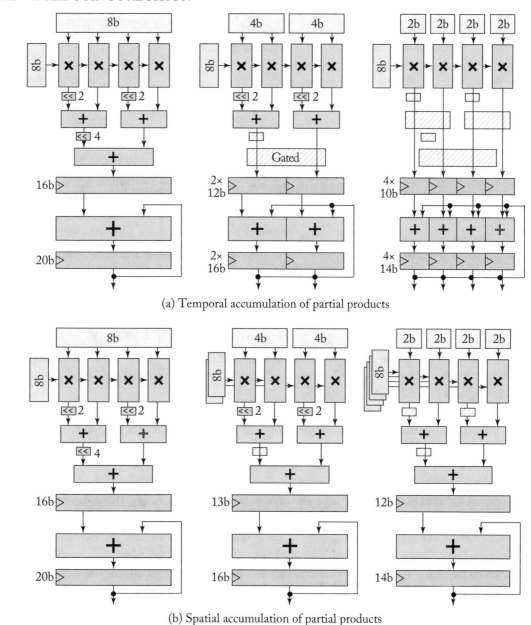

(a) Temporal accumulation of partial products

(b) Spatial accumulation of partial products

Figure 7.12: Examples of spatial precision-scalable MACs. For simplicity, in this example only one input is precision scalable (weights). Note that the designs with spatial accumulation take more inputs but keep all units busy and do not partition the final adder. (Figures adapted from [212].)

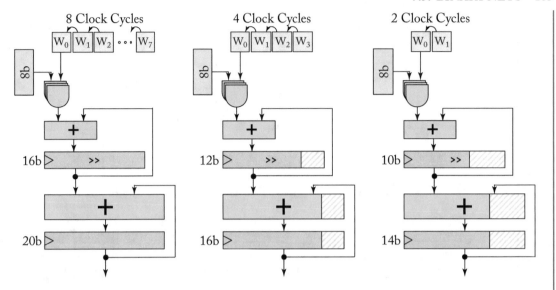

Figure 7.13: Example of temporal precision-scalable MACs. For simplicity, in this example only one input is precision scalable (weights). (Figure adapted from [212].)

critical path compared to a spatial multiplier.[7] Furthermore, since bit-serial multipliers have a smaller area than spatial multipliers, and several of them can be used in parallel to increase throughput [96, 216, 217].

Bit-serial multipliers also provides an opportunity for exploiting sparsity at the bit level.[8] Only the non-zero bits in the operand(s) require a computational cycle; the rest can be skipped [218]. At the extreme, the operand only has one non-zero bit, which makes it a multiplication by a power of two, and can be implemented with just a shift (i.e., "multiplier-free") as previously described in the context of log domain quantization.

Camus et al. [212] evaluates the energy, area, bandwidth, and throughput of many variations of precision-scalable MAC architectures including whether one or both inputs are scalable (i.e., 1D or 2D scalable) and multi-bit serial designs. It shows that the hardware overhead to support precision-scalable MACs beyond the simple data-gated MAC can be quite substantial and significantly reduces the benefits; this is a clear example of a trade-off between flexibility and efficiency. The overhead can be reduced by amortizing it across multiple MACs [219].

7.5 BINARY NETS

At the most extreme, there have been works looking at reducing the number of bits for weights and/or activations down to a single bit (i.e., limit each input operand to only two unique values);

---

[7]However, this may also increases the clock tree power.
[8]In Chapter 8, we will discuss exploiting sparsity at the operand (value) level (i.e., in the weights and activation).

these types of DNN are often referred to as *binary nets*. In addition to reduce the memory requirements, binary nets also significantly reduce the cost of the computation. If one of the operands is binary, then the multiplication turns into an addition. If both of the operands are binary, the multiplication operation can be turned into an XNOR, and multiple multiplications can be performed with a bit-wise XNOR. Following the bit-wise operation, the accumulation is performed with a popcount.[9] Binary nets are widely used in conjunction with processing in memory accelerators, discussed in Chapter 10, where the compute is performed using the storage element, which typically stores a binary weight.

One of the main drawbacks of binary nets is its impact on accuracy. Directly using binary weights and activation results in a 29.8 percentage point degradation in accuracy when applied to AlexNet for image classification on the ImageNet dataset [220, 221]; with such a large drop in accuracy, one might ask if traditional handcrafted approaches might offer a better accuracy versus energy trade-off [48].

Accordingly, binary nets are often combined with many of the previously discussed techniques to recover the accuracy. For instance, the first and last layers in the DNN are often kept at full precision (32-bit float), which is a form of varying precision that requires flexible hardware. Another example is XNOR-Net [221], which uses a different scale factor for the weights at each layer, and a different scale factor for the activations at each spatial location (i.e., $H \times W$) in the feature map; this form of non-uniform quantization reduces the degradation in accuracy to 11 percentage point for AlexNet on ImageNet.

While binary nets limit the weights and/or activations to two values, there may be benefits to allowing for a third value, specifically, the value of zero. Although this requires an additional bit per operand, the sparsity of the operand can be exploited to reduce computation and storage cost as discussed in Chapter 8, which can potentially cancel out the cost of the additional bit. DNNs that allow two unique values plus zero are often referred to as *ternary nets* [222].

Recent works have also explored mixed precision approaches, where the weights and activations use different forms of quantization, to achieve higher accuracy. For instance, QNN [210], DoReFA-Net [211], and HWGQ [223] use 1-bit with a scale factor for weights, and 2-bits for activations, with various forms of quantization.

Using binary/ternary weights offer additional opportunities for optimizations. For instance, there tends to be more redundancy between filters since the number of unique weights is reduced to two (binary) or three (ternary). This can be exploited by reducing memory access or by transforming the filters to further increase sparsity as explored in Yin et al. [224]; more discussion on exploiting weight sparsity can be found in Section 8.1.2.

Hardware implementations for binary/ternary nets have been explored in recent publications. YodaNN [225] and Yin et al. [224] uses binary weights, while BRein [167] uses binary weights and activations. Binary weights are widely used in the processing-in-memory accelera-

---

[9]Popcount (short of population count) is method to count the number of non-zero binary values in a vector. It is also often referred to as binary accumulation. Many CPU architectures today have a dedicated instruction for popcount.

tors discussed in Section 10.2. Hardware implementation of these low precision networks have also been explored in FPGAs, where if binarizing the network sufficiently reduces the number of weights, then all the weights in the DNN can be "hard coded" via synthesis and stored on a single FPGA [226]. Finally, the nominally spike-inspired TrueNorth chip can implement a reduced precision neural network with binary activations and ternary weights using TrueNorth's quantized weight table [14]. Most of the hardware implementations for binary/ternary nets are demonstrated on small DNN models such as LeNet, with a few on larger DNN models such as AlexNet [224, 225].

7.6   INTERPLAY BETWEEN PRECISION AND OTHER DESIGN CHOICES

Up to this point, we have discussed many design choices related to reduced precision both from the algorithm and hardware perspective. Exploring reduced precision does not have to happen in isolation. In fact, it can be combined with other design decisions both at the algorithm and hardware levels for an even larger design space. This can often lead to improved trade-offs between efficiency and accuracy.

Reduced precision can be explored in conjunction with the shape of the DNN model. For instance, the accuracy loss due to reduced precision can potentially be recovered by increasing the number of filters (i.e., widening the network) as explored in Wide Reduced Precision Network (WPRN) [227].

Reduced precision can also be accounted for in the design of the hardware dataflows discussed in Chapter 5. For instance, if weights are only 1-bit, but the activations use a higher 16-bit, an input-stationary dataflow may be more favorable than a weight-stationary dataflow as demonstrated in UNPU [96].

7.7   SUMMARY OF DESIGN CONSIDERATIONS FOR REDUCING PRECISION

First and foremost, it is important to consider and carefully evaluate the impact of reduce precision techniques on accuracy. This must include factors such as the difficulty of the dataset, task, and DNN model, as previously discussed in Section 3.1. For instance, while it might be possible to reduce precision for an easy task (e.g., digit classification) without impacting accuracy, applying the same approach to a more difficult task may result in a significant drop in accuracy.

It is also critical to ensure that the hardware cost to support reduced precision does not exceed the benefits. This is particularly true when considering variable precision, as it requires addition hardware, as discussed in Section 7.4. Another important factor to consider is the granularity of the variable precision as more overhead is required to support fine-grained variability (e.g., 2b, 4b, 8b) than coarser-grained variability (e.g., 4b, 8b); however, having finer granularity can provide support for DNNs that require fewer bits.

Finally, when evaluating reduced precision approaches, it is important to compare to the correct baseline. As mentioned in Section 7.2.2, 8-bit fixed point precision for inference and 16-bit floating point precision for training has already become common practice and hardware support for these precisions are readily available in commercial products; thus, using 32-bit floating point is considered a weak baseline.

CHAPTER 8

Exploiting Sparsity

A salient characteristic of the data used in DNN computations is that it is (or can be made to be) sparse. By saying that the data is sparse, we are referring to the fact that there are many repeated values in the data. Much of the time the repeated value is zero, which is what we will assume unless explicitly noted. Thus, we will talk about the sparsity or density of the data as the percentage of zeros or non-zeros, respectively in the data. The existence of sparse data leads broadly to two potential architectural benefits: (1) sparsity can reduce the footprint of the data, which provides an opportunity to reduce storage requirements and data movement. This is because sparse data is amenable to being compressed, as described in Section 8.2;[1] and (2) sparsity presents an opportunity for a reduction in MAC operations. The reduction in MAC operations results from the fact that $0 \times$ *anything* is 0. This can result in either savings in energy or time or both. In Section 8.3, we will discuss how the dataflows for sparse data can translate sparsity into improvements in energy-efficiency and throughput. However, first in Section 8.1 we discuss the origins and ways that one can increase sparsity in the data used in DNN computations.

8.1 SOURCES OF SPARSITY

Efficient processing of feature map activations becomes increasingly important as the size of the input to the DNN model grows (e.g., increased image resolution), while efficient processing of filter weights becomes increasingly important as the size of the DNN model grows (e.g., increased number of layers).

This section will discuss various approaches that can exploit properties such as redundancy and correlation in the feature maps and filters to increase their activation sparsity (Section 8.1.1) and weight sparsity (Section 8.1.2), respectively. The requirements for these approaches may differ as activation sparsity is often data dependent and not known a priori, while weight sparsity can be known a priori. As a result, methods to increase sparsity for weights can be performed off-line (as opposed to during inference) and can be more computationally complex than methods applied to increase activation sparsity. For instance, increasing weight sparsity can be incorporated into training.

---

[1]Note: We use the words sparsity or density to refer to a statistical property of the data, while we use the words compressed or uncompressed to describe the characteristics of a representation of the (typically sparse) data.

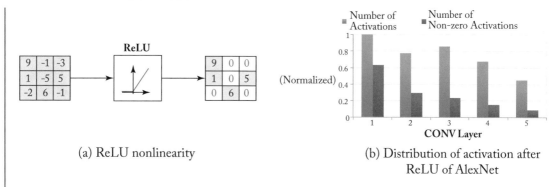

(a) ReLU nonlinearity

(b) Distribution of activation after
ReLU of AlexNet

Figure 8.1: Sparsity in activations due to ReLU.

8.1.1   ACTIVATION SPARSITY

Sparsity in the activations can come from several sources. The most obvious and commonly exploited source is from the use of ReLU as a nonlinearity. Other sources include exploiting correlation in the input data or the upsampling of the feature maps in a up-convolution layer. In this section, we will discuss how design choices of the DNN model or additional processing of the input data or feature maps can reveal additional sparsity.

Sparsity Due to ReLU

As discussed in Section 2.3.3, ReLU is a popular form of nonlinearity used in DNNs that sets all negative values to zero, as shown in Figure 8.1a. As a result, the output activations of the feature maps after the ReLU have fewer non-zero values, i.e., are *sparse*; for instance, the feature maps in AlexNet have a percentage of non-zero values, i.e., *density*, between 19 to 63%, as shown in Figure 8.1b. This activation sparsity gives ReLU advantages over other nonlinearities such as sigmoid, etc.

The activations can be made even more sparse by setting values below a certain threshold to zero; this is often referred to as *pruning* and can also be applied to weights, as discussed in Section 8.1.2. Activation pruning can be implemented by increasing the threshold in the ReLU or reducing the bias in the filter. Such pruning of small-valued activations can be translated into an additional 11% speed up [228] for image classification on ImageNet with little impact on accuracy. Aggressively pruning more activations (i.e., increasing the threshold) can provide additional throughput improvement at the cost of reduced accuracy, as shown in Figure 8.2.

The fact that ReLU effectively discards negative output activations opens up the possibility of using approximate computing to reduce the number of MAC operations. Specifically, we can terminate the computation of the output activation value if we can predict early on that the output will be negative. The main challenge in this approach is how early and how accurately can we predict that the output will be negative. Early prediction will allows us to terminate earlier

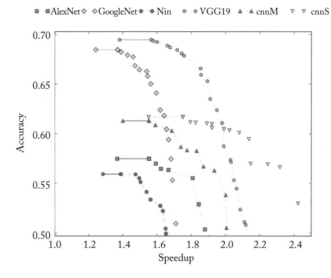

Figure 8.2: As the threshold for pruning activations increases, the accuracy drops as amount of sparsity, and consequently speed, increases. (Figure adapted from [228].)

and avoid more operations, while accurate prediction helps reduce the impact on accuracy (i.e., an incorrect prediction may result in a drop in accuracy). At the same time, it is desirable to use simple logic to perform the prediction, as it adds overhead in terms of energy and area.

Various works have explored different methods to predict that value of the output activation with the goal of giving an early and accurate prediction with minimum additional compute overhead. For instance, PredictiveNet [229] and Song et al. [230] propose computing the most significant bits (MSBs) of the partial sum to predict whether the output will be negative and only compute the remaining bits (i.e., least significant bits (LSBs)) if the output is positive; this can translate into improvements in energy-efficiency and/or throughput. This can be implemented with a precision-scalable multiplier (e.g., bit-serial), or an n_{MSB}-bit multiplier that computes the MSBs and additional hardware that is conditionally invoked to compute the remaining bits. SnaPEA [231] proposes reordering the weights based on their sign and then terminating the accumulation when the partial sum drops below a threshold, as shown in Figure 8.3. This is feasible since the order of accumulation does not affect the final result.[2] It also relies on the fact that the input to the filter will be zero or positive; this is true if the input activations were processed with a ReLU in the previous layer, and the input data to the first layer is zero or positive (e.g., pixels in an image). Finally, the cost of sorting and reordering of the weights can be amortized across multiple inputs and can happen offline if the weights are known in advance.

---

[2]This is not exactly true for floating point, but should be sufficiently accurate.

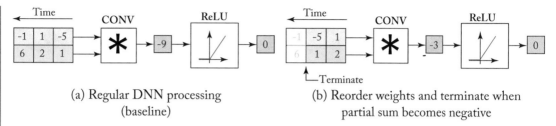

(a) Regular DNN processing
(baseline)

(b) Reorder weights and terminate when
partial sum becomes negative

Figure 8.3: Example of how we can reduce number of MAC operations if we can predict that the output will be zero. (Figure adapted from [231].)

Correlation in Input Data

Depending on the application, there may be correlation in the input data that can translate into the correlation in the activation values of the feature maps, both of which can be exploited. In terms of image and video data, this can manifest itself in terms of spatial or temporal correlation, where the values of neighboring pixels within the image, or pixels in consecutive frames, tend to be similar, respectively. In fact, it is this type of correlation that is exploited by popular and widely used image and video compression standards such as JPEG [232], H.264/AVC [233], or HEVC [234]. Correlation can also be found in other types of data, such as speech; however, in this section we will use image and video as driving examples.

One approach to exploiting this correlation is to process the *difference* between the correlated pixels (activations) rather than each pixel (activation) independently. If the pixels (activations) are well correlated, then it is likely that difference between them will be *zero* or close to zero; thus we can compute a sparse difference map between correlated pixels (activations), where the degree of sparsity depends on the degree of correlation between the pixels (activations).

We can examine this approach using the following toy example. Assume we would like to multiply two pixels (activations), a_1 and a_2, with the weight w. We could compute these products separately as follows:

$$y_1 = a_1 \times w$$
$$y_2 = a_2 \times w. \tag{8.1}$$

This would require two multiplications. Alternatively, we could compute the difference between a_1 and a_2, and compute

$$y_1 = a_1 \times w$$
$$y_2 = a_1 \times w + (a_2 - a_1) \times w = y_1 + \Delta_a \times w. \tag{8.2}$$

We can reuse y_1 in the computation of y_2. If a_1 and a_2 are the same, then Δ_a is zero, and the above calculation requires one multiplication rather than two.

The main challenge in this type of approach is the computational overhead for generating the difference map (i.e., Δ_a). On the one hand, it would be desirable to minimize the effort

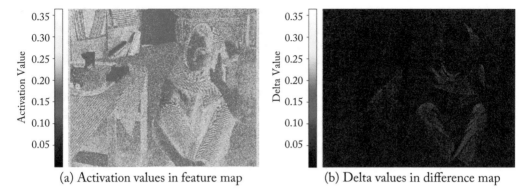

(a) Activation values in feature map   (b) Delta values in difference map

Figure 8.4: Delta values (i.e., Δ_a) in difference map (left) generated by exploiting correlation between immediate horizontal spatial neighbors in feature map (right). Notice that delta values tend to be either zero or close to zero as compared with activations in feature map. (Figure from [235].)

to compute the difference map; on the other hand, it would be desirable to find the pixels (activations) that are the *most* correlated, and thus maximize the sparsity of the difference map. Furthermore, additional storage is required to store the previous product (i.e., y_1 in the toy example). For temporal correlation, this can be costly as it may require storing the entire previous frame and/or all the intermediate feature maps of the previous frame.

A simple approach for computing the difference map is to compute the difference between immediate neighboring pixels. For instance, Diffy [235] generates a difference map based on the difference between immediate horizontal spatial neighbors within an image, as shown in Figure 8.4. Another example is Riera et al. [236], which generates the difference map between pixels (activations) at the same spatial position in consecutive frames (feature maps) to exploit temporal correlation, as shown in Figure 8.6b; these pixels can be thought of as immediate temporal neighbors. To increase the correlation between consecutive frames (and their feature maps), additional quantization can be applied to reduce the number of unique values, as discussed in Chapter 7; however, this may come at the cost of reduced accuracy.

When considering temporal correlation, particularly in video, it can be beneficial to account for the moving objects within the video (i.e., change in the location of objects across frames). This can be represented in the form of assigning motion vectors to different objects or pixels in the video, which indicates the correlation between pixels across frames, as shown in Figure 8.5. While using motion vectors can help identify highly correlated pixels, and thus enable increased sparsity in the difference map, the amount of computation needed to generate these motion vectors can be very expensive. One solution is to build specialize hardware to perform the motion estimation necessary to generate the motion vectors as proposed in EVA^2 [237].

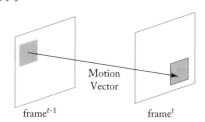

(a) Motion vector shows movement between frames

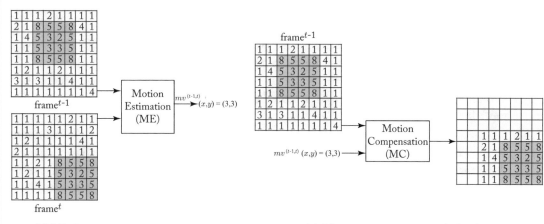

(b) Motion estimation compute
the motion vector

(c) Motion compensation shifts the input
based on motion vector

Figure 8.5: (a) The motion vector indicates how pixels move between frame $t-1$ and t. It can also be used to indicate which pixels (or activations) are temporally correlated between consecutive frames. (b) Motion estimation computes the motion vector $mv^{(t-1,t)}$ based on shift of pixels (or activations) within frame. (c) Motion compensation warps (shifts) the pixels in frame $t-1$ based on motion vector $mv^{(t-1,t)}$ to align them with frame t. The different map can be computed between the motion compensated frame $t-1$ and original frame t.

Motion estimation is a popular form of computation used in video compression and image processing, and thus the motion vectors might be freely available if one considers the interaction of the DNN processing with other parts of the system. For instance, if the DNN accelerators is part of a larger System-on-Chip (SoC), it may be possible to obtain the motion vectors from other blocks in the system such as the Image Signal Processor (ISP) as proposed in Euphrates [238]. Another option might be to consider the format of the incoming data. If the incoming data is in compressed form (e.g., compressed video using H.264/AVC or HEVC), which is usually the case for streaming or stored video, the motion vectors are already embedded in the syntax elements of the compressed video itself and can be directly accessed at low cost as proposed in FAST [239].

There can also be advantages of processing the difference map even if the delta values (i.e., Δ_a) are not zero, but close to zero. For instance, for the values close to zero, most of the higher-order bits will be zero and bit-serial processing, as discussed in Chapter 7 can be used to reduce the cost of the MAC operations [235].

Finally, the approaches described in this section can be selectively applied to the input data and the feature maps, depending on the amount of correlation and thus sparsity, as well as the computation and storage costs. For instance, when processing images, exploiting spatial correlation can be applied to the input data and all the intermediate feature maps, as the amount of additional storage is limited to only the previous pixel or activation [235].

On the other hand, exploiting temporal correlation is more expensive in terms of storage cost. As previously discussed, the entire previous frame or feature map may need to be stored, and the number of feature maps to be stored can increase with the number of layers in the DNN model.[3]

Accordingly, there tends to be more variation in temporal correlation based approaches, as shown in Figure 8.6. For instance, EVA^2 [237] uses temporal correlation to generate the difference map for the feature map of an intermediate DNN layer; the feature maps of the subsequent DNN layers are directly processed (not their difference maps), as shown in Figure 8.6c. Alternatively, in FAST [239], temporal correlation is selectively applied to different regions of the frame; highly correlated regions exploit temporal correlation in the *final* output feature map (i.e., the DNN outputs are copied from the previous frame and no DNN layer processing needs to be performed) as shown in Figure 8.6d, while the other regions need to undergo full DNN processing.

Up-Convolution Layers

Another source of sparsity comes from the use of *up-convolution layers*, as previously discussed in Chapter 2 [60].[4] Up-convolution layers and its variants are often used in DNN models that generate dense output predictions (e.g., assign a label to each pixel in an image) for applications such as semantic segmentation [92, 240, 241], depth estimation [64, 242], super-resolution [243–245], and style-transfer [246]. Specifically, they are typically used in the applications that generate high-resolution outputs, such as generative adversarial networks (GANs) and image segmentation, which differ from applications that generate a single value output, such as image classification.

To generate the dense output, the up-convolution layers and its variants increase the size of the feature map by producing an output feature map whose spatial dimensions (i.e., P and Q in

---

[3]The storage cost for exploiting spatial correlation tends to be less than temporal, thus designs such as [235] apply this approach to all intermediate feature maps.

[4]Note variants of the up-convolution layer with different types of upsampling include deconvolution layer, sub-pixel or fractional convolutional layer, transposed convolutional layer, and backward convolution layer [69].

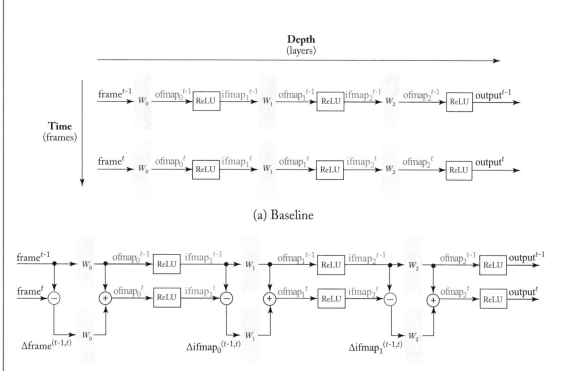

(a) Baseline

(b) Exploit correlation of immediate temporal neighbors for all feature maps

Figure 8.6: Various ways to exploit temporal correlation. (a) Baseline where all feature maps for sequential frames are processed separately, and no temporal correlation is exploited. (b) Exploit correlation of immediate temporal neighbors for all feature maps [236]. The difference map (Δ) is computed by directly subtracting the frame or feature map of time $t - 1$ from t. After processing the difference map (Δ) by the filters (W_i), the frame or feature map of time $t - 1$ is added back, which follows Equation (8.2). (*Continues.*)

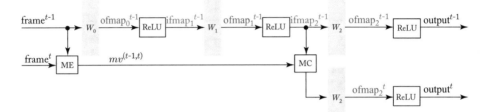

(c) Exploit temporal correlation to skip early layers

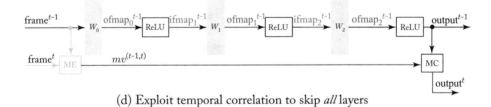

(d) Exploit temporal correlation to skip *all* layers

Figure 8.6: (*Continued.*) Various ways to exploit temporal correlation. (c) Motion estimation (ME) is used obtain the motion vector ($mv^{(t-1,t)}$) between the frames at time $t-1$ and t. Processing of the earlier layers is skipped for frame t. The final output (*output^t*) is generated by only applying the later layers (in this example, layer 2) to the motion compensated (MC) intermediate feature map (in this example, the input feature map of layer 2, *ifmap$_2^{t-1}$*). The trade off between how many layers are skipped and accuracy is explored in [237]. (d) Processing of *all* layers is skipped for frame t. The final output (*output^t*) is generated by only applying motion compensation (MC) to the final output of frame $t-1$ (*output^{t-1}*). In [238, 239], the motion estimation (ME) can also be skipped, since the motion vector ($mv^{(t-1,t)}$) can be obtained directly from the data or another part of the system.

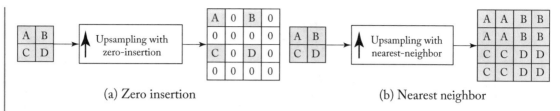

(a) Zero insertion   (b) Nearest neighbor

Figure 8.7: Example upsampling methods in decoder layer. (Figures adapted from [64].)

the DNN model dimensions) are larger than its input feature map (i.e., H and W in the DNN model dimensions). This is in contrast to approaches such as strides and pooling that reduce the size of the feature map. Increasing the size of the feature map can be done by upsampling the input feature map *before* processing it with the filter weights, as discussed in Section 2.3.4. Common forms of upsampling used in up-convolution layers and its variants include inserting zeros between the activations, as shown in Figure 8.7a, interpolation using nearest neighbors as show in Figure 8.7b, and interpolation with bilinear or bicubic filtering.

Upsampling introduces sparsity and correlation into the input feature map that is structured and known a priori. For zero-insertion, the sparsity is coming from the zeros being inserted between rows and columns of input values, leading to around 75% sparsity. For nearest-neighbors upsampling, correlation is coming from a pixel value being copied into adjacent pixel locations, resulting in windows of pixels that are known to have identical values. Accordingly, the cost of detecting sparsity or computing the difference map can be significantly less than for the sources of sparsity described in the previous sections, which tend to be unstructured or data-dependent. In fact, in some cases the input feature map and/or filter weights can be restructured such that input feature map can be processed in a dense form and avoid the overhead of sparse processing all together. For instance, the filter can be decomposed into a set of smaller filters such that the input feature map is processed in a dense form, and the outputs of those smaller filters can be interleaved to form the larger output feature map, as shown in Figure 8.8 [242]; this makes the processing of the decoder layer similar to the convolutional layer [69].

8.1.2 WEIGHT SPARSITY

Sparsity in the weights can come from several sources. Repeated weights in a filter naturally occur if the number of weights in the filter exceed the number of unique weights, which is typically set by the number of bits per weight (e.g., a $3 \times 3 \times 128$ filter will have repeated weights if each weight is 8-bits, since 8-bits can only represent 256 unique values). Repeated weights can also be enforced in the DNN model through the design of the filter or during training by setting weights to zero (referred to as pruning) or by reducing the number of unique weights

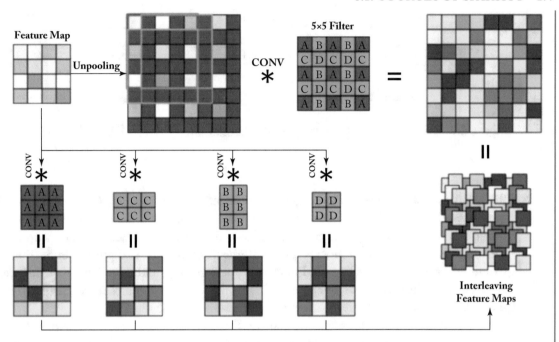

Figure 8.8: Fast upsampling by exploiting structure in activation feature map. (Figure from [242].)

(e.g., with reduced precision, as discussed in Chapter 7).[5] In this section, we will discuss how various design choices for the DNN model can be used to reveal additional weight sparsity including weight repetition for both zero and non-zero values. As previously discussed, unlike activation sparsity, which in most cases is data dependent and needs to be computed online (e.g., generating difference map during inference as discussed in Section 8.1.1), the approaches used to increase weight sparsity can be done offline, and thus can be more computationally intensive (e.g., can be incorporated into training).

Weight Reordering and Reuse
Repeated weights can appear in a filter for a variety of reasons as previously discussed. Reordering the weights, and thus the activations and operations, can help reduce the number of weight accesses from memory as well as the number of multiplications. For instance, the weight can be multiplied with the *sum of the activations* rather than each activation as demonstrated in UCNN [247].

---

[5]Enforcing repeated weights can also be thought of as a form of weight sharing, since the same weight is shared across many locations in the filter.

We can illustrate this approach using a toy example. If the filter weights are $[a, b, a, c]$ and the input activations are $[w, x, y, z]$, rather than computing the dot product as $aw + bx + ay + cz$, which would require four weight reads, four multiplications, three additions, the weights and subsequently the activations can be reordered to be $[a, a, b, c]$ and $[w, y, x, z]$, and the dot product can be computed as $a(w + y) + by + cz$, which would require three weight reads, three multiplications, and three additions. This approach is also referred to as dot product factorization; it was inspired by the re-association used by the transforms (e.g., Winograd) described in Section 4.3, which were used to reduce the number of multiplications.

Weight reordering does not impact accuracy. The main challenge with weight reordering is the hardware overhead due to: (1) the indirection tables required to fetch the activations and the weights in the desired order, which increase storage and memory accesses;[6] and (2) the increase in the number of bits in the adder and the input operand to the multiplier.

Network Pruning

To make network training easier, DNN models are usually over-parameterized. Therefore, many of the weights in a DNN model are believed to be redundant and can be removed (pruned, i.e., set to zero) without reducing accuracy. In fact, it has been shown that sparse DNN models (i.e., DNN models with high weight sparsity) tend to have higher accuracy than dense DNN models for a fixed number of effectual (i.e., non-zero) weights [248]. *Network pruning* refers to the process of removing weights in the DNN model.

The concept of network pruning dates back to the late-1980s to mid-1990s [249–252]. The popularity of DNNs has renewed interest in this topic resulting in a significant amount of research over the past few years. Many of today's approaches outperform random pruning when generating a sparse DNN model [248, 253–255].

Network pruning algorithms tend to follow the same process, which is illustrated in Figure 8.9. They begin with a large pre-trained dense DNN model and undergo the following two stages: (1) *weight removal* to determine which weights to remove or set to zero; and (2) *fine tuning* to update the values of the weights (typically the remaining non-zero ones). These two stages are typically iteratively applied several times to gradually increase the weight sparsity in the DNN model. The process of assigning the number of weights to prune per iteration is referred to as *scheduling*. We will now describe the various stages in detail and give examples of how they can vary across different network pruning algorithms.

Weight removal can be broken down into three main steps, as shown in Figure 8.9:

- *Scoring*: Assigns a score to each weight or a group of weights based on a given criterion. The most common criterion is the impact of the weight(s) on accuracy.

---

[6]In the previous example, the activation indirection table would store $[0, 2, 1, 3]$ while the weight indirection table would store $[0, 0, 1, 2]$. This overhead also exists for the weight reordering approach discussed in Section 8.1.1, which proposed early termination of the output activation calculation when using ReLU.

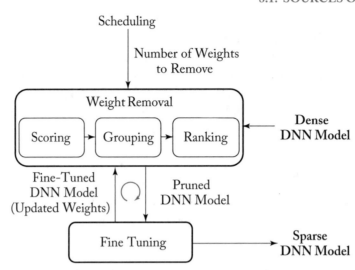

Figure 8.9: The typical process of network pruning. Network pruning begins with a large pre-trained dense DNN model. A two-stage iterative process is then applied involving *weight removal* followed by *fine tuning* to remove and update the weights, respectively. The *scheduling* process determines the number of weights that are pruned for each iteration. The output of network pruning is a sparse DNN model.

- *Grouping*: Weights can be grouped based on a pre-defined structure to allow groups of weights to be removed rather than a single weight.

- *Ranking*: The weights are ranked based on their scores. Depending on grouping, each weight can be ranked individually, or each group of weights is ranked relative to other groups. The likelihood that each weight or group of weights is removed is based on its rank.

There are various approaches for *scoring* the weights. The goal of many of these approaches is to use a score that can help minimize the impact on accuracy while maximizing weight sparsity and/or minimizing the number of MAC operations. For instance, early works such as Optimal Brain Damage [256] assigned scores to the weights based on the impact of each weight on the training loss (discussed in Section 1.2), referred to as weight saliency. However, weight saliency is expensive to compute for today's large DNN models, since it requires the second derivative to be determined for each weight.

Currently, the most popular approach of scoring the weights uses the magnitude of the weights as the scores, which is often referred to as *magnitude-based pruning* [250, 257]. The motivation behind this approach is that removing weights with small magnitudes will minimize changes to the filters and thus potentially minimize the impact on the feature maps and hence

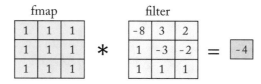

(a) Original without pruning

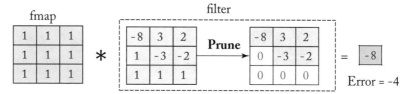

(b) Pruning based on the magnitude of weights (i.e., magnitude-based pruning)

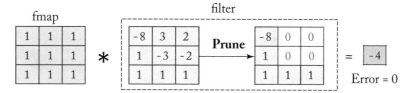

(c) Pruning based on the impact on the output feature map
(i.e., feature-map-based pruning)

Figure 8.10: Example showing how pruning based on the impact on the output feature map (i.e., feature-map-based pruning) can result in a higher accuracy (lower error) than magnitude-based pruning while achieving the same sparsity. Values in red indicate pruned weights.

the accuracy; furthermore, evaluating the magnitude of the weights is easy to do. Several works have also explored the idea of scoring the weights based on their impact on the corresponding output feature maps to maximize the accuracy, called *feature-map-based pruning* [253]. Feature-map-based pruning can achieve higher accuracy with the same sparsity than magnitude-based pruning since it minimizes the change to the feature maps directly rather than indirectly through minimizing the change to the filters; for instance, weights with large magnitudes but opposing signs could both be removed if their effect on the output feature map cancels each other out, as shown in Figure 8.10. However, the cost of evaluating the impact of the weights on the output feature map can be more complex than magnitude-based pruning, since they also require knowledge of the input data to the filters.

　　In addition to considering the impact of each weight on the accuracy when scoring the weights, it is also important to consider the impact that each weight has on other metrics (Chapter 3) such as energy efficiency and throughput. In other words, we want to remove the weights

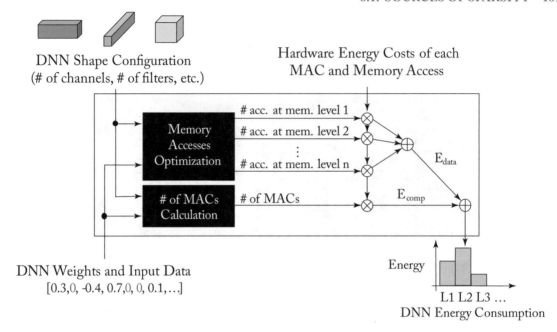

Figure 8.11: Energy estimation methodology [259] used for energy-aware pruning in [253]. It estimates the energy based on data movement from different levels of the memory hierarchy, number of MACs, and data sparsity. This tool is available at https://energyestimation.mit.edu/.

that achieve not only the smallest decrease in accuracy but also the largest increase in energy efficiency or throughput after being pruned. For instance, *energy-aware pruning* considers both of the energy and the accuracy while scoring the weights, which results in a 1.7× increase in energy efficiency compared to magnitude-based pruning for the same accuracy [253]. Alternatively, the impact on latency can also be considered as proposed in NetAdapt [258]. These approaches are often referred to as *hardware-aware* or *hardware in the loop* as they incorporate hardware metrics such as energy and latency into the DNN model design process; in other words, they are mechanisms for hardware and DNN model co-design. These hardware metrics can be obtained using throughput and energy estimation tools [187, 259, 260] or using empirical measurements directly obtained from the hardware itself [258]. Figure 8.11 shows an example of the energy estimation methodology [259] used for energy-aware pruning in [253].

There are also various approaches for *grouping* the weights. The weights can be pruned individually[7] (referred to as *fine-grained* pruning) or in groups with a pre-defined structure (referred to as *coarse-grained* pruning). For coarse-grained pruning, the weights in the same group can be restricted to specific locations with different degrees of granularity, such as the same col-

---

[7]This can be viewed as having only one weight in a group.

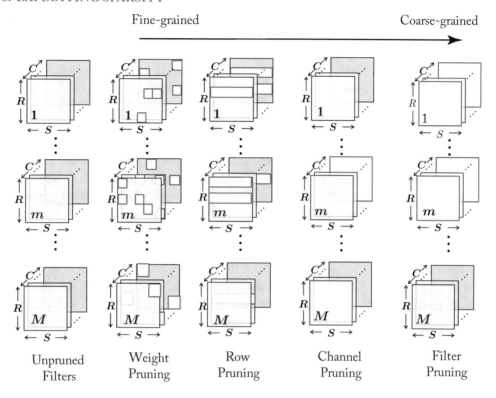

Figure 8.12: Pruning across various degrees of granularity.

umn, row, channel, or filter [258, 261–264]. Figure 8.12 illustrates the differences between these different forms of grouping.[8] It should be noted that pruning channels or filters can be thought of as changing the layer shape of the DNN network architecture (e.g., changing C or M of the DNN model dimensions in Figure 2.2b); therefore, tools that perform automatic channel or filter pruning, e.g., NetAdapt [258], can also be viewed as a form of network/neural architecture search (NAS), which is discussed in Chapter 9.

There are several things to consider when selecting the amount of granularity. On one hand, fine-grained pruning often results in higher degrees of sparsity than coarse-grained pruning (for a given accuracy) since fine-grained pruning is less constrained. On the other hand, it is more costly for hardware to exploit fine-grained pruning than coarse-grained pruning since it must check for non-zero weights more frequently and at any location. In addition, there is more

---

[8]In the literature, *coarse-grained* and *structured* pruning are often used interchangeably when referring to pruning channels or filters, while *fine-grained* and *unstructured* pruning are often used interchangeably when referring to pruning individual weights.

overhead for signaling the location of non-zero weights, which reduces the benefits of using compression; this overhead is discussed in Section 8.2.

Due to the challenges of handling DNN models with fine-grained sparsity, they are typically best suited for processing on custom sparse DNN accelerators, which are discussed in Section 8.3. While general-purpose platforms such as CPUs and GPUs do have libraries that handle coarse-grained sparse data (e.g., Sparse BLAS, cuSPARSE), they often require extremely high sparsity (e.g., >90%) to achieve throughput benefits. Therefore, DNN models with coarse-grained sparsity are typically preferred for platforms such as CPUs and GPUs. The granularity of pruning can also be customized to the hardware platform to increase the impact on throughput for the same amount of weight sparsity. For instance, Scalpel [265] matches the pruning granularity to the underlying hardware parallelism, specifically, the SIMD width, and achieves a 2 to 3× speedup compared to fine-grained pruning. This is yet another example of bringing hardware in the loop of DNN model design.

Weight reordering can be used on top of pruning to increase the coarseness of the sparsity, as shown in Figure 8.13b [266, 267]. Specifically, non-zero weights can be grouped together to form a dense weight matrix, which can then be processed using hardware that efficiently supports dense matrix multiplications, including general purpose platforms such as CPUs and GPUs, as well as dense DNN accelerators, such as the processing-in-memory based DNN accelerators discussed in Chapter 10.

There are also various approaches for *ranking* the weights. The weights can be ranked at a local scale (e.g., per layer) or at a global scale (e.g., across the entire DNN model). Global ranking of all the weights within the DNN model often provides better performance (e.g., higher sparsity for a given accuracy) than local ranking. However, global ranking is often more computationally complex than local ranking because global ranking involves more comparisons and requires the use of scoring methods that can fairly compare weights in different layers, such as weight saliency [256], which are often computationally expensive. In contrast, local ranking typically only ranks the weights within each of the layers, which requires fewer comparisons and allows the use of simple scoring methods, such as magnitude-based pruning [250, 257] or feature-map-based pruning [253], which makes it a popular choice. One of the main challenges of local ranking is that per-layer sparsity needs to be specified and the optimal specification usually difficult to determine.

Fine tuning updates the values of the remaining weights to restore accuracy. There are several important design decisions that are considered in the various approaches used for fine tuning. The first is how to initialize the weights at the beginning of fine tuning. For instance, the weights can be reinitialized each iteration, they can start from the state at the end of the previous iteration (most popular), or they can rewind to an earlier state [268]. Another important decision is whether the weights are updated by performing a global or local optimization. Performing a global optimization is similar to typical network training approaches, where all weights across all layers are jointly optimized and updated simultaneously. Performing a local optimization,

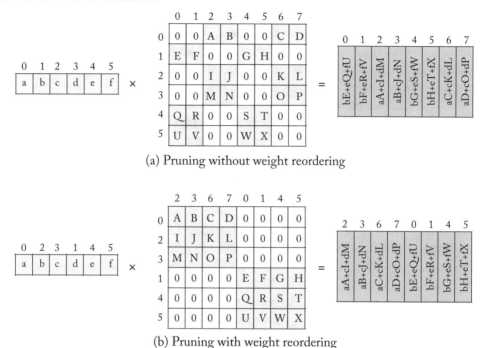

(a) Pruning without weight reordering

(b) Pruning with weight reordering

Figure 8.13: Applying weight reordering on top of pruning to increase the coarseness of the sparsity. In this example: (a) before weight reordering, the granularity of the sparsity is 1×2; and (b) after weight reordering, the granularity of the sparsity is 3×4. Note that the input activations and output activations are also reordered.

where only a subset of weights are jointly optimized and updated simultaneously, can speed up the time required for fine tuning. For instance, fine tuning can be performed per layer, where the weights are updated by minimizing the difference between the output feature maps before and after pruning [253]. This minimization has a closed-form solution, so local optimization can be performed faster than global optimization. Finally, since fine tuning updates the weights, previously pruned weights might become important at a later iteration. Accordingly, some recent works explore the idea of restoring some of the pruned weights during fine tuning [269–271].

Scheduling involves determining how many weights to prune in each iteration in order to achieve a target weight sparsity in the final sparse DNN model. We can set the number of weights per iteration and desired weight sparsity explicitly [257] or infer them from energy [253] or latency [258]. The common scheduling methods include: (1) pruning all the weights required to achieve the target number of weights in a single iteration (one shot) [272]; (2) pruning a fixed fraction of weights at each iteration across multiple iterations [257]; or (3) pruning different fraction of weights at each iteration across multiple iterations [254]. Scheduling becomes more challenging when ranking the weights locally since the number of weights to prune per iteration needs to be further specified at a finer granularity (e.g., on a per-layer basis rather than for the entire network). One approach to tackling this challenge is to apply pruning to only one layer per iteration, specifically the layer that achieves the best accuracy versus latency (or energy consumption) trade-off [258] with pruning.

It should be noted that there is an interplay between network pruning and the shape and size of the DNN model (i.e., the network architecture of the DNN model). While pruning can improve the efficiency of a given DNN model, switching to a DNN model with a more efficient network architecture, as described in Chapter 9 can often result in better efficiency, as shown in Figure 8.14, where efficiency is defined as number of weights or MAC operations versus accuracy [248]; this has yet to be explored for energy efficiency or throughput versus accuracy. Furthermore, network pruning is more effective on inefficient DNN models than DNN models that already have an efficient network architecture [248]. For instance, pruning on AlexNet can reduce the number of weights by 10.6×, the number of MACs by 6.6×, and the energy consumption by 3.7×, whereas pruning on GoogLeNet only reduces the number of weights by 2.9×, the number of MACs by 3.4×, and the energy consumption by 1.6× [253]. Finally, pruning is often more effective in reducing the number of weights on fully connected layers than convolutional layers. For instance, several of the fully connected layers in AlexNet can be pruned to over 90% weight sparsity, while the convolutional layers only reach around 60% weight sparsity [253]; however, this could also be because AlexNet has three fully connected layers, whereas more modern DNN models (e.g., GoogLeNet, ResNet, MobileNet) only use one fully connected layer, and thus the three fully connected layers of AlexNet are extremely over-parameterized.

Network pruning continues to be a very active field in both research and industry. Unfortunately, the field currently lacks standardized benchmarks and metrics to properly compare and evaluate the numerous methods that have been developed, making it difficult to measure the progress in the field. To address this, Blalock et al. [248] identifies issues with current practices based on a survey of over 80 papers (e.g., lack of controlled comparisons, imprecise and incomplete specification of experimental setup and metrics), provides various solutions to those issues (e.g., report results as multiple points on a trade-off curve, with multiple dataset and DNN model combinations), and proposes the use of an open-source framework called ShrinkBench, which provides a standardized collection of pruning primitives, model, datasets, and training routines to enable standardized evaluation of pruning methods.

Dilated Convolutions

Increasing the receptive field of a filter (i.e., the spatial dimensions of the filter R and S of the DNN model dimensions in Figure 2.2b) can sometimes help increase accuracy, since each output activation is a combination of a larger region of inputs (e.g., in an image this could mean seeing the entire face versus just the nose, or seeing more of the environment around an object, which

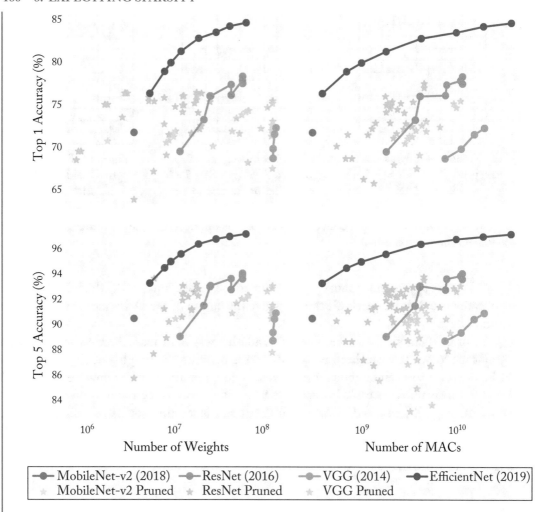

Figure 8.14: Pruning versus network architecture of DNN model. Using a DNN model with a more efficient network architecture (e.g., EfficientNet or MobileNet) can result in a better trade off between accuracy and number of weights and accuracy and number of MAC operations. (Figure adapted from [248].)

could provide more context to help identify the object). However, increasing the receptive field increases the number of weights in the DNN model and the number of MAC operations that need to be performed per output activation.

Various approaches can be applied to achieve the same large receptive field while reducing the additional cost. One approach is stacking multiple filters with smaller receptive fields; this was explored in DNN models such as VGGNet [73], and is discussed in more detail in Chap-

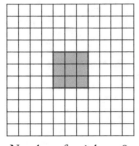 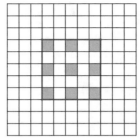 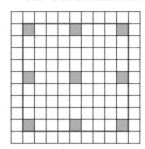

Number of weights = 9   Number of weights = 9   Number of weights = 9
Receptive Field (R × S) = 3 × 3   Receptive Field (R × S) = 5 × 5   Receptive Field (R × S) = 9 × 9
Dilation Rate: 1   Dilation Rate: 2   Dilation Rate: 4

Figure 8.15: Filter for dilated convolution at various dilation rate (i.e., number of zeros inserted between non-zero weights).

ter 9. Another approach is spatial pooling (described in Section 2.3.4), which is often used for applications such as image classification, where the output of the DNN model is reduced down to one label.

For DNN models with dense outputs, another way to increase the receptive field without significant complexity overhead is to insert zeros between the weights of the filter, as shown in Figure 8.15; this can be thought of as upsampling the filter, similar how we upsampled the input feature map, as discussed in Section 8.1.1. This approach is commonly referred to as *dilated* convolutions [273] or *atrous* convolution [240] and is another source of weight sparsity. As mentioned in Section 8.1.1, the sparsity from upsampling is structured and known a priori and thus the cost of detecting sparsity can be significantly reduced compared to the unstructured sparsity described in the previous section.

8.2   COMPRESSION

The existence of sparsity in both weights and activations, as just described in the previous sections, inspires the application of techniques to compress the data to reduce storage space, data movement, and/or computation to save time and/or energy in DNN accelerators. To explore this opportunity, recall that in Chapter 2 it was noted that the multi-dimensional operands used in DNN computations can be viewed as tensors. For instance, input activations can be represented as 4-D tensors with dimensions N, H, W, and C and filter weights can be represented as 4-D tensors with dimensions R, S, C, and M. Since the data in these tensors is often sparse, we will discuss compression in the context of compressing tensors.

There have been a large number of formats proposed for representing sparse tensors as nicely summarized by the TACO project in [274]. From a hardware design perspective, each of these formats have characteristics that distinguish them along one of several axes. For example, they can vary in terms of the size in memory, the cost of accessing the elements of the tensor, and the cost of operators that modify the tensor (e.g., add an element) or combine it with other tensors (e.g., intersection). These metrics can also vary with the sparsity of the tensor or other statistical characteristics (e.g, clustering along a diagonal). In conjunction with a design that uses a specific format, these characteristics will be reflected in many of the operational metrics described in Chapter 3.

Rather than just enumerate compression techniques, however, we will present an abstraction for the representation of tensors in memory and the operations that can be performed on them. Within this abstraction the various attributes of different compression techniques can be classified and characterized, and the opportunities for mixing and matching techniques should be evident.

Providing a common abstraction for the operations that can be performed on a tensor has a further benefit in understanding the hardware (e.g., its dataflow) used to process sparse tensors. We find that often design descriptions get bogged down in the detailed manipulations required by a specific data representation. These details can obscure the high-level principles behind the design. Having a common abstraction for any sparse tensor representation and a common set of operations on that abstraction allows for a separation between the details of the exact manipulations required to operate on a specific representation of the tensor in memory and the essence of the dataflow. Hopefully, this will provide an opportunity to more clearly compare and contrast different designs and gain insights into the tradeoffs between them.

Note, however, that although there is a conceptual partitioning between the algorithmic activity (e.g., dataflow) on the tensors and their representation, design decisions will need to take both into account. A design that relies heavily on a particular operation probably should not be paired with a representation for which that operation is especially expensive. Instead, this should be viewed as an opportunity for a co-design process, where the design's algorithmic activity and the representation for the tensors used in the design are jointly selected for overall optimal metrics. Furthermore, a specific design may employ *cross-layer optimization* breaking the strict boundary between the implementation of the algorithm and the implementation of the data representation.

8.2.1   TENSOR TERMINOLOGY

Figure 8.16 shows a matrix (i.e., a 2-D or rank-2 tensor) with the terminology we are going to use to describe tensors. Specifically, the axes of the matrix are referred to as *ranks* and labeled with a *rank-id*. In this case, the matrix has two rank-ids (H and W), corresponding to the ranks of a channel of input activations. The individual elements of a rank are also labeled with a *coordinate* and thus the value at an individual cell (or *point*) in the matrix can be identified by a tuple of coordinates - one for each rank. Thus, the value "f" at coordinate 1 in rank H and coordinate 2 in rank W is at point $(h, w) = (1, 2)$.

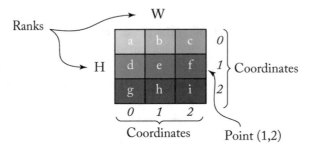

Figure 8.16: A rank-2 tensor (i.e., a matrix) showing the key terms for describing a tensor: *rank*, *coordinate* and *point*. Assuming the ranks are ordered by their ids as H, W, the value at the point (1, 2) is "f".

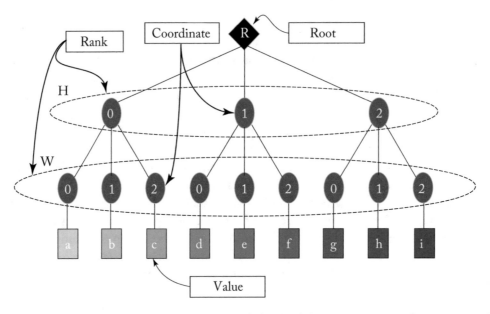

Figure 8.17: A tensor represented as a tree. Each level of the tree corresponds to a *rank* of the tensor, and contains a node for each *coordinate* in that rank. The leaves of the tree are the values at each *point* in the tensor.

To discuss the scope of compression opportunities, we are going to explore an expanded version of the tree-based tensor representation described in [275] and used in [276]. Figure 8.17 shows an abstraction for a tensor as a tree with a root node (black diamond) and intermediate nodes (brown circles), leaf nodes (blue squares), and edges (black lines). Below the root node (black diamond "R") each level (dashed oval) corresponds to a *rank* (or dimension) of the tensor.[9] The first rank is the dashed oval "H". Each intermediate node of the tree corresponds to a coordinate in some rank of the tensor. For intermediate ranks, a node's children are the coordinates of the next lower rank (e.g., dashed oval "W" of the tensor). At the lowest rank, the child of a coordinate (i.e., a leaf node) is the value of a specific point of the tensor (blue square). Note, one can determine the point a value is at from the sequence of coordinates passed at each rank while traveling down the tree from the root to the value. Thus, the value "c" is at point $(h, w) = (0, 2)$.

An important and common operation on a tensor is finding the value at a particular point (i.e., ordered set of coordinates) in the tensor. In this tree representation, it should be clear that finding a point in the tensor involves traversing the tree looking for the desired coordinate at each rank in turn. Although abstractly the order of the ranks have no meaning, in most concrete tensor representations the order of the ranks is of great importance, because the order can significantly affect the cost of accesses and therefore is often a salient characteristic of any algorithm that operates on the tensor. Thus, even in our abstract representation of a tensor the order is manifest in the representation. We discuss this in more detail later, but first we are going to consider sparsity.

Representing a sparse tensor in this tree is straightforward, as illustrated in Figure 8.18. In the figure each element of a rank is manifest simply as a coordinate node having fewer children (i.e., just the coordinates of non-empty elements in the next rank). Furthermore, if we ignore the details of the implementation of the nodes and edges of the tree, this representation abstracts most of the details of the way a tensor is represented in storage, giving the accelerator designers the opportunity to optimize the representation for the storage and operational characteristics desired in their design.

Although viewing a tensor as a tree where each node has edges emanating from it connecting to each of its multiple children is not an unreasonable model, to improve the relationship between our model and actual design considerations we are going to use an alternative model. In specific, we are going to assume that each coordinate node has a single *payload* that is either (1) a collection of coordinates each with their own payload (at intermediate levels in the tree) or (2) a single value at the lowest level (i.e., rank) of the tree.

For case 1, the collection of coordinates that are children of a single coordinate will be called a *fiber*, as shown in Figure 8.19.[10] A rank will therefore contain one fiber for each coordinate in the rank above, and each fiber will be the payload of one of those coordinates.

Figure 8.19 illustrates this *fiber-tree* representation. The figure is interpreted as follows: the root of the tree (black diamond marked "R") points at the single fiber (solid oval) in the top (or highest) rank (dotted oval "H") of the tensor. Rank H's fiber holds the non-empty coordinates at that rank. Each coordinate in rank H's fiber has a payload that is a reference (black line) to a fiber (solid oval) at the next lower rank (dotted oval "W"). In this case, rank W contains two fibers.

---

[9]Later we will discuss optimizations that relax this assumption.

[10]This roughly corresponds to the mathematical notion of a fiber where all coordinates of a tensor are fixed except one. In our case, that would map to a tensor consisting of the ranks up to and including the rank of the fiber of interest.

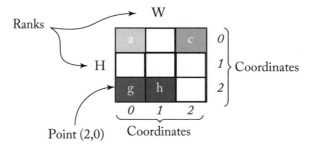

(a) Sparse tensor-matrix. Blank cells represent an empty point in the tensor

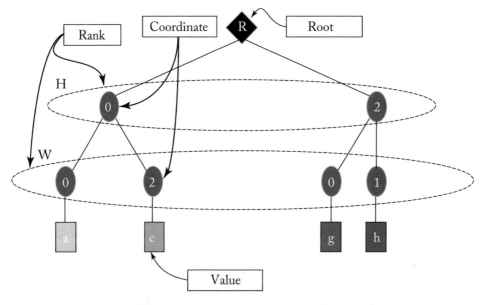

(b) Sparse tensor-tree. All coordinates with no payload are dropped from the tree. Thus coordinates in rank W corresponding to empty cells are dropped and, if all children of a coordinate in rank H have been dropped, that coordinate is dropped as well

Figure 8.18: Sparse tensor as matrix and tree.

For a higher dimension tensor, the interpretation would continue recursively in that pattern until the lowest rank (rank W in this case), where the payload of each coordinate is a value (blue square) marked with letters that represent scalars or other terminal data values. Using this abstraction, we can consider a variety of concrete representations of a tensor and their efficiency in space and time.

In summary, the terminology we will use is as follows:

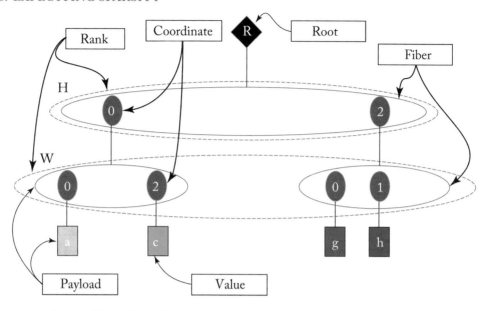

Figure 8.19: Sparse Fiber Tree. Each level of the tree corresponds to a rank *rank* that contains one or more *fibers*. Each fiber contains a set of *coordinates*, whose *payload* is either another fiber at the next lower rank or a *value* at the bottom of the tree. Note, fibers at the lower rank only have coordinates for non-empty values, and empty fibers are dropped from all levels of the tree.

- **Rank** - an axis, such as a row, column, or higher dimension, of a tensor. In our representation, ranks are labeled with a name, called a *rank-id* (e.g., H and W in Figure 8.19). Furthermore, in our abstract tensor representation, ranks are ordered from top (highest rank) to bottom (lowest rank).

- **Coordinate** - an identifier associated with each element (or item) contained in *rank* of a tensor. For example, the numbers in the brown circles in Figure 8.19.

- **Payload** - the value associated with a particular *coordinate* in a *rank*. In Figure 8.19, references to a payload are indicated by the black lines (edges) in the tree. For intermediate *ranks*, the *payload* of a *coordinate* will be a *fiber* that corresponds to a sub-tensor (of the lower ranks). For example in Figure 8.19, the coordinate 2 in rank H has a fiber in rank W as its payload, so coordinate 2's payload corresponds to 1-D sub-tensor whose top fiber has coordinates 0 and 1. The *payloads* of the coordinates in the lowest *rank* will be a simple value (e.g., a number).

- **Point** - a set of *coordinates* (one for each *rank* of the tensor) that localizes a single value in the tensor. The order of the coordinates match the order of the ranks (from top to bottom in the fiber tree). For example, the value "c" (blue square) in Figure 8.19 is at point $(h, w) = (0, 2)$.

- **Fiber** - a set of *coordinates* and their associated *payloads*. All of the *coordinates* in a fiber are children of a single *coordinate* in the rank above. Therefore, a *rank* will contain one or more *fibers* and each *fiber* in a *rank* will be the *payload* associated with a *coordinate* of the *rank* above.

8.2.2 CLASSIFICATION OF TENSOR REPRESENTATIONS

In general, there are a variety of characteristics of the concrete representation (or format) in memory of a tensor that will affect its desirability for use in a particular DNN accelerator design. Since the fundamental component of our tensor abstraction is a fiber, our focus will initially be on concrete representations of fibers, and on the tradeoffs among representations based on factors such as how much space they occupy and the complexity of the operations that access or operate on them.

In practice, a key operation on a fiber is to *lookup* a payload associated with a specific coordinate. That will generally mean finding the address (in physical storage) of the payload associated with the given coordinate. Therefore, finding the value at a point in the tensor will nominally require a series of such lookups traveling down the tree. That series of lookups will locate (e.g., by address) the fibers at the intermediate ranks and ultimately the final payload, which will contain the value of the desired point.

Obviously, the efficiency of a specific fiber representation, as characterized by metrics such as size and speed of payload lookup, will depend on the in-storage layout of the fiber and the algorithmic choices available to traverse that layout. Hardware DNN implementations need to consider the implementation costs and tradeoffs among these metrics in order choose the representation that best optimizes the design. From this perspective, a wide variety of fiber representations have been proposed for use in software and hardware, and can be classified in a variety of ways.

Explicit versus Implicit Coordinate Representations
Fiber representations can be classified by whether the values of the coordinates are manifest explicitly in the representation or not. These will be referred to as *explicit* or *implicit* coordinate representations, respectively.

An example of an *implicit coordinate representation* is an implementation of a fiber as a standard 1-D array data structure. An example is shown in Figure 8.20a.

For fibers represented in an *array-style representation*, the coordinate never appears directly in memory, but are determined directly as an offset to a specific payload in the list of payloads. In general, we will refer to the offset (or relative location in memory) of each element in a fiber as the element's *position*. Payload lookup by coordinate generally needs to find the position of a payload with the given coordinate. In an array-style representation, payload lookup by position would be

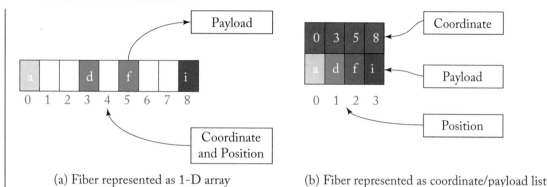

(a) Fiber represented as 1-D array   (b) Fiber represented as coordinate/payload list

Figure 8.20: Uncompressed and compressed fiber representations. Subfigure (a) shows an array-style representation, where the payloads (blue rectangles) and zero values (blank rectangles) occupy all the elements of the array. Below each element is a number indicating its *position* in the array and the same number corresponds to the element's (implicit) *coordinate*. Subfigure (b) shows a coordinate/payload representation where each position holds a tuple containing each non-zero element's coordinate (brown rectangle) and value (blue rectangle).

very efficient, since the position can be determined directly from the coordinate. Unfortunately, for sparse fibers the space required to hold the fiber could be quite large because even empty elements of the fiber occupy space.

An *explicit coordinate representation*, where the coordinate values actually appear in memory, can be useful to save space for very sparse fibers. For example, a list of coordinate/payload tuples for non-empty coordinates would occupy much less space than an uncompressed array of payloads. An example of a fiber represented as such a *coordinate/payload list* is shown in Figure 8.20b.

When using a coordinate/payload list representation, one can see that if a coordinate in a high-level fiber is empty (i.e., indicative that an entire tile of the tensor is zero) a considerable amount of space at all the lower levels is saved. Such an explicit coordinate representation, however, would have more costly payload lookups because, although each coordinate has a *position* in the fiber, there is not a direct mapping from coordinate to position. Thus, a more complex lookup (e.g., via a binary search) would be required. This creates a tradeoff between space (large for the array-style representation) and speed/energy (more memory accesses for the binary search needed by lookup in a coordinate/payload representation). However, sometimes the space savings can be worthwhile, and such a *coordinate/payload list* representation is used as a component of a variety of well-known tensor representations, such as compressed sparse row (CSR) [277], compressed sparse column (CSC) [278], and compressed sparse fiber (CSF) [275]. Additional characterizations of these tensor formats will be presented in Section 8.2.5.

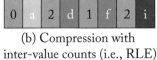

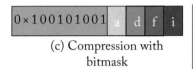

(a) Original (uncompressed) array-style fiber | (b) Compression with inter-value counts (i.e., RLE) | (c) Compression with bitmask

Figure 8.21: Examples of implicit coordinate-style methods to compress a fiber. Blue boxes are non-zero values, white boxes are zero values, and grey boxes are encoding metadata. In subfigure (b), the metadata contains the number of zero values between two non-zero values. In subfigure (c), the metadata is a bitmask where each one in the bitmask indicates that the corresponding coordinate is non-zero and counting ones to its left indicates the value's position.

Compressed versus Uncompressed Representations

Fiber representations can also be classified by whether the size of the fiber in memory is a function of its sparsity. Or, in other words, is the fiber *compressed* or not.

Clearly, the (implicit coordinate) 1-D array-style representation of a fiber described above is an *uncompressed representation* because the space occupied is not a function of sparsity. However, one can enhance it to create a *compressed representation* by adding metadata that records where there are empty values. This includes any of the myriad *run-length encoding* (RLE) schemes,[11] where payloads are interspersed with information about the number of empty values between non-empty values (see Figure 8.21b). Empty values can be identified in other ways, such as with a bit-mask (see Figure 8.21c). These compressed representations can reduce storage and data movement costs as was done in the Eyeriss design [101]. In Eyeriss, the volume of partial sum data transmitted to/from DRAM and the size of the data stored there was reduced using an RLE compression scheme.

Unfortunately, compression schemes are not a panacea as there are tradeoffs that need to be considered when applying a compression scheme. For example, consider the number of bits allocated to describe gaps between non-empty values in a RLE scheme. In a very sparse fiber, one would like a large number of bits to express a large gap between non-zero values (or *span*) to always be able to jump directly to the next value. Otherwise, one would have to introduce zeros as concrete values to serve as a stepping stone to the next non-zero value. On the other hand, if the fiber is not so sparse, then the bits allocated to hold a large span could be a significant overhead. Such choices are a function of statistical properties of the data, which may or may not be known a priori. Thus, these choices can have a significant impact on the flexibility metrics of the design, as discussed in Section 3.5. Adding these kinds of compression also impacts lookup because there is a more complex relationship between a value's coordinate and its position.

Explicit coordinate schemes can also be a compressed representation. For example, one could implement coordinate to payload mappings by putting the payloads in a hash table. In such a scheme, a coordinate could be used as the key into the hash table that returns the payload for

---

[11]These schemes date back to very early inspiration by Shannon [279].

that coordinate. Because hash table sizes are a function of the number of values contained (i.e., the non-empty coordinates), a compression benefit will accrue. Furthermore, note that within a rank each fiber could have its own hash table or there could be a single table associated with all the fibers in the entire rank. Between those two schemes the keys needed for the hash table lookups differ. The former scheme just needs a coordinate since each fiber has its own hash table, while the latter needs a fiber-id and a coordinate because the single table has the coordinates for multiple fibers. Therefore, depending on the hash table organization used by a rank, the information returned by a lookup operation on the rank above will differ.

8.2.3   REPRESENTATION OF PAYLOADS

The hash table alternatives for representing the fibers in a rank described in Section 8.2.2 illustrate the fact that the information in the payloads used in a rank is dependent on the fiber representation used in the next rank lower in the tree. In the fiber-tree abstraction in Figure 8.19, this corresponds to implementation of the edges (black lines) that connect a coordinate to its payload.

Like coordinates, these payloads can be explicit or implicit. In the case of the *hash table per fiber* representation, the *explicit payload* at the rank above in the tree is just a direct reference to the hash table for the appropriate fiber. Such a direct reference is a very common form of payload, but differs from what is needed for the *hash table per rank* representation, where the (explicit) payload is a fiber-id that is used as part of the hash table lookup.

A payload can be *implicit* such as when the position in a coordinate/payload list can be interpreted directly as an offset into information at the next lower rank. In that case, that position constitutes an *implicit payload* that occupies no space.

In some cases, either coordinates, payloads, or both can be compressed by using an information theoretic compression scheme, such as a Hamming code. Note, however, that this can introduce complexities for implicit payload schemes or complicate lookup operations.

8.2.4   REPRESENTATION OPTIMIZATIONS

So far, we have exclusively considered representations where each level of the tree corresponds to exactly one rank of the tensor. In this case, the cost of finding a value (i.e., a payload at a leaf of the tree) is a direct function of the number of ranks in the tensor. Fortunately, it is possible to consider representations with multiple ranks combined together in a level.

Conceptually, having two ranks in a single level means combining two consecutive ranks into one. After this combination, the coordinates of the new rank are tuples of the coordinates from the two original ranks. This processes is referred to as *flattening* ranks and is illustrated for a fiber-tree in Figure 8.22.

The implementation of flattening varies with the representations chosen for the individual ranks. For fibers represented in uncompressed array-style, it should be clear that two ranks can easily be *flattened* into a single level. The resulting representation is still uncompressed and

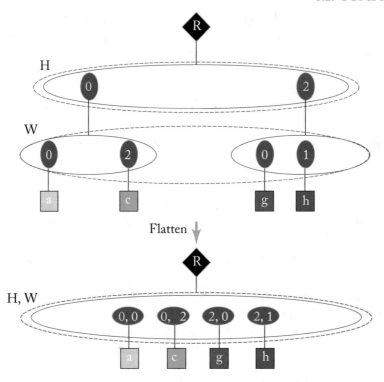

Figure 8.22: Effect of flattening ranks. Two ranks H and W are flattened into one rank and the coordinates of the new rank (H,W) are tuples of the original coordinates.

lookup involves two coordinates and is implemented with a simple arithmetic calculation. For a 2-D tensor with ranks H and W, the calculation for lookup by coordinates h and w would be $h \times |W\_rank| + w$. This is a standard way that multi-dimensional arrays are represented in software. Flattened compressed representations are also possible, although lookup by coordinate might be serial. Efficient (binary search) lookup can, however, be maintained when combining a pair of ranks in coordinate/payload list format where the two ranks can be combined into a single level and the coordinates become a tuple of the coordinates from the two original ranks. Similarly, two ranks in hash table format can be combined into a single level with a key lookup by coordinate tuples.

8.2.5 TENSOR REPRESENTATION NOTATION

A small table of fiber representation schemes described in the preceding sections is shown in Table 8.1. The table includes a short label and description for each scheme, its characteristics and an example design that uses it. Other (possibly flattened) fiber representations can be assigned

Table 8.1: Fiber representations

Label	Description	Coordinates	Compressed	Example
U	Uncompressed array	Implicit	No	DianNao [150]
R	Run-length-encoded (RLE) stream	Implicit	Yes	Eyeriss [98]
B	Bitmask of non-zero coordinates	Implicit	Yes	SparTen [277]
C	Coordinate/payload list	Explicit	Yes	SCNN [156]
H_f	Hash table per fiber	Explicit	Yes	
H_r	Hash table per rank	Explicit	Yes	

their own unique labels. However, when two levels of fiber with known representations are flattened, we use a notation like U^2, P^2, or (RU) to describe the flattened combination.

Given the above fiber representations, the combination of all the choices for representing a fiber's coordinates and payloads leads to a large number of implementation choices. This is multiplied by the fact that each rank of a tensor might use a different representation.

To provide a specific example, we will show how the well-known compressed sparse row (CSR) format [277] can be represented as a concrete representation of a fiber tree. Figure 8.23 illustrates this by showing the matrix of Figure 8.18 in CSR format. The figure shows CSR as a two rank tree that uses an uncompressed array as it top rank fiber (H), and a coordinate/payload list as its bottom rank (W). Thus, the rows are compressed.

In CSR, each position in the upper rank (which is also its coordinate since it is uncompressed) has a payload consisting of a open range that points at a fiber in the bottom rank.[12] And each fiber in lower rank consists of a list of explicit coordinates each of whose position is an implicit payload that is the position of the value in the value array.

So in the figure, if we want to find the value at coordinate (2,1) we start by looking at position (and coordinate) 2 of the upper rank and find its payload is the open range [2, 4). Note how the open range is cleverly encoded with information from two successive positions in the array. Then looking at the fiber in the bottom rank at positions 2 and 3 (i.e., in the open range [2,4)), we search for the coordinate 1. Finding that it is at position 3, we know that the value for coordinate (2,1) is the value "h" at position 3 in the value array.

The CSC representation is the dual of CSR and is basically the same just with the rank order reversed. These schemes statically pick a representation per rank, but an even more complex approach would be to dynamically choose the representation used for each fiber. That choice could be made at the rank level, so that the rank would have a tag indicating the representation for all its fibers or the choice could be made at the individual fiber level.

[12]Note, that the CSR representation depends on the fibers in the lower rank being consecutive in memory so the payloads of the upper rank can point to a position in the lower rank.

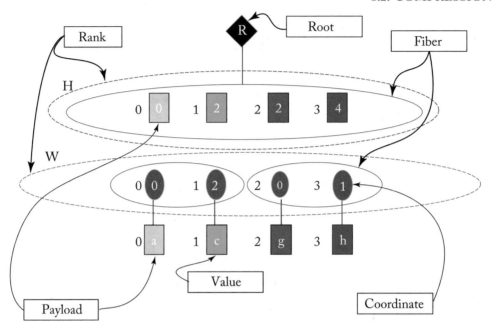

Figure 8.23: Compressed Sparse Row (CSR) format as a concrete fiber-tree representation. CSR is implemented as two ranks, the top rank is an *uncompressed array* in memory with payloads consisting of a pointer to a fiber in the next rank (in open range form). The fibers in the lower rank are *coordinate/payload lists* that are concatentated together in memory. In the diagram, *positions* in memory for each element of a fiber are indicated with black numbers next to a shape.

Noticing that many previously proposed tensor representations have been created by selecting a fiber representation for each rank (or flattened set of ranks) of the tensor leads to the idea of a more generic specification of a tensor representation. With this objective in mind, a full specification of tensor would require a selection for each rank of (1) a fiber representation and (2) a rank-id. Combining concepts from [274] and [276] we use the following notation to represent the specification of a tensor:

$$Tensor < \textbf{FIBER-REPRESENTATION-REGEX} > (\textbf{RANK-ID}...), \qquad (8.3)$$

where **FIBER-REPRESENTATION-REGEX** is a regular expression consisting of a sequence of labels of fiber representations from Table 8.1, and **RANK-ID** is the name of a rank. Table 8.2 lists some common tensor representations in this notation and their common name. Note, the actual rank-ids are only used for the rows for CSR and CSC because those representations only differ in the names they assign to the ranks.

DNN accelerators need to make design decisions on what representations they will employ based on the how they affect the metrics of the design. However, those decisions must

Table 8.2: Tensor representations

Specification	Name
Tensor<U+> (…)	Standard multi-dimensional array
Tensor <UC > (H,W)	Compressed sparse row (CSR) [274]
Tensor <UC > (W,H)	Compressed sparse column (CSC) [275]
Tensor <C +> (…)	Compressed sparse fiber (CSF) [272]
Tensor <C^n> (…)	Coordinate format (COO) [274]

be considered in conjunction with the computation sequencing, which is described in the next section.

8.3   SPARSE DATAFLOW

In Section 8.2, we discussed the opportunity that sparsity presents to compress the tensors used in DNN computation. This provides obvious benefits savings in storage space, access energy costs and data movement costs by storing and moving compressed data. However, we also recognize that sparsity means that individual values of activations or weights (or entire multi-dimensional tiles) are zero, that the multiplication of anything by zero is zero, and furthermore the addition with zero simply preserves the other input operand. Such operations therefore become *ineffectual* (i.e., doing the operation had no effect on the result). As a consequence, when performing the pervasive sum of product operations (i.e., dot products) in DNN computations there is an opportunity to exploit these ineffectual operations.

The simplest way to exploit ineffectual operations is to save accessing operands and avoid executing the multiplication when an operand is zero. The Eyeriss design saved energy by avoiding reading operand values and running the multiplier when an activation was zero [160]. That eliminated activity (in energy) includes the accesses to input operands, writes/updates to output operands, and arithmetic computation so it can enhance the energy efficiency/power consumption metrics discussed in Section 3.3. However, this only saved energy, not time, for ineffectual operations.

When the hardware can recognize the zeros in the products terms, the amount of time spent performing the dot product can be reduced by eliminating all the time spent on ineffectual activity related to product terms with any zeros as an operand. This will not only improve energy, but improve operations per cycle, as described in Equation (3.2).

Figure 8.24 gives an indication of the potential for reducing computation time by showing the density (proportion of non-zeros or 1 - sparsity) in both weights and input activations for the layers of VGGNet. From those statistics, an architecture that could optimally exploit weight or activation sparsity could provide speedups of 2 to 5×. If one assumes that weight and activation

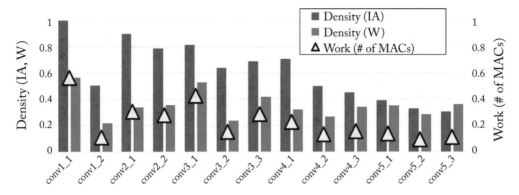

Figure 8.24: Distribution of weight and activation density. (Figure adapted from [159].)

sparsity are statistically independent, a savings would be accrued whenever at least one operand is zero, resulting in potential speedups of over 10×.

In Section 8.1.2, we noted that specialized DNN hardware is generally applied when trying to exploit fine-grained, unstructured sparsity in weights or any sparsity in activations. So this section will generally focus on techniques targeted at fine-grained, unstructured sparsity, but the concepts described and notation used are generally applicable to any type of sparsity. In any case, supporting sparsity means that additional hardware will be used to identify non-zero values and for looking up payloads, which adds energy and area overheads. Therefore, the challenge is creating an efficient design that achieves savings and does so across a range of sparsities, such that the benefits of exploiting sparsity exceeds the cost of the additional hardware required to identify the non-zero data; achieving gains is particularly challenging when the amount of sparsity is low, fine-grained, and unstructured.

There are two major aspects involved in the challenge of exploiting sparsity to reduce computation time in DNN computation: (1) choosing an optimal representation for the sparse data, and (2) choosing a computation flow. These two aspects must be co-designed in order to achieve an overall optimal design. However, considering them simultaneously can be confusing, so we will use the fiber-tree notation presented in Section 8.2.1 to allow their consideration separately.

The challenge of choosing a computation flow for a sparse DNN accelerator is somewhat analogous to the challenge of maximizing reuse, as discussed in Chapter 5. Just as Chapter 5 described the opportunities for achieving different forms of reuse, and used the notion of different dataflows to explore that domain of reuse opportunities, there are a variety of options and often tradeoffs in the amount of sparsity that can be exploited. And again dataflows can be described that allow one to express those options.[13] In the next several sections, we will begin with a re-

---

[13]Note that in Chapter 5 on dataflows, the focus was on the fundamental ordering of computations, while important issues such as tiling (Chapter 4) and mapping (Chapter 6) were left to be considered independently. This chapter will largely do the same.

Design 8.4 1-D Weight-Stationary Convolution Dataflow with Dense Weights & Activations

```
1    i = Array(W)        # Uncompressed  input    activations
2    f = Array(S)        # Uncompressed  filter   weights
3    o = Array(Q)        # Uncompressed  output   activations
4
5    for s in [0, S):
6        for w in [s, W – s):
7            q = w – s
8            o[q] += i[w] * f[s]
```

---

view of a dataflow for dense convolution and proceed to describe dataflows for exploiting sparse weights, sparse activations, and then both for convolutions. Finally, we will present dataflows for fully-connected computations with both sparse weights and activations.

In Chapter 5, the different dataflows were described as different loop nests. A standard output-stationary dataflow is shown in Design 8.4. In lines 1-3, the `Array(...)` declarations correspond to a tensor declared as *Tensor $< U^n >$* (...) in the notation of Section 8.2.5. Lines 5 and 6 show the `for` loops that create an index variable used to traverse the DNN tensors (weights, input and output activations). Those `for` loops traverse an open range of values (e.g, [0, S) iterates over the values 0 to S-1). Those index variables are used to directly index into the arrays containing the tensors. In the terminology of Section 8.2, those index variables hold coordinates and the arrays correspond to an uncompressed (often flattened—see Section 8.2.4) data representation for the tensor. Note that we follow the convention that a small letter (e.g., s) is a coordinate in the rank of the corresponding capital letter (e.g., S) and that same capital letter indicates that the coordinates occupy the open interval (e.g., [0,S)). For these uncompressed representations, it is easy to express payload lookups using standard array access notation (e.g., o[q]), since the coordinates equal the position in the array (line 8).

For sparse data in a compressed representation, the behavior for a particular dataflow (i.e., loop nest) can be replicated by replacing all the array accesses with a *getPayload()* lookup by coordinate (or set of coordinates). A weight-stationary example of this approach is shown in Design 8.5. In that dataflow the `Tensor()` declarations follow the notation of Section 8.2 except the specific fiber representations are omitted, but would need to be selected by co-design with the dataflow for an actual implementation.

This approach, however, has two significant drawbacks: (1) it involves iterating over *all* the coordinates in each fiber of the tensors, and therefore there will be no time or energy savings with respect to operand accesses; and (2) the cost of individual payload lookup given a point (i.e., set of coordinates) can be very expensive, since it might involve a traversal of the fiber tree - one

Design 8.5 1-D Weight-Stationary Convolution Dataflow with lookup of Sparse Weights & Activations

```
1    i = Tensor(W)        # Compressed  input   activations
2    f = Tensor(S)        # Compressed  filter  weights
3    o = Array(Q)         # Uncompressed output  activations
4
5    for s in [0, S):
6        for w in [s, W - s):
7            q = w - s
8            o[q] += i.getPayload(w) * f.getPayload(s)
```

level per coordinate. Fortunately, there are better dataflows that create considerable regularity in the access pattern, and therefore that inefficiency can be ameliorated.

The cost of looking up a single payload in a tensor by its coordinates, which was described in Section 8.2, can be quite expensive. Fortunately, this is often not the key metric of interest, because it is very common for the payloads in a fiber to be traversed in *coordinate order*, i.e., in order of monotonically increasing (non-empty) coordinates. As we will see below, this can be true for DNN accelerators that attempt to exploit sparsity.

Coordinate order traversal for many fiber representations (e.g., the array-style representation and the coordinate/payload list representations) can be very efficient, often by employing a mechanism that remembers the current position in the fiber. For a multi-rank tensor (i.e., a tree with fibers as the payloads in the intermediate levels), a traversal of the tree in the order of the ranks in the tree would be correspondingly efficient. Such a traversal follows a depth-first traversal of the tree. We will refer to this highly desirable path through the tree as a *concordant* traversal of the tensor. Conversely, the less desirable path through the tensor where ranks are traveled in a different order than they appear in the tree is referred to as a *discordant* traversal.[14]

In many cases, a discordant traversal would require many distinct (possibly indirect) payload lookup operations, and its cost would be a function of the details of the implementation (e.g., caching could help). However, some representations support reasonably efficient discordant traversal, such as a flattened uncompressed representation, which sacrifices spatial locality, but does not require indirect references. The balancing of the tensor representation, efficiency of traversal, and the application of concordant or discordant traversal are critical design considerations for sparse DNN accelerators.

Most proposed DNN accelerators designed to exploit sparsity employ concordant traversal of the sparse input activation and/or weight tensors. Such a traversal generates a series of points (i.e., coordinate tuples) and the value at each of those points. The sequencing and use of the

---

[14]This terminology was coined by Michael Pellauer as part of the Symphony sparse computation accelerator project as words that have connotations that correspond to the characteristics of the traversals and have a musical allusion.

Design 8.6 Sparse 1-D Summation

```
1    t = Tensor(H)
2    sum = 0
3
4    for (h, t_val) in t:
5        sum += t_val
```

---

values generated by such traversals for all the tensors in a DNN computation correspond to a dataflow of the accelerator. Although important for understanding the ultimate design metrics of a design, we believe the detailed representation of the tensor is not necessary to understand the dataflow of the accelerators. One can get rough behavioral characterizations by counting the lengths of each concordant traversal and number of times each type operation is performed on each tensor. Therefore, we can use traversals of (and operations on) the sparse tensor abstraction presented in Section 8.2 to express dataflows of sparse DNN accelerators.

Given the desire to express dataflows that operate on sparse data, we employ a *Python-like* language with iteration operators to traverse a sparse tensor. As an example, in Design 8.6 we illustrate summing all the elements of a sparse 1-D tensor. The first line is a declaration of a 1-D tensor with a rank named H. Since the sum only needs to consider the non-zero elements of the tensor, we will express iteration over just the non-zero elements of a 1-D tensor using a `for` loop. This is shown in line 4 by a loop that iterates over tensor t. In reality, an iteration over a tensor implies an iteration over the elements in the topmost (and in this case, only) *fiber* of the tensor. Each step of the iteration returns a tuple consisting of the *coordinate* of the next non-empty element, h, and its *payload*, t_val, which for a 1-D tensor is a non-zero scalar value at coordinate h. Accumulating those values is performed in line 5. Assuming the concordant traversal of the fiber is efficient, this code represents the accumulation with efficient accesses to only the non-empty elements of the tensor, and its performance should be proportional to the length of the tensor t.

Design 8.7 extends the above simple example to a summation of all the elements of a 2-D tensor, which is declared in line 1 with ranks named "H" and "W". Now the payloads of the first traversal are fibers from the second rank, i.e., the W rank. Therefore the `for` loop in line 4 returns a coordinate h from the H rank and a fiber (i.e., the payload at that coordinate). The fiber is named t_w to represent the fact that is a fiber of the W rank of the tensor t. The `for` loop in line 5 traverses the non-empty elements of that fiber returning a coordinate w from the W rank and a non-zero value at the coordinate (h,w) (i.e., the payload at that coordinate). The value is named t_val to indicate that it is a value of the tensor t, and is combined into the sum in line 6.

By considering the `for` loops in this dataflow, the performance can be estimated to be proportional to the product of the length of the fiber in the top rank (H) of t and the average

Design 8.7 Sparse 2-D Summation

```
1    t = Tensor(H, W)
2    sum = 0
3
4    for (h, t_w) in t:
5        for (w, t_val) in t_w:
6            sum += t_val
```

Table 8.3: Sparse dataflow roadmap

Section	Dataflow Description	Examples
8.3.1	Convolution with sparse weights	Cambricon-X [278]
8.3.2	Convolution with sparse activations	Cnvlutin [225]
8.3.3	Convolution with sparse weights and activations	SCNN [156], SparTen [277], Eyeriss V2 [158]
8.3.4	Fully connected with sparse weights and activations	EIE [279], ExTensor [273]

length of the fibers in the bottom rank (W) of t. Similar performance estimates can be made for the other dataflows presented in this section.

Given this notation, which focuses on the dataflow without the complexity of dealing with a specific tensor representation, we can succinctly express a variety of dataflows that exploit sparsity. The following sections explore various dataflows that are designed to exploit sparsity in weights and/or activations for both CONV and FC layers. In those sections, we will both describe designs in terms of loop nests and block diagrams whose structure is implied by the loop nests.

Figure 8.25 shows a key for the components used in the design block diagrams. The block diagrams generally illustrate a single storage-level design that process untiled computations. Therefore, the storage elements are assumed to hold the entire data sets. In actual designs, those storage elements would be implemented using one of the hierarchical buffering schemes described in Section 5.8, such as caches, scratchpads, or explicit decoupled data orchestration (EDDO) units, like buffets.

Table 8.3 gives a roadmap of the dataflows that will be explored.

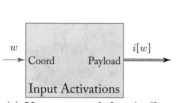

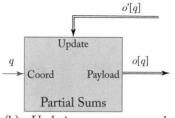

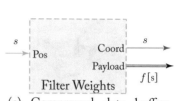

(a) Uncompressed data buffer: Buffer that accepts a *coordinate* via a network link (brown line) and returns via a network link (black double line) the value (i.e., the payload) for that coordinate by reading the value at that coordinate. Recall that *position* == *coordinate* when the tensor is uncompressed).

(b) Updating uncompressed data buffer: Enhanced uncompressed data buffer that also updates the value at the specified coordinate with a new value when that value arrives. Note, these designs wait for the update before accepting the next coordinate. Deeper pipelining optimizations are feasible, but ignored here.

(c) Compressed data buffer: Buffer that accepts a *position* via a network link (orange line) and returns both the *coordinate* and a value (i.e., the *payload*) at that coordinate. The dotted border of the box indicates that the data is sparse.

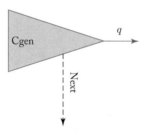

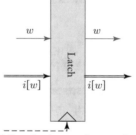

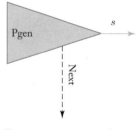

(d) Coordinate generator (Cgen): Finite state machine configured to generate a sequence of *coordinates* as well as a control signal (dashed black line) that indicates a breakpoint in a sequence of outputs. The sequences generated by Cgen correspond to the index variables in the **for** loops in a loop nest. The configuration of the coordinate generators is not expressed in the diagrams.

(e) Latch: A one element buffer that holds a value or coordinate/value pair. There is a control signal (dashed line) that tells when to latch a new value. A "stationary" value in a dataflow will typically be held in a latch.

(f) Position generator (Pgen): Finite state machine programmed to operate like the Cgen, but generates positions (orange line) instead of coordinates.

Figure 8.25: Key for structures in block diagrams of sparse architecture dataflows. (*Continues.*)

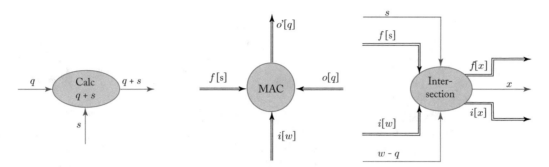

(g) Coordinate calculator: A small arithmetic unit for doing coordinate calculations, such as simple additions and subtractions.

(h) Multiply accumulate unit (MAC): Arithmetic unit that takes in three values (double black lines) multiplies two of the values together and adds the product to the third value and returns the result.

(i) Intersection unit: Unit that takes in two streams of coordinate/value pairs and generates a stream of coordinates and two streams of values only when the coordinates from the two input streams match.

Figure 8.25: (*Continued.*) Key for structures in block diagrams of sparse architecture dataflows.

8.3.1 EXPLOITING SPARSE WEIGHTS

A very natural opportunity to exploit sparsity in DNN accelerators is to save time for zero weights, such as those provided by weight pruning, as described in Section 8.1.2. An advantage of dealing with sparse weights (as opposed to activations) is that they can be known statically and their compressed representation can be generated once (possibly offline in software) alleviating the hardware of the responsibility and costs of doing that compression.

Just as with dense computations described in Chapter 5, there are different dataflow options for processing a convolution with sparse weights. Design 8.8 illustrates a weight-stationary dataflow that exploits sparse weights. In that dataflow, the input and output feature maps are assumed dense and uncompressed, so they are represented as a standard `Array` and lookup by coordinate uses the standard array access operator (`[]`). On the other hand, the weights are assumed sparse and compressed, so are represented by the fiber-tree tensor abstraction, but are assumed to be in a design-specific compressed representation that allows efficient concordant traversal of its fibers.

Computation of the weight-stationary 1-D convolution in Design 8.8 proceeds by accessing each non-zero filter weight via the `for` loop in line 5 and holding it stationary, while the `for` loop in line 6 traverses the output locations. Line 7 calculates the coordinate of the input activation needed for this step using the current output coordinate, q, and the filter weight coordinate, s. The computation in line 8 performs the computation of partial sums by direct accesses to the

---

Design 8.8 1-D Weight-Stationary Convolution Dataflow - Sparse Weights

---

```
1    i = Array(W)        # Uncompressed input  activations
2    f = Tensor(S)       # Compressed filter  weights
3    o = Array(Q)        # Uncompressed output  activations
4
5    for (s, f_val) in f:
6        for q in [0, Q):
7            w = q + s
8            o[q] += i[w] * f_val
```

Design 8.9 1-D Output-Stationary Convolution Dataflow - Sparse Weights

---

```
1    i = Array(W)        # Uncompressed input  activations
2    f = Tensor(S)       # Compressed filter  weights
3    o = Array(Q)        # Uncompressed output  activations
4
5    for q in [0, Q):
6        for (s, f_val) in f:
7            w = q + s
8            o[q] += i[w] * f_val
```

---

input activations (`i[]`) and partial sums (`o[]`) by coordinate. The filter weight itself, `f_val`, is provided by the traversal of the sparse filter weight tensor.

A block diagram of the 1-D weight-stationary convolution dataflow with sparse weights is shown in Figure 8.26. A notable feature of the diagram are the latch that holds the "stationary" weight and its coordinate, while successive non-zero weights are blocked until the latch grabs a new value. That is controlled by the "Next" signal from the partial sum coordinate generator, which is sent on each of the |`f[]`| passes through the output partial sums.[15]

An advantage of the weight-stationary dataflow for sparse weights is that any complexity in accessing the compressed weights can be buried under the multiple iterations through the output activations in the inner loop. On the other hand, the output-stationary dataflow for sparse weights in Design 8.9 allows the hardware to accumulate partial sums into a single register, thus avoiding the access costs required for repeated reads and writes into a larger storage array.

---

[15]We are using the absolute value notation |`f[]`| to represent the number of valid positions (i.e., non-zero values) in the tensor `f`.

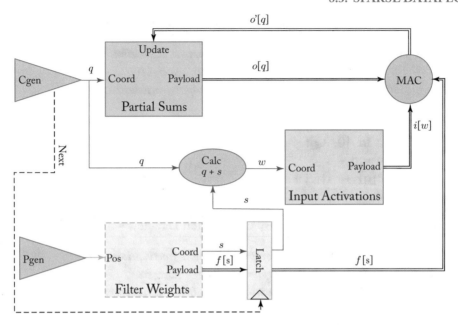

Figure 8.26: 1-D Weight-Stationary Convolution Dataflow – Sparse Weights: in this design, the filter weight Pgen (outer loop) is configured to generate the sequence of positions $[0, |f[]|)$ exactly once. The partial sum Cgen (inner loop) is configured to generate the coordinate sequence $[0, Q)$ one time for each non-zero filter weight (i.e., $|f[]|$) times). The weights (f[s]) and their coordinates (s) are held "stationary" in the latch, which is filled each time the partial sum Cgen starts a sequence with the "Next" control line from partial sum Cgen. That weight and the appropriate input activation are multiplied and added to the sequence of partial sums (o[q]), which are updated with the output of the MAC.

A block diagram of the 1-D output-stationary convolution dataflow with sparse weights is shown in Figure 8.27. Notable features are the partial sum latch which holds partial sums "stationary" until the final sum is sent to the partial sum buffer.

The performance of either dataflow can be estimated to be proportional to the product of the number of output activations (Q) and the number of non-zero weights (i.e., the length of the rank S fiber in the filter weight tensor (f)). This highlights the fact that neither dataflow provides any time savings benefits when an activation (as opposed to a weight) is zero.

Cambricon-X

A DNN accelerator architecture that attempts to exploit weight sparsity is Cambricon-X [281]. Cambricon-X is a weight-stationary dataflow for sparse weights enhanced from the above simple examples in a variety of ways. One way it is enhanced is through additional parallelism, which

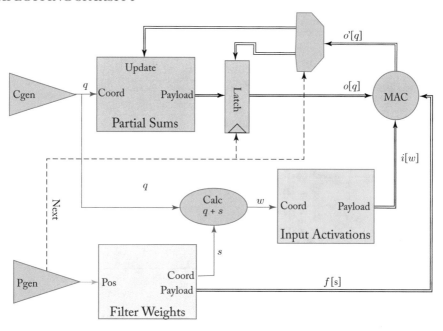

Figure 8.27: 1-D Output-Stationary Convolution Dataflow – Sparse Weights: in this design, the partial sum Cgen (outer loop) is configured to generate the coordinate sequence $[0, Q)$ exactly one time. The filter weight Pgen (inner loop) generates the sequence of positions $[0, |f[]|)$ repeatedly Q times. The partial sums (o[q]) are held "stationary" in the latch, which is latched with the initial partial sum each time the Pgen starts a sequence with the "Next" control line from filter weight Pgen. That control signal also sends the final partial sum back to the buffer through the demultiplexer. MAC operations are performed for each non-zero weight (f[s]) and the appropriate input activation (i[q+s]) and added to the current partial sum o[q] creating a new partial sum (o'[q]).

is manifest by both working on multiple weights in parallel and working on multiple output activations in parallel.

A sample dataflow for a 1-D convolution illustrating parallelism for weights and outputs is shown in Design 8.10. In line 5, the weight fiber is split into subfibers of length 2 (i.e., fibers each with 2 coordinates) by the splitEqual() method. The effect of such splitting is illustrated in Figure 8.28 where a fiber with six sparse coordinates is split into groups of two by grabbing coordinate/payload pairs from consecutive *positions* in the original fiber. Because the criteria for splitting is based on positions, this is referred to as *position-space splitting*.[16] After the split, the original coordinates from the original rank (S) are preserved in the lower rank (S0), and

---

[16]A variety of other splitting semantics exist, such as splitting in the original coordinate space and/or splitting into unequal pieces.

Design 8.10 1-D Parallel Weight-Stationary Convolution Dataflow - Sparse Weights

```
 1      i = Array(W)          # Uncompressed  input  activations
 2      f = Tensor(S)         # Compressed  filter  weights
 3      o = Array(Q)          # Uncompressed  output  activations
 4
 5      for (s1,  f_split ) in  f. splitEqual (2):
 6          for q1 in  [0,  Q/4):
 7              parallel-for  (s0,  f_val ) in  f_split
 8                  parallel-for  q0 in  [0,  4)
 9                      s = s0
10                      q = q1*4 + q0
11                      w = q + s
12                      o[q]  += i [w] * f_val
```

coordinates in the upper rank (S1) just indicate the group number. Note, there are many choices for representing a split fiber; for instance, adding a small amount of meta-data referring to the original fiber may be a superior design choice compared to adding a completely new rank and creating new split fibers.

Returning to the dataflow, after the split a `parallel-for` in line 7 operates on each of the elements in that split sub-fiber (`f_split`) in parallel. In lines 6 and 8, the indices of the uncompressed output activations are partitioned (in position space) into groups of 4 for additional parallel execution. Therefore, in total this dataflow provides eight-way (2×4) parallelism.

An actual sparse DNN accelerator, such as Cambricon-X, would need to consider a number of additional factors. Factors in common with dense DNN accelerators include the need to expand each of the input and output operands to the full multi-dimension tensors of standard DNN computations (e.g., multiple input and output channels). It also needs to consider multi-level buffering and the sizing of those buffers to accommodate various problem sizes. However, for sparse DNN accelerators there is an additional wrinkle, because they need to cope with the fact that the space used in a buffer is now a function of sparsity. Since weight sparsity is known statically, the mapping can take those sizes into account, otherwise conservative assumptions or an exception handling mechanism to cope with buffer overflows would be needed.

A final consideration is the actual data representation to be used for the sparse data. This can have a significant impact on the overall efficiency of a design and needs to be considered as a co-design with the dataflow selection. For example, the Cambricon-X team considered two *compressed, implict-coordinate* schemes for holding filter weights. They evaluated a representation using a bit-mask indicating non-empty elements in the weight fiber and an RLE-type scheme. They found the RLE-type scheme more efficient for their dataflow.

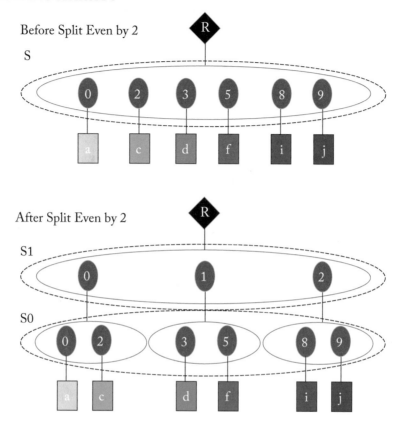

Figure 8.28: Effect of splitting a fiber into subfibers where each subfiber contains the same number of coordinates, i.e., evenly. This is referred to as splitting the fiber evenly in *position* space.

8.3.2   EXPLOITING SPARSE ACTIVATIONS

As described in Section 8.1.1, activations can also be sparse, so it should be unsurprising that analogous to the dataflows that exploit sparse weights in Section 8.3.1 one can find dataflows that exploit sparse activations.

A 1-D weight-stationary convolution dataflow that exploits sparse activations is shown in Design 8.11. Iterating over the coordinates of the uncompressed weights in the outermost **for** loop (line 5) highlights that this is a weight-stationary dataflow. While in the inner **for** loop (line 6) the dataflow only iterates over the non-zero activations. Note, however, that depending on the current weight's coordinate, some input activations do not contribute to any output activations and so these *edge effects* are skipped with the **if** condition on the **for** statement on line 6. The hardware would need to account for such edge effects in this design and later designs. This hardware ideally would be able to save time for coordinates outside the constraints, but may

---

Design 8.11 1-D Weight-Stationary Convolution Dataflow - Sparse Activations

---

```
1    i = Tensor(W)        # Compressed input activations
2    f = Array(S)         # Uncompressed filter weights
3    o = Array(Q)         # Uncompressed output activations
4
5    for s in [0, S):
6        for (w, i_val) in i if s <= w < Q+s:
7            q = w - s
8            o[q] += i_val * f[s]
```

Design 8.12 1-D Output-Stationary Convolution Dataflow - Sparse Activations

---

```
1    i = Tensor(W)        # Compressed input activations
2    f = Array(S)         # Uncompressed filter weights
3    o = Array(Q)         # Uncompressed output activations
4
5    for q in [0, Q):
6        for (w, i_val) in i if q <= w < q + S:
7            s = w - q
8            o[q] += i_val * f[s]
```

---

need to spend a cycle when it goes outside the constraint. Such considerations are beyond the scope of this book.

Hardware that implements this dataflow could have logic that overlaps the determination of skipped activations with other processing and not actually waste cycles. The designer would also need to pick a fiber representation for the input activations that makes traversal of the non-zero elements and generation of their coordinates efficient. Thus, for example, either a coordinate/payload or an RLE-style fiber representation could be a good choice.

A block diagram of the 1-D weight-stationary convolution dataflow with sparse activations is shown in Figure 8.29. Note that there is a latch that holds the current weight and its coordinate "stationary" through a series of input activations.

An alternative dataflow for exploiting sparse activation in a 1-D convolution is shown in Design 8.12. In the outer **for** loop (line 5), the dataflow iterates over all the coordinates of the output tensor, which confirms that this is an output-stationary dataflow. The inner **for** loop (line 6) iterates over a restricted set of the non-zero input activations that contribute to the current output. That restriction is represented by the condition in the **if** statement on line 6

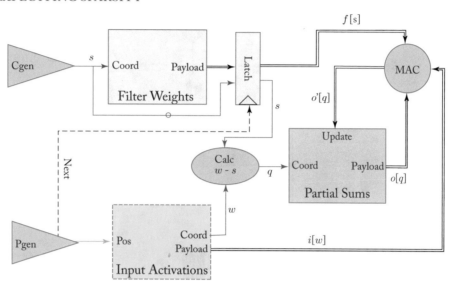

Figure 8.29: 1-D Weight-Stationary Convolution Dataflow – Sparse Activations: in this design, the filter weight Cgen (outer loop) is configured to generate the coordinate sequence $[0, S)$ exactly one time. The input activation Pgen (inner loop) generates the data dependent sequence of positions between $[0, |i\,[\,]|)$, where $s \leq w < Q + s$ for each weight (i.e., S times). That Pgen signals the beginning of each sequence on the "Next" control line to latch the current weight (f[s]) and its coordinate (s). MAC operations are performed for each non-zero input activation (i[w]) and "stationary" weight (f[s]) and added to the appropriate partial sum (o[w-s]) creating a new partial sum (o'[w-s]), which updates the partial sum buffer.

such that the iterator over the input activation fiber only returns the coordinates and values for the appropriate non-zero input activations in the fiber.

Most of the hardware to cope with sparsity in Design 8.12 is concentrated in line 6. Here, we see the iteration over the non-zero elements of the fiber containing the input activations, i. Thus, that fiber can be in a compressed representation (e.g., a coordinate/payload list). Line 6 also shows the restriction to input activations that contribute to the current output activation. Careful examination of the pattern of coordinate/value pairs shows that the loop will traverse a sliding window of input activations, where each window has a variable number of coordinates, but will cover a constant distance in coordinate space. An example sequence of windows for a sparse set of input activations and a 3-wide filter is shown in Figure 8.30. Again, the logic to calculate that window can be overlapped with fetching values in the window.

A block diagram of the 1-D output-stationary convolution dataflow with sparse activations is shown in Figure 8.31.

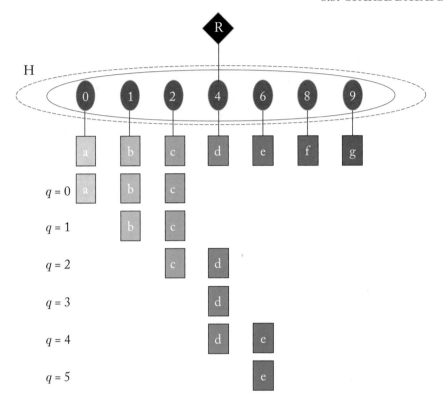

Figure 8.30: Input activation sparse sliding window – filter size S= 3.

Cnvlutin

The Cnvlutin sparse DNN accelerator is an augmentation of the output-stationary sparse activation dataflow in Design 8.12 [228]. The augmentation includes processing input activations from multiple input channels simultaneously to allow a spatial accumulation (see Section 5.9). Cnvlutin also combines each input activation with weights from distinct output channel filters to simultaneously generate output activations for multiple output channels. It also needs to cope with the fact that the feature maps and filters have both a height and a width. Thus, the code for the dataflow for Cnvlutin would require additional `for` loops for the additional ranks of the weights, and input and output activations.

Finally, Cnvlutin needs a representation for the input activations, and is described as having a relative coordinate (or offset)/payload representation.

8.3.3   EXPLOITING SPARSE WEIGHTS AND ACTIVATIONS

The preceding two sections (Sections 8.3.1 and 8.3.2) described dataflows that only exploit the fact that one of the operands of the convolution, either weights or input activations, were sparse.

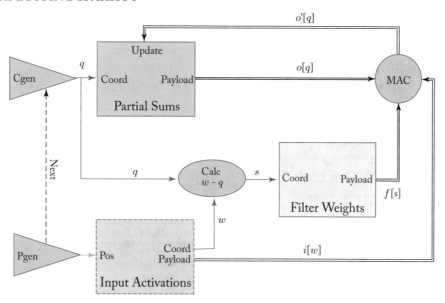

Figure 8.31: 1-D Output-Stationary Convolution Dataflow – Sparse Activations: in this design, the partial sum Cgen (outer loop) is configured to generate the coordinate sequence $[0, Q)$ exactly one time. The input activation Pgen (inner loop) generates the data dependent sequence of positions between $[0, |i[]|)$, where $q \leq w < q + S$ for each partial sum (i.e., Q times). That Pgen signals the beginning of each sequence on the"Next" control line to move the Cgen to the next coordinate. MAC operations are performed for each non-zero input activation (i[w]) and weight (f[w-q]) and added to the appropriate partial sum (o[q]) creating a new partial sum (o'[q]), which updates the partial sum buffer. Note this design could easily be extended to use a single "stationary" partial sum latch to hold the accumulating partial sum.

Naturally, since a multiplier has to do no work when either of the input operands, weights or input activations, are zero, it would be attractive to try to implement a dataflow that saves time when either operand is zero.

Exploiting sparsity in both weights and activations in a dataflow is more challenging than exploiting sparsity in only one datatype. Therefore, some works have combined a dataflow that exploits sparsity in one datatype (weights) with hardware that exploits bit-level sparsity in the other datatype (input activations) to reduce the cost of a multiplication [283]. However, a number of works have successfully created dataflows that exploit sparsity in both datatypes, which we will describe in this section.

A 1-D input-stationary convolution dataflow that exploits both weight and input activation sparsity is presented in Design 8.13. Here, we see that every input activation is multiplied by every weight and accumulated into some output activation (at coordinate q = w-s). Note, that

---

Design 8.13 1-D Input-Stationary Convolution Dataflow - Sparse Weights & Activations

```
1    i = Tensor(W)         # Compressed input   activations
2    f = Tensor(S)         # Compressed filter  weights
3    o = Array(Q)          # Uncompressed output  activations
4
5    for (w, i_val) in i:
6        for (s, f_val) in f if w-Q < s <= w:
7            q = w - s
8            o[q] += i_val * f_val
```

Design 8.14 1-D Sparse Input-Stationary Convolution Dataflow – Cartesian Product

```
1    i = Tensor(W)         # Compressed input   activations
2    f = Tensor(S)         # Compressed filter  weights
3    o = Array(Q)          # Uncompressed output  activations
4
5    for (w1, i_split) in i. splitEqual (2):
6        for (s1, f_split) in f. splitEqual (2):
7            parallel-for (w0, i_val) in i_split :
8                parallel-for (s0, f_val) in f_split  if w0-Q<s0<=w0
9                    w = w0
10                   s = s0
11                   q = w - s
12                   o[q] += i_val * f_val
```

---

this is with the exception of edge effects that try to generate values for invalid output activation coordinates. Those edge effects are controlled by the if statement on line 6.

A block diagram of the 1-D input-stationary convolution dataflow with sparse weights and activations is shown in Figure 8.32.

SCNN

The dataflow in Design 8.13 has interesting ramifications when parallelism is added. Specifically, consider the dataflow in Design 8.14. Following the splitting pattern from Design 8.10, we generate fibers with two input activations (i_split in each iteration of line 5) and two weights (f_split in each iteration of line 6). Those pairs of values are delivered in parallel (in lines 7 and 8) to the multipliers in line 12.

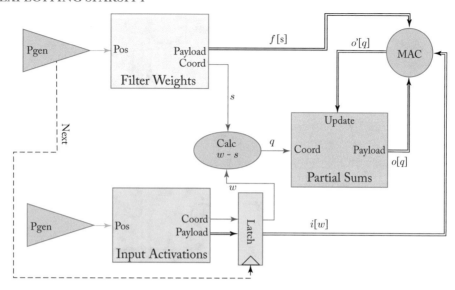

Figure 8.32: 1-D Input-Stationary Convolution Dataflow – Sparse Weights and Activations: in this design, the input activation Pgen (outer loop) is configured to traverse the positions of the non-zero inputs exactly once in the with the sequence $[0, |i[]|)$. The filter weight Pgen (inner loop) generates a data dependent sequence of positions $[0, |f[]|)$, where $w - Q < s \leq w$ repeatedly for each input activation (i.e., $|i[]|$). Each time it starts a sequence it signals the latch via the "Next" control line to latch a new input. MAC operations are performed for each non-zero input activation (i[w]) and non-zero weight (f[s]) and added to the appropriate partial sum (o[w-s]) creating a new partial sum (o'[w-s]), which updates the partial sum buffer.

Examination of the pattern of operand delivery to line 12 reveals that two inputs activations (`i_val`) and two weights (`f_val`) are delivered to the four multipliers in a cross-product pattern. Figure 8.33 illustrates the flow of the information from the `i_split` and `f_split` fibers to the multipliers. The attractive feature of this topology is that reads of just $2 \times N$ values provide operands for N^2 multipliers. Note that while the values must flow to the multipliers, the coordinates must also flow to those units to calculate the coordinate of the output activation that needs to be accumulated into. So at the right side of the diagram the resultant spray of operands are shown. This cross (or Cartesian) product is a integral part of the SCNN design [159], which uses clusters with 16 multipliers (i.e., $N = 4$).

The input activation and weight fibers in a 1-D convolution would undoubtedly be too small to produce high utilization in the Cartesian product unit. So the SCNN design *flattens* (see Section 8.2.4) various ranks of both the input activations and weights. Specifically, the H and W ranks of the input activations are flattened (as illustrated in Figure 8.22) and the R, S, and M ranks of the weights are similarly flattened. This produces larger fibers that improve

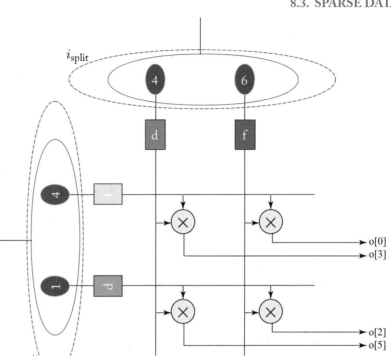

Figure 8.33: Sparse Cartesian product.

multiplier utilization in the sparse input-stationary Cartesian product dataflow. The tuples of coordinates in the flattened rank are used to compute the output activation coordinate for each product. In SCNN, distinct input channels (rank C) are processed serially and additional parallelism is achieved by having distinct Cartesian product complexes work on different tiles of input activations (split in coordinate space on the H and W ranks).

The loops in Design 8.13 could easily be reversed to create a weight-stationary dataflow, however, as argued in the SCNN paper, that would provide no performance gain and would result in more frequent accesses to the larger input activation storage array resulting in lower energy efficiency.

The energy cost associated with SCNN's spray of outputs into the output feature map tensor has led to a consideration of output-stationary dataflows that exploit sparsity in both weights and input activations. Design 8.15 illustrates such a dataflow. The outer loop (line 5) is a standard traversal of the coordinates of the output feature map tensor. The inner loop (line 6), however, show some new tricks: intersection and projection.

The ampersand (&) operator returns the intersection of two fibers. This operator scans the coordinates of each of its input operands and returns a new fiber with only the coordinates

Design 8.15 1-D Output-Stationary Convolution Dataflow - Sparse Weights & Activations

```
1    i = Tensor(W)       # Compressed input   activations
2    f = Tensor(S)       # Compressed filter  weights
3    o = Array(Q)        # Uncompressed output  activations
4
5    for q in [0,Q):
6        for (w, (f_val, i_val)) in f.project(+q) & i:
7            o[q] += i_val * f_val
```

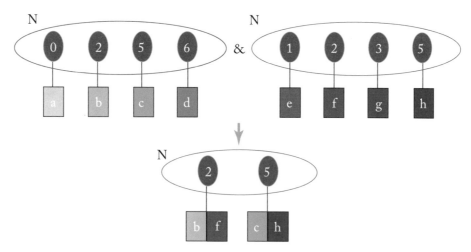

Figure 8.34: Fiber intersection.

that appear in both and also returns payloads that are a combination of the payloads from the original fibers. Figure 8.34 illustrates this action on two 1-D fibers. The intersection between the two fibers returns a new fiber with the common coordinates of the original fibers (2 and 5) and payloads that are tuples of the original fiber's payloads ((b, f) and (c, h)). An implementation of a fiber intersection unit for coordinate/payload-style fibers was explored in the ExTensor design [276].

In order for the intersection to not repeatedly act on the same elements of both fibers (f and i), the filter weights involved in the intersection need to move as a *window* over the input activations for each distinct partial sum. To achieve this effect, we employ a projection method (project(<offset>)) that shifts the coordinates in a fiber by a specified offset.[17] The operation

---

[17]Although we use a simple offset for projection here, in general one would need something more powerful, like a lambda function, that can more generally calculate a new coordinate.

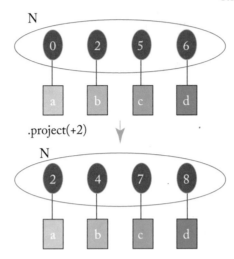

Figure 8.35: Fiber projection.

of the `project()` method on a 1-D fiber is illustrated in Figure 8.35, where all the coordinates of the input fiber are shifted up by two.

A block diagram of the 1-D output-stationary convolution dataflow with sparse weights and activations is shown in Figure 8.36. That diagram shows that separate streams of input activations and (sliding window of) filter weights being intersected to create the stream of operand pairs for the MAC.

SparTen

The SparTen DNN architecture uses a core dataflow similar to Design 8.15 optimized with specialized tensor representations for the input feature map and filter weights in order to implement intersection efficiently [280]. SparTen uses a bitmask with zeros for zero coordinates and ones for non-zero coordinates, so the position in the bitmask corresponds to a coordinate. In such a representation, the `getPayload()` method is implemented by counting the number of ones in the bitmask up to the desired coordinate to find the position of the desired payload. Finding intersected coordinates uses a simple Boolean AND operation on the bitmasks. Finding the positions of the payloads that survive the intersection is performed by counting ones in the original bitmask up to the desired coordinate (i.e., invoking `getPayload()`).

Eyeriss V2

As described in Chapter 3, a key challenge in the design of DNN accelerators is finding the right balance between flexibilty and other metrics, primarily energy and performance. The Eyeriss V2 DNN accelerator strove to be efficient both across a wide range of DNN network shapes (see Chapters 2 and 9), but also work efficiently across a range of sparsities [161]. Like the

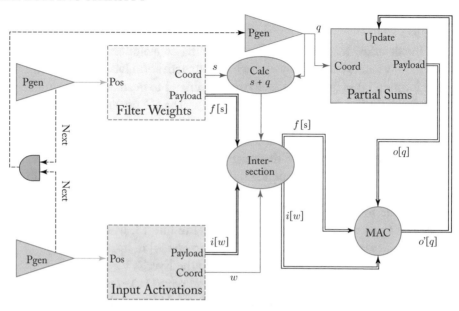

Figure 8.36: 1-D Output-Stationary Convolution Dataflow – Sparse Weights and Activations: in this design, the filter weight and input activation Pgens are each configured to traverse the positions of the non-zero values in their respective buffers repeatedly for each partial sum (i.e., Q times). Ideally, the input activation Pgen will only traverse the active part of the sliding window of positions. In any case, the resultant stream of input activation coordinates and values; and the filter weight's (projected) coordinates and values will be intersected generating a stream of values pairs when the same coordinate exists in both streams. Those resulting input activations and filter weights are multiplied together in MAC unit and added to the current partial sum from the partial sum buffer. The same partial sum (at coordinate q) is accumulated to until both the filter weight and input activation Pgens both have started a new sequence and informed the partial sum Pgen (via the ANDed combination of "Next" signals) that it should move to the next partial sum coordinate (q).

SCNN and SparTen accelerators described above, Eyeriss V2 exploits both weight and input activation sparsity. But in contrast to SCNN, which loses significant efficiency when the data is dense, Eyeriss V2 strives to provide good efficiency irrespective of sparsity. It achieves this with a dataflow that provides modest parallelism per PE and good utilization across a diverse set of workloads.

The Eyeriss V2 dataflow is a derivative of the original Eyeriss row-stationary dataflow (see Section 5.7.4). Each PE handles the convolution for one row of input activations, and partial sums are combined in a column of PEs to create an output activation. Eyeriss V2 is extended, however, to support sparse input feature maps and filter weights. The PEs also provide 2-way parallelism.

The dataflow for Eyeriss V2 for performing convolution[18] is shown in Design 8.16. The PE's dataflow works successively on each output activation location (q) in line 5 and each filter weight location (s) in line 6. It uses those coordinates to get a fiber (i_c) holding the input activations for all the input channels using the getPayload() operation in line 8. Those accesses will follow a sliding window pattern through the input activations that the hardware must generate. Then for each input channel with a non-zero activation (i.e., traversing i_c), the getPayload() in line 10 will access a fiber of filter weights (f_m) containing weights for each output channel with a non-zero weight. The concrete representation of f is uncompressed for the "C" and "S" ranks, which makes that getPayload() operation efficient. The filter weight tensor is in coordinate/payload form for the "M" rank, which allows for splitting the weights into groups of two (line 11) and processing two weights in parallel in line 13, where each weight is multiplied by the same input activation and accumulated in to $o[m,q]$ in line 14.

To improve throughput, the multiplies in the Eyeriss V2 PE are performed with two-way parallelism as indicated by the splitEven() and **parallel-for** in lines 11 and 12. The typical number of filter weights for a specific input channel is generally large enough to achieve good utilization of the multipliers. The two-way spray of parallel accumulates (line 13) also means that the design requires a two-port register file for output activations. This degree of parallelism is a compromise between the larger amount of regular parallelism that would be available for dense data, and both the amount of parallelism available and the complexity of a register file with more ports that would be needed to support more parallelism for sparse data.

8.3.4   EXPLOITING SPARSITY IN FC LAYERS

The previous subsections have considered the fundamental dataflows of DNN accelerators that targeted convolutional layers, but there also has been some work targeting FC layers. As described in Chapter 4, FC layers are essentially matrix-matrix or vector-matrix multiplications, so these accelerators target that calculation.

EIE

EIE is a DNN accelerator for FC layers that processes an input activation vector into a M-channel output activation vector [282]. A simplified representation of the EIE DNN accelerator for FC layers is depicted as Design 8.17. There are two things to note about this dataflow. First, as described in Section 8.2.4, the input channel (C), height (H), and width (W) ranks can be flattened into one rank (CHW). Second, recall that for FC layers the range of filter weights is equal to the range of input activations (i.e., CRS == CHW). So we see that for each input activation, this input-stationary dataflow selects a row of weights (f_m) and multiplies the

---

[18]Eyeriss V2 also supports FC layers with a different dataflow.

Design 8.16 Eyeriss V2 Convolution Dataflow - Sparse Weights & Activations

---

```
1   i = Tensor(C,W)          # Compressed input   activations
2   f = Tensor(C,S,M)        # Compressed  filter   weights
3   o = Array(M,Q)           # Uncompressed output   activations
4
5   for q in [0, Q):
6     for s in [0,S):
7         w = q+s
8         i_c = i_w.getPayload(w)
9         for (c, i_val) in i_c:
10            f_m = f_c.getPayload(c, s)
11            f_m_split = f_m.splitEqual(2)
12            for (_, f_m) in f_m_split:
13                parallel-for (m, f_val) in f_m:
14                    o[m, q] += i_val * f_val
```

Design 8.17 1-D Input-Stationary Fully Connected Dataflow – Sparse Weights & Activations

---

```
1   i = Tensor(CHW)          # Compressed input   activations
2   f = Tensor(CHW, M)       # Compressed  filter   weights
3   o = Array(M)             # Uncompressed output   activations
4
5   for (chw, i_val) in i:
6       f_m = f.getPayload(chw)
7       for (m, f_val) in f_m:
8           o[m] += i_val * f_val
```

"stationary" input activation by each of the weights in the row, and contributes the product to the partial sum for the appropriate output channel (m).

In order to have efficient execution of the getPayload() method on the filter weights' CHW rank and concordant traversal of the filter weights' M rank, the fiber-tree should have a rank order of CHW and M. Furthermore, it is extremely likely that there is at least one weight for each coordinate in CHW (i.e., that rank is dense), so an uncompressed format is appropriate. Thus, EIE uses a format that is a representation for a 2-D tensor using an uncompressed upper rank and an RLE-style lower rank because only a fraction of the output channels will

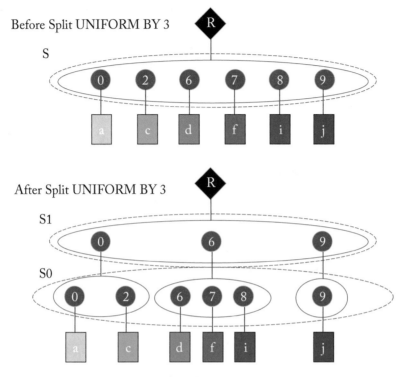

Figure 8.37: Effect of splitting a fiber uniformly in coordinate space by 3.

have a weight for any particular input activation coordinate.[19] It appears that EIE also achieves concordant traversal of the input activations by employing an uncompressed format.

To provide parallelism, EIE partitions the M rank of the weight tensor in position space among multiple PEs using a filter weight tensor where the output channels (M) are split equally in position space into two ranks (M1 and M0) and then the ranks are re-ordered M1, CHW, and M0.

Design 8.18 shows EIE's parallelism. Since this splitting is done on weights, which are known statically, the partitioning need not be done at runtime. However, note further that the subtrees that result from this split may not all be the same size. This can result in load imbalance, which can cause under-utilization of the MAC units and is a pervasive issue in sparse dataflows. One can also see that the same input activations' coordinate (chw) and value (i_val) are used by multiple parallel units, so they must be broadcast to all the PE units.

One final attribute of the EIE design is that it uses quantized weights (see Chapter 7). This could be added to this (or most any) dataflow in a straightforward fashion by treating the weight (f_val) as a index into another array containing the actual weights.

---

[19]The EIE papers calls this a variant of CSC, but our taxonomy would make it a distinct format.

Design 8.18 Parallel Input-Stationary Fully Connected Dataflow - Sparse Weights & Activations

```
1    i = Tensor(CHW)            # Compressed input    activations
2    f = Tensor(M1,CHW, M0)     # Compressed filter   weights
3    o = Array(M)               # Uncompressed output activations
4
5    for (chw, i_val) in i:
6        parallel-for (m1, f_chw) in f:
7            f_m = f_chw.getPayload(chw)
8                for (m, f_val) in f_m:
9                    o[m] += i_val * f_val
```

ExTensor

FC layers that exploit both sparse input activations and weights can also use other dataflows. For example, although not specifically designed for DNN acceleration, the ExTensor accelerator implements multi-level tiled sparse matrix multiplication [276]. Therefore, it can execute an arbitrary batch size (N) for fully connected layers. Furthermore, it selectively uses either a weight-, input-, or output-stationary dataflow at each level of the storage hierarchy and its performance is enhanced with optimized intersection units.

Tiling the operands of a multiply requires that the corresponding coordinates exist in the tiles of both operands, because the arithmetic units need operands with the same coordinates. This involves splitting fibers in coordinate space.

Figure 8.37 displays the effect of splitting a fiber uniformly in coordinate space. The figure shows that the coordinates of the original fiber divided into groups of three by coordinates. Therefore, the groups in coordinate space are 0-2, 3-5, 6-8, and 9-11. The newly created upper fiber (S1) has a coordinate that matches the first coordinate of each group, and has a payload that is a fiber in the lower rank (S0) with the coordinate/payload pairs that existed in the original fiber with coordinates of that group. Note that the fibers in S0 are of different sizes. In fact, since there were no coordinates in the range 3-5, i.e., the fiber in S1 would be empty, and there is no coordinate 3 in the upper (S1) fiber. This can lead to load imbalance between parallel units.

A sample weight-stationary single-tile-level ExTensor dataflow is shown in Design 8.19. Like the output-stationary sparse convolution dataflow (Design 8.15), this dataflow also employs an intersection. However, note that the intersection is between the coordinates of weight *values* (f_val) and the coordinates of input activation *fibers* (i_n). This implies that when an intersection drops a coordinate a considerable amount of work might be being saved (i.e., the entire traversal of the input activation fiber).

Design 8.19 ExTensor Weight-Stationary Fully Connected Dataflow – Sparse Weights & Activations and batch size N

```
1    i = Tensor(CHW, N)        # Compressed  input   activations
2    f = Tensor(M, CHW)        # Compressed  filter   weights
3    o = Array(M)              # Uncompressed output   activations
4
5    for (m, f_chw) in f:
6        for (chw, (f_val, i_n)) in f_chw & i:
7            for (n, i_val) in i_n:
8                o[n,m] += i_val * f_val
```

8.3.5   SUMMARY OF SPARSE DATAFLOWS

In this section, we have surveyed various dataflows used to exploit sparsity in both filter weights and input feature maps for both convolutional and FC computations. These dataflows form the core of a variety of DNN accelerator designs, each of which is augmented with additional layers of buffering and higher-level parallelism. The dataflows were represented in a uniform notation that separated the dataflow from the details of the representation (and manipulations) of the input and output tensors. This allows for a comparison of the core computation flow of the individual designs, by illuminating the data accesses and the operations needed to implement each dataflow. This can be used to characterize their behavior on various workloads and to infer the hardware necessary to implement the dataflows. At present, however, there is no comprehensive comparative analysis of the alternative sparse dataflows across a wide range of design parameterizations (e.g., buffer sizes) and workloads.

8.4   SUMMARY

This chapter explored the origins, explicit creation, and exploitation of sparsity in DNN computations, where sparsity refers to the fact that there are many repeated values, usually zeros, in the data. This chapter presents various sources of sparsity (e.g., the ReLU nonlinearity that sets negative values to zero in the feature map activations, or repeated values in the weights of a filter) as well as methods that can increase sparsity (e.g., exploiting correlation in the data or removing weights using pruning).

Two potential architectural benefits of sparsity were also discussed: (1) compression of sparse data can reduce its footprint, which provides an opportunity to reduce storage requirements and data movement; and (2) sparsity presents an opportunity for a reduction in MAC operations. The reduction in MAC operations results from the fact that $0 \times anything$ is 0. This can result in either savings in energy or time or both.

To explore the potential architectural benefits on energy-efficiency and throughput of sparsity, this chapter presented a sampling of sparse architectures as a composition of two elements: (1) a dataflow that operated on an abstract representation of a sparse tensor; and (2) implementation choices for the concrete data formats for the tensors.

CHAPTER 9

Designing Efficient DNN Models

The previous two chapters discussed the use of DNN model and hardware co-design approaches, such as reducing precision (Chapter 7) and exploiting sparsity (Chapter 8), to reduce storage requirements, data movement, and the amount of MAC operations required for processing a DNN model. In this chapter, we will discuss how designing DNN models with efficient "structures," often referred to as the *DNN network architecture*,[1] can also help enable efficient processing of DNNs. The network architecture is defined by the layer types, the layer shapes, the number of layers, and the connections between layers, as defined in Chapter 2. Designing efficient network architectures involves applying techniques to these different aspects to enable efficient processing. As with the other co-design approaches, the main challenge is to improve the efficiency of the network architecture as evaluated by the metrics described in Chapter 3, such as energy consumption and latency, without sacrificing the accuracy.

In earlier works, improving the network architecture relied on the researchers' expertise to manually design layers and figure out the optimal connections between them; however, this is often a tedious and challenging task. Consequently, in recent years, the use of machine learning to automatically design the network architecture, referred to as *Neural Architecture Search (NAS)*, has become an increasingly popular research area. While many efficient network architecture design approaches focus on reducing the number of weights, activations, and MAC operations to improve efficiency, this does not necessarily translate to reduced energy consumption and latency. As previously discussed, one must also account for factors, such as utilization and the cost of data movement, which depend on the dataflow and memory hierarchy; in other words, one must consider how the DNN model is mapped onto the hardware in order to evaluate efficiency. Accordingly, recent research efforts have proposed methods that directly target hardware metrics such as energy consumption and latency when designing efficient network architectures.

In this chapter, we will first discuss common methods that are widely used in manual network design. Second, we will explain how NAS can be used for network architecture design, describe the key components of NAS, and discuss the associated design considerations. In this context, we will also discuss how to bring hardware into the design loop to directly target metrics such as energy consumption and latency. Third, we will discuss a class of methods called *knowl-*

---

[1]Note: The term DNN network architecture differs from the *hardware* architecture or *network-on-chip* architecture described in the other chapters.

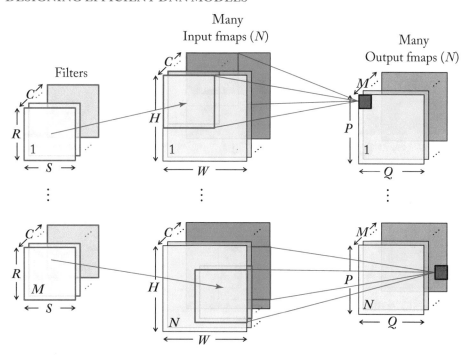

Figure 9.1: Dimensions of a CONV layer.

edge distillation, which can be combined with network architecture design to further increase the accuracy. Finally, we will discuss design considerations when choosing amongst the various techniques for designing efficient DNN models.

9.1 MANUAL NETWORK DESIGN

The CONV and FC layers account for most of the computation and data movement of a DNN model. Therefore, manual network design focuses on improving the efficiency of these two types of layers. Manual design approaches typically focus on reducing the number of weights, activations, and/or MAC operations to indirectly reduce energy consumption and reduce latency.

9.1.1 IMPROVING EFFICIENCY OF CONV LAYERS

Various methods have been proposed to improve the efficiency of CONV layers by reducing the number of weights in the filters, which may help reduce storage requirements, data movement, and number of MAC operations. As shown in Figure 9.1, the filters in the CONV layer are parameterized by the spatial size (R and S) of the filter, the number of input channels (C), and the number of output channels (M). Recall that the number of output channels (M) also

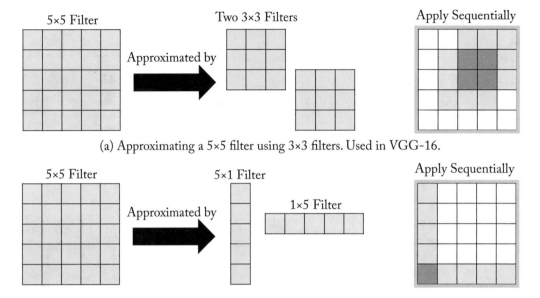

(a) Approximating a 5×5 filter using 3×3 filters. Used in VGG-16.

(b) Approximating a 5×5 filter using 1×5 and 5×1 filters. Used in GoogLeNet/Inception v3 and v4.

Figure 9.2: Approximating larger filters using smaller filters.

corresponds to the number of filters. These methods can be categorized based on how they reduce each of these dimensions.[2]

The spatial size (R and S) of the filter can be reduced by replacing a single filter with a large spatial size (large filter) by several filters with small spatial sizes (small filters). Several small filters can emulate the receptive field of a large filter but with fewer operations and weights, as shown in Figure 9.2. For example, one 5×5 convolution can be approximated by two 3×3 convolutions [73], as shown in Figure 9.2a. Alternatively, one R×S convolution can be approximated by two 1-D convolutions, one 1×R and one S×1 convolution [76], as shown in Figure 9.2b. A similar idea has been widely used in image processing for decades and achieved great success in improving algorithm efficiency when the filters are separable [137].

The number of input channels (C) can be reduced by using *1×1 CONV layer insertion* and *grouped convolutions*. *1×1 CONV layer insertion* [24, 74, 75] involves inserting a 1×1 CONV layer before a "large" CONV layer to reduce the number of input channels (C) in the "large" CONV layer; a "large" CONV layer typically refers to a CONV layer where R and S are greater than one. Specifically, the inserted 1×1 CONV layer has fewer output channels than input channels ($M < C$), which reduces the number of input channels in the next layer from C to M; this is often referred to as a "bottleneck," as discussed in Section 2.4.1. For instance, Figure 9.3 shows

[2]Note that one potential downside of reducing the number of weights in a filter is that it reduces the number of unique filters that can be represented, which may impact accuracy.

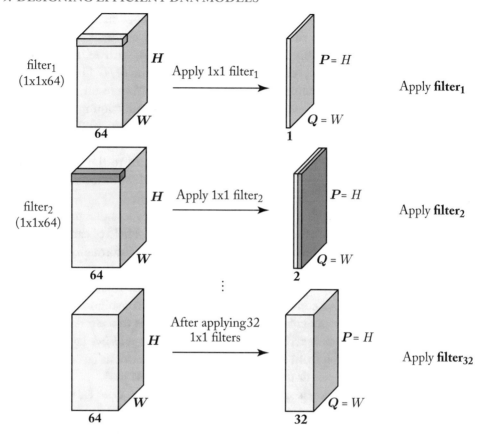

Figure 9.3: The 1×1 CONV layer insertion can be used to reduce the number of output channels (M) of the current layer, and consequently the number of input channels (C) in the next layer. In this figure, the number of output channels is reduced from 64 to 32 by using a 1×1 CONV layer with 64 input channels and 32 filters. Therefore, the next CONV layer only needs to have 32 input channels instead of 64.

how a 1×1 CONV layer with 64 input channels and 32 filters can transform an input with 64 input channels into an output of 32 channels, which reduces the number of input channels in the next layer to 32. SqueezeNet is an example of a network architecture that extensively uses 1×1 CONV layer insertion to reduce the number of weights [284]. It proposes the use of a *fire module* that first reduces the number of input channels with a 1×1 CONV layer and then increases its number of output channels with multiple 1×1 and 3×3 convolution layers. This approach results in a network architecture that has 50× fewer weights than AlexNet, while still achieving the same accuracy.

Grouped convolutions (GC) divide the filters and the input channels into multiple groups. For conventional convolution, each of the M filters, with C channels, is applied to all C input channels. For GC, the M filters are divided into G groups, with M/G filters per group and C/G input channels per group; within each group, each of the M/G filters, with C/G channel, is applied to the C/G input channels within the same group. As a result, the number of channels in the filter is reduced by G times, which leads to G times reduction in the number of weights and MAC operations.[3] Figure 9.4 gives an example of GC with $G = 2$ and $M = 4$, where the first half of the filters are applied to the first half of the input channels to generate the first two output channels, and the second half of the filters are applied to the second half of the input channels to generate the last two output channels. GC was first introduced by AlexNet to fit a layer onto two GPUs, as discussed in Section 2.4.1.

A downside of GC is that there maybe be an accuracy loss due to the reduced receptive field in the channel dimension (see Section 2.1). The receptive field of each group is restricted to a subset of input channels, which prevents the cross-channel information from being fully exploited in a similar manner as for a conventional convolution. In other words, each output channel no longer contains information from all the input channels. This limitation can be addressed by adding an additional 1×1 convolution (called point-wise convolution) [183] to combine the information from all the input channels. Another approach is to use a *shuffling* operation [285] to shuffle the output channels across groups, such that after *multiple* layers each output channel will contain information from all input channels in the previous layers. Figure 9.5 shows an example of the shuffling operation for GC with $G = 2$ and $M = 4$.

Depth-wise convolution is an extreme case of GC, where $G = C$, as shown in Figure 9.6. As a result, each group has one filter and one input channel, and each filter has only one channel and thus performs 2-D filtering. Depth-wise convolution is typically followed by point-wise convolution with C channels to combine the different output channels from the depth-wise convolution. MobileNets [183] use the combination of depth-wise convolution followed by point-wise convolution to significantly reduce the number of weights and MAC operations.

Squeeze and excitation (SE) [286] can be viewed as depth-wise convolution with *dynamic* weights, where the weights change based on the input feature map rather than being fixed (*static*) after training, to increase the accuracy. The idea of SE is to put more attention on the certain channels of the feature map (determined from training) by increasing the magnitude of their activations and decreasing the magnitude of activations in the other channels. Figure 9.7 illustrates the SE operation. It first applies global pooling (see Section 9.1.2) to reduce the spatial resolution of the input feature map to 1×1. This feature map will then be processed by multiple

---

[3]Another way to think about this is that the number of input channels that contribute to each output channel is reduced from C to C/G.

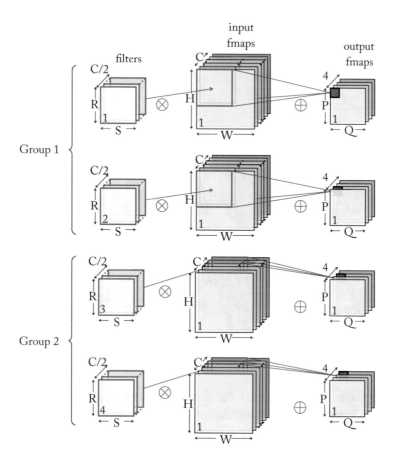

Figure 9.4: This figure illustrates grouped convolution with two groups ($G = 2$). The first half of the filters are applied to the first half of the input channels to generate the first two output channels, and the second half of the filters are applied to the second half of the input channels to generate the last two output channels. Note that the number of channels of filters is reduced from C to $C/2$, which leads to $2\times$ reduction in the number of weights and MAC operations. Note that this example only uses a batch size of 1 ($N = 1$) and thus each feature map is labeled at "1". For illustrative purposes, we repeat the input and output feature maps so that the reader can see which channels of the feature map are being processed by each filter.

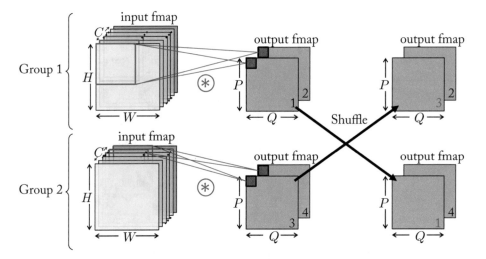

Figure 9.5: This figures illustrates the shuffling operation for grouped convolution. In this example, $G = 2$ and $M = 4$, where the receptive field of the first group is restricted to the first half of the input channels, and the receptive field of the second group is restricted to the second half of the input channels. By swapping the first and third output channels of the layer across the different groups, the output channels within each group contain information from all input channels.

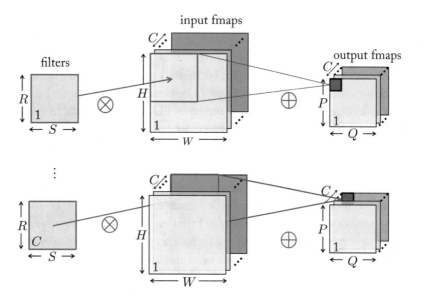

Figure 9.6: Depth-wise convolution is an extreme case of GC, where $G = C$. As a result, each group has one filter and one input channel, and each filter has only one channel and thus performs 2-D filtering. Note that this example only uses a batch size of 1 ($N = 1$) and thus each feature map is labeled at "1". For illustrative purposes, we repeat the input and output feature maps so that the reader can see which channels of the feature map are being processed by each filter.

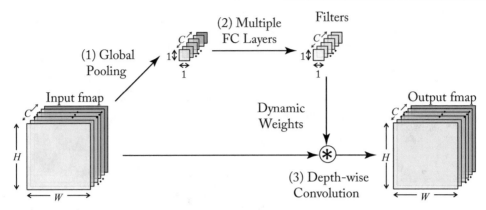

Figure 9.7: Squeeze and excitation can be viewed as a 1×1 depth-wise convolution with dynamic weights and consists of three steps: (1) global pooling is applied to reduce the resolution of the input feature map to 1×1; (2) multiple FC layers with nonlinearity is used to process resulting feature map; and (3) 1×1 depth-wise convolution, which uses the output of the FC layers as dynamic weights in the C 1×1 filters, is used to process the original input feature map.

FC layers with nonlinearity. The results are used as the dynamic weights of a 1×1 depth-wise convolution[4] to process the original input feature map.

The number of output channels (M) that needs to be generated per layer can be reduced by reusing the output channels of output feature maps generated by previous layers, as shown in Figure 9.8. Reusing output channels has been shown to provide a good trade-off between accuracy and efficiency. *Feature map aggregation* defines how output channels from previous layer(s) can be combined (e.g., concatenated) and reused; this is one of the key distinguishing properties between different output channel reuse approaches. For instance, DenseNet [84] proposes the *Dense Block* where each layer concatenates the output channels from all the previous layers as the input. In comparison, Yu et al. [287] explores different ways to hierarchically combine the output channels rather than combining all the previous output channels in a single step.

Note that output feature maps with different spatial resolutions cannot be directly combined (reused). This happens frequently in dense prediction applications, such as image segmentation. One common way to address this is to upsample the lower resolution feature map to match the higher resolution feature map. However, this method increases the memory requirements quadratically with respect to the up-sampling factor. Rather than up-sampling the low resolution feature map, the *space-to-depth (S2D)* operation [241] can be used to reduce the spatial resolution of the high-resolution feature map without losing information, as shown in Figure 9.9. Specifically, the S2D operation moves input activations from the spatial dimension

---

[4]Recall that 1×1 depth-wise convolution differs from conventional 1×1 convolution (also referred to as point-wise convolution) in that 1×1 depth-wise convolution performs 2-D convolution while 1×1 point-wise convolution performs 3-D convolution. In both cases, 1×1 refers to the spatial dimension of the filter, specifically, R=1 and S=1.

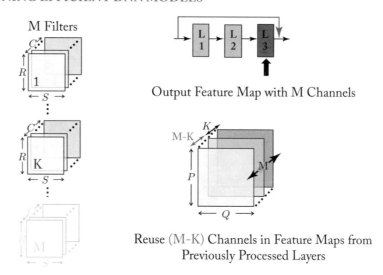

Figure 9.8: Reusing the output channels of feature maps from previous layers can help reduce the number of filters in a layer.

to the channel dimension. After filtering the combined feature map, the high resolution can be restored using depth-to-space (D2S) operation, which is the inverse of S2D, as shown in Figure 9.9. With the S2D and D2S operations, the spatial resolution of output feature maps can be changed without substantial increase in memory requirements.

9.1.2   IMPROVING EFFICIENCY OF FC LAYERS

One of the main challenges of FC layers is the large number of weights, since each filter needs to have an associated weight for each activation in the feature map (i.e., $R = H$, $S = W$). As a result, the number of weights grows quadratically with the resolution of the input feature map (i.e., $H \times W$). Therefore, the number of weights can be significantly reduced by decreasing the resolution of the feature map. The trend in image classification is to keep only one FC layer at the end of the network and insert a *global pooling* (also referred to as adaptive pooling) layer right in front of it. The global pooling layer, as shown in Figure 9.10, is the same as a regular pooling layer, except that its receptive field (window size) is always the same as the resolution of the input feature map. This layer reduces the resolution of the feature maps to 1×1 while keeping the number of channels the same and hence decreases the required number of weights in the following FC layer.

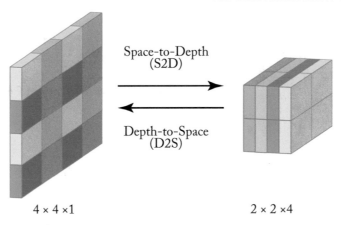

Figure 9.9: An example of the space-to-depth (S2D) and depth-to-space (D2S) operations. The S2D operation moves input activations from the spatial dimension to the channel dimension, and the D2S operation is the inverse.

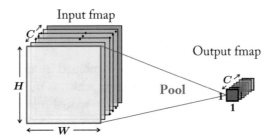

Figure 9.10: The global pooling layer is the same as a regular pooling layer, except that its window size is always the same as the resolution of the input feature map. This layer reduces the resolution of the feature maps to 1×1 while keeping the number of channels the same.

9.1.3 IMPROVING EFFICIENCY OF NETWORK ARCHITECTURE AFTER TRAINING

The previous sections described methods of designing the network architectures *before* training. The DNN model can also be made efficient *after* training. Specifically, the method of approximating large filters by a series of small filters (Section 9.1.1) can be applied after training the weights in the DNN model through a process called *tensor decomposition*. It treats the weights in a layer as a 4-D tensor and decomposes it into a combination of smaller tensors (i.e., several layers), which jointly approximate the original 4-D tensor. Low-rank approximation can then be applied to further increase the compression rate at the cost of accuracy degradation, which may be restored by fine-tuning the weights. This approach has been demonstrated using Canonical

Polyadic (CP) decomposition, which is a high-order extension of singular value decomposition that can be solved by various methods, such as a greedy algorithm [288] or a nonlinear least-square method [289]. Combining CP-decomposition with low-rank approximation can achieve 4.5× speed-up on CPUs [289]. However, CP-decomposition cannot be computed in a numerically stable way when the dimension of the tensor, which represents the weights, is larger than two [289]. To alleviate this problem, Tucker decomposition can be used instead [290].

9.2 NEURAL ARCHITECTURE SEARCH

Neural architecture search (NAS) aims to automatically find a network architecture that achieves good performance, where performance for network architecture typically refers to a good trade-off between accuracy and latency, or accuracy and energy. Manual network design has been an effective approach for designing network architecture with reasonable performance, but determining the network architecture is a tedious process. For example, there are a large number of hyperparameters to tune, such as the number of layers, the connections between layers, and the type and shape of each layer. Due to the highly complex relationship between these hyperparameters and the performance of the resulting network, determining the optimal network architecture usually relies on trial and error, which makes network architecture design very complicated and time consuming. To address this problem, NAS leverages the advancement in machine learning to automate the design process.

Figure 9.11 illustrates the general flow of NAS algorithms. NAS is generally carried out in an iterative manner. At each iteration, the optimization algorithm samples several network architectures (i.e., samples) from a predefined search space, which consists of all discoverable network architectures. The performance of each network sample will then be evaluated. Based on the evaluation results, the optimization algorithm samples the next set of network architectures from the search space. This process continues until a termination criterion (e.g., the maximum number of search iterations) is met and generates the searched network architecture.

In summary, the three main components of NAS are as follows:

- **Search space:** what is the set of all samples.

- **Optimization algorithm:** where to sample.

- **Performance evaluation:** how to evaluate samples.

The two main metrics for gauging the performance of a NAS algorithm are (1) the achievable network performance and (2) the required search time. Ideally, the network with the best performance can be found using an exhaustive search, which evaluates all the possible networks and selects the best one. However, this is impractical given the large number of possible networks. For example, exhaustively searching for the optimal network architecture with up to 10 layers and 100 filters per layer involves evaluating 10^{100} networks. Therefore, NAS research focuses on reducing the search time with minimal loss in the achievable network performance.

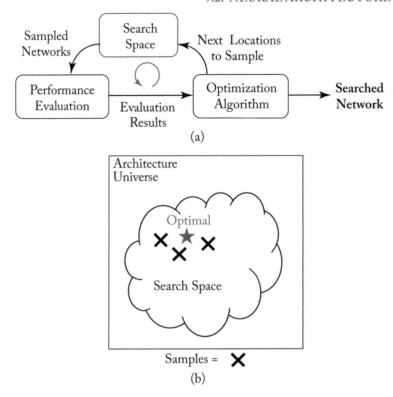

Figure 9.11: The illustration of the general flow of neural architecture search.

The search time of NAS (time_{nas}) can be estimated by the following equation:

$$\text{time}_{nas} = \text{num}_{samples} \times \text{time}_{sample}, \tag{9.1}$$

where $\text{num}_{samples}$ is the number of samples explored and time_{sample} is the time required for evaluating a sample. Each term can be further decomposed to reveal the following factors:

$$\text{time}_{nas} \propto (\frac{\text{size}_{search\_space} \times \text{num}_{alg\_tuning}}{\text{efficiency}_{alg}}) \times (\text{time}_{eval} + \text{time}_{train}). \tag{9.2}$$

The number of samples ($\text{num}_{samples}$) is determined by the size of the search space ($\text{size}_{search\_space}$), the efficiency of the optimization algorithm (efficiency_{alg}), and the number of times it takes to tune the optimization algorithm (num_{alg_tuning}).

The design and size of the search space offer a trade-off between performance versus search time. A larger search space can allow for more discoverable networks, which can improve performance; however, it may require more samples in order to find the optimal network architecture with improved performance.

Table 9.1: This table summarizes which terms in Equation (9.2) can be improved by improving each of the three main components of NAS

	num$_{samples}$			time$_{samples}$	
	size$_{search\_space}$	efficiency$_{alg}$	num$_{alg\_tuning}$	time$_{eval}$	time$_{train}$
Search Space	✓				
Optimization Algorithm		✓	✓		
Performance Evaluation				✓	✓

The selection of the optimization algorithm also influences performance and search time. A more efficient optimization algorithm (efficiency$_{alg}$) can better utilize the samples and hence reduce the number of required samples. However, it is also important to consider the difficulty of tuning the hyperparameters of the optimization algorithm itself (e.g., network architecture of the reinforcement learning agent). Optimization algorithms that are difficult to tune may require multiple iterations before they can enable effective search (num$_{alg\_tuning}$). This critical factor is often overlooked.

The time required for evaluating a sample (time$_{sample}$) includes the time required for evaluating the network performance (time$_{eval}$). Once a given network is sampled, it may need to be trained to get the precise accuracy numbers, which leads to the training time (time$_{train}$).

Researchers improve NAS algorithms by introducing innovations in the three main components, where each improves different terms in Equation (9.2) (summarized in Table 9.1):

- **Shrinking the search space**, which reduces size$_{search\_space}$.

- **Improving the optimization algorithm**, which increases efficiency$_{alg}$ and reduces num$_{alg\_tuning}$.

- **Accelerating the performance evaluation**, which reduces time$_{eval}$ and time$_{train}$.

It is important to note that these three components are not independent of each other, and a change in one component may involve a change in another component. For example, some optimization algorithms cannot support hardware metrics (e.g., latency and energy consumption) and thus can only use proxy metrics (e.g., number of MACs and number of weights).

9.2.1   SHRINKING THE SEARCH SPACE

Shrinking the search space increases the search speed by limiting the discoverable network architectures. The idea is to only search a subset of the network architectures in the network architecture universe out of all the possible network architectures. Although this class of methods can effectively reduce the required number of samples, it may irrecoverably limit the achievable network performance and needs to be carried out carefully. For example, the optimal network

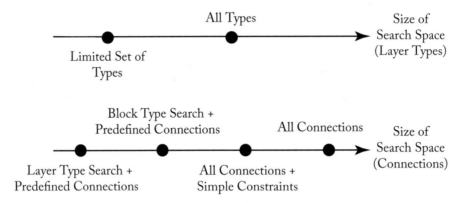

Figure 9.12: Different methods for shrinking the search space and the relative sizes of the corresponding search spaces.

architecture may fall outside of the search space, so it is impossible for the optimization algorithm to sample it. Therefore, the domain knowledge learned from manual network design plays an important role in properly guiding the reduction of the search space.

Figure 9.12 illustrates different methods for shrinking the search space, and the relative sizes of the corresponding search spaces. The search space is determined by the possible layer types and the possible connections between layers.

It is a common practice for NAS to reduce the possible layer types to a set of widely used layer types, such as convolution with different filter sizes and strides, and pooling with different pooling functions. These layer types have shown to be effective in manual network design and have become key components of modern network architectures. However, fixing the layer types also prevents NAS from discovering new layer types.

There is a wider variety of methods for reducing the possible connections between layers. Searching arbitrary connections provides the maximum flexibility, but it is intractable ("all connections" in Figure 9.12). A few works [291, 292] add some simple constraints, such as setting the maximum depth of the network, and show that a new network architecture can be discovered ("all connections + simple constraints" in Figure 9.12). However, the resulting search space is still too large in practice. This requires a significant amount of computational resources.

Motivated by the modular design strategy in manual network design, other methods [293–303] first search the connections between a few layers, define them as a block, and then connect these blocks in a predefined way ("block type search + predefined connections" in Figures 9.12 and 9.13a). Because blocks contain significantly fewer layers than the whole network, the connections in blocks are easier to search.

The search space can be further reduced by pre-defining the connections in the blocks and only searching for the layer types to use for each layer [258, 304–308] ("layer type search + predefined connections" in Figures 9.12 and 9.13b). With more constraints, we further reduce

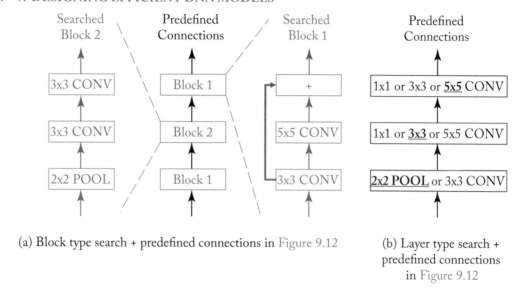

(a) Block type search + predefined connections in Figure 9.12

(b) Layer type search + predefined connections in Figure 9.12

Figure 9.13: Illustration of two ways to shrinking the search space.

the search space and hence the search time. However, more domain knowledge is required to guarantee that there exist network architectures with good performance in the reduced search space, and NAS becomes more similar to manual network design. Therefore, increasing the search speed and the performance of the searched network architecture while minimizing the required domain knowledge is a key challenge of NAS.

9.2.2   IMPROVING THE OPTIMIZATION ALGORITHM

These optimization algorithms differ in several aspects, such as how they use the previous samples to determine the next set of samples, which leads to different computational complexity, restrictions to the search space and performance metrics, and the ease of hyperparameter tuning. Popular optimization algorithms for NAS include random search, coordinate descent, gradient descent, evolutionary algorithm, reinforcement learning, and Bayesian optimization. Each has its own benefits and drawbacks, and which algorithm to select depends on the target application and other factors, such as the search space and the performance metrics used. We will briefly introduce these six optimization algorithms in their canonical form.

Random search [296] is one of the simplest optimization algorithms of NAS. It randomly samples the entire search space and chooses the sample with the best performance. Although this algorithm is simple, it can find the networks with similar performance to those found by more complicated optimization algorithms [302]. Random search does not use the samples from the previous iterations to determine the next set of samples, which reduces the search efficiency.

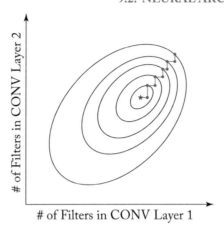

Figure 9.14: An example of the coordinate descent optimization algorithm on a network with two CONV layers and an FC layer.

In contrast, coordinate descent [258] does use the samples from the previous iterations. At each iteration, it starts from the previous best sample and greedily samples the nearby region along a few predefined "directions" in the search space. Searching along a "direction" can involve sweeping a hyperparameter and keeping the other hyperparameters fixed. Figure 9.14 illustrates an example of applying coordinate descent on a network with two CONV layers and an FC layer. In this example, the search space is defined by the number of filters in each CONV layer, and the two pre-defined directions are moving leftward (i.e., reducing the number of filters in the CONV layer 1), and downward (i.e., reducing the number of filters in the CONV layer 2). After evaluating the samples along the two directions, it adopts the sample with the best performance and then performs the same process again. Because it greedily samples the networks along all the predefined directions, the number of directions is limited.

Gradient descent [293, 303, 305, 307, 308] also starts from the previous best sample like coordinate descent, but searches along the direction indicated by the gradient. The gradient can be computed analytically and efficiently without explicitly exploring multiple directions. However, it requires that the gradient can be computed, which is not always feasible. For example, certain performance metrics (e.g., latency) are not differentiable.

Evolutionary algorithms [292, 300, 302, 306] typically keep several samples (e.g., a thousand) from the previous iterations. At each iteration, evolutionary algorithms randomly remove samples with the worse performance and sample the neighborhoods of the remaining networks in the search space. Keeping multiple samples allows multiple positions in the search space to be searched in parallel; however, it also increases the requirements on computation resources and introduces new hyperparameters, such as the number of samples to keep.

Reinforcement learning [291, 295, 297–299] proposes a better way to use the previous samples. At each iteration, instead of keeping the previous best sample and starting from it, reinforcement learning trains an agent to learn from the previous samples and uses the agent to determine the next set of samples. Therefore, the new samples do not need to be in the neighborhood of the previous best sample. This provides higher flexibility for the search, but designing the agent and tuning its hyperparameters can be challenging and more complicated than other optimization algorithms.

Bayesian optimization [301, 304] takes a very different route from the previously discussed approaches. It aims to model the distribution of the entire search space so that it can pick the sample with the best performance according to the distribution model. Starting from an initial guess of the distribution model of the search space (i.e., the prior), it updates the model using the samples at each iteration and generates the next set of samples based on the updated model. With the prior, this method can better incorporate the knowledge learned from manual network design. However, the main issue is that precisely modeling a large search space can be difficult and requires many samples.

9.2.3   ACCELERATING THE PERFORMANCE EVALUATION

Accelerating the performance evaluation is another effective way to speed up NAS. Because NAS requires only the rank of the performance values instead of the exact values, this leaves room for approximation. There are at least three items that can be approximated: accuracy, weights, and hardware metrics.

The accuracy is one of the most important metrics for network performance and can be approximated using (1) proxy task, (2) early termination, and (3) accuracy prediction. The proxy task is a simpler task that can be used to approximate the target task, such as using a simpler and smaller dataset [291, 297, 298, 300] and reducing the resolution of the images [303]. For example, CIFAR-10 [105] is usually used as the proxy task for ImageNet [23]. Early termination [258, 292, 295, 296, 298, 301–303] (Figure 9.15a) terminates network training before the training converges and uses the accuracy at this point as the approximation. For instance, if a network requires one million iterations to converge, we would only train it for 10,000 iterations. Accuracy prediction [306] (Figure 9.15b) takes this one step further by extrapolating the early terminated training curve to predict the converged accuracy.

In addition to the accuracy, we can also approximate the weights to speed up training if there exists a trained network with a similar network architecture as the current sampled network. For example, if we reduce the number of filters in a layer of the trained network to generate the current sample, we can either (1) transfer the weights or (2) estimate the weights. For transferring weights [258, 292–294, 299, 302, 305, 307, 308] (Figure 9.16a), because the current sample shares a similar network architecture of a trained network, we can directly initialize the weights of the sample by using part of the weights in the trained network. For estimating the weights [253] (Figure 9.16b), the weights can be quickly approximated by solving an $\ell2$-

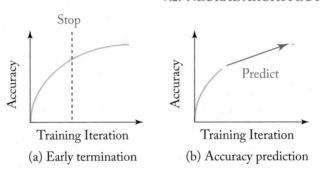

Figure 9.15: Two methods for approximating the accuracy during neural architecture search.

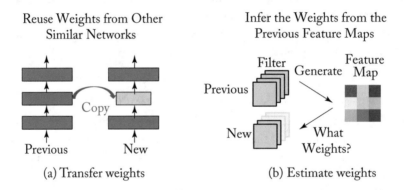

Figure 9.16: Two methods for approximating the weights during neural architecture search.

minimization problem to minimize the difference in the output feature map between a layer in the sample network and the corresponding layer in the trained network. Since there is a close-form solution for the $\ell 2$-minimization problem, this step can be carried out efficiently.

Approximating the hardware metrics (e.g., latency, energy consumption) is another important topic if hardware metrics are involved. Evaluating hardware metrics can be slow and difficult to parallelize due to the limited number of available hardware devices. One popular method is using proxy metrics [305]. For example, the number of MAC operations is commonly used for approximating the latency. However, the proxy metrics typically fail to consider the properties of the hardware, which results in inaccurate approximation and thus inferior performance of the searched network. Instead, building look-up tables of the hardware metrics and performing fast table lookup during NAS can be a promising choice [258]. Figure 9.17 illustrates an example of using layer-wise look-up tables for fast latency evaluation. These layer-wise look-up tables contain pre-measured latency of each layer with various shapes. During NAS, we can look up the table of each layer, and sum up the layer-wise latency to estimate the latency of

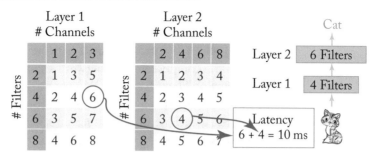

Figure 9.17: This figure illustrates how layer-wise look-up tables are used for fast hardware metric evaluation.

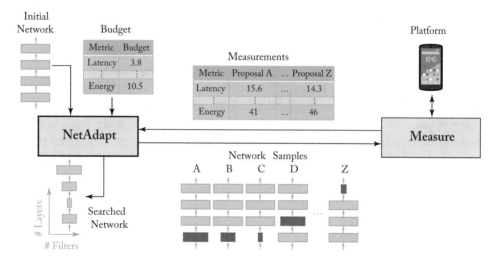

Figure 9.18: NetAdapt is a NAS algorithm that has been shown to be an effective approach for improving the Pareto frontier of accuracy-latency trade-off [309], requiring only small number of hyperparameters to tune. It is directly guided by hardware metrics and generates a family of networks with different accuracy-latency trade-offs in a single run. The code is available at https://netadapt.mit.edu/.

a network. This approach has been widely used with various optimization algorithms, such as coordinate descent [258], evolutionary algorithm [306], and gradient optimization [307].

9.2.4   EXAMPLE OF NEURAL ARCHITECTURE SEARCH

We will use NetAdapt [258] as an concrete example of an NAS algorithm. NetAdapt has been shown to be an effective approach for improving the Pareto frontier of accuracy-latency trade-off [309]. Figure 9.18 illustrates the algorithm flow of NetAdapt. We will now highlight the three main components of NetAdapt.

Search space (Section 9.2.1): NetAdapt uses an initial network architecture to define the initial search space. The search space is composed of the network architectures derived from the initial network architecture by removing filters (i.e., reducing channels) of various layers. Since the network architectures in the search space are a subset of the initial network architecture, they are referred to as sub-networks. At each iteration, the search space consists of only the sub-networks of the best performing network from the previous iteration; this effectively shrinks the search space at each iteration.

Optimization algorithm (Section 9.2.2): NetAdapt proposes using the coordinate-descent optimizer, which has only a few optimizer-related hyperparameters (e.g., the target latency per iteration). At each iteration, the optimizer samples the sub-networks of the best network architecture from the previous iteration with the same target latency for this iteration. The sampled network architecture with the best accuracy-latency trade-off will then be used to generate the samples in the next iteration. The best samples from each of the iterations form a network family with different accuracy-latency trade-offs, which enables the support of use cases with different latency requirements without the need to run NetAdapt multiple times.

Performance evaluation (Section 9.2.3): while training the sampled network architectures to evaluate their accuracy numbers, NetAdapt uses the early termination technique to reduce the training time. Moreover, to further improve the performance of the searched network architecture, NetAdapt is guided by latency instead of number of MAC operations, and uses look-up tables to significantly speed up the performance evaluation.

9.3 KNOWLEDGE DISTILLATION

Using a DNN model with many layers or averaging the predictions of different models (i.e., ensemble) typically gives a better accuracy than using a single DNN model with only few layers. However, the computational complexity is also higher. To get the best of both worlds, a method called *knowledge distillation* is used to transfer the knowledge learned by one or more complex DNN models (teacher) to a simpler DNN model (student) to increase the accuracy of the student network without increasing its complexity. More precisely, knowledge distillation involves first training the complicated teacher network and using the outputs of the teacher network as the labels to train the simpler student network. The student network can therefore achieve an accuracy that would not be achievable if it was directly trained with the labels in the same dataset [310, 311]. For example, Hinton et al. [312] shows how using knowledge distillation can improve the speech recognition accuracy of a student network by 2%, which is similar to the accuracy of a teacher network that is composed of an ensemble of ten networks.

Figure 9.19 shows the simplest knowledge distillation method [310]. The softmax layer is commonly used as the output layer in networks for image classification to generate the class

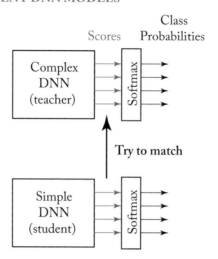

Figure 9.19: Knowledge distillation matches the class scores of a simple network (student) to those of a complex network (teacher).

probabilities from the class scores;[5] it normalizes the class scores into values between 0 and 1 that sum up to 1, where large and small values are pushed toward the two extremes, 1 and 0, respectively. For this knowledge distillation method, the class scores of the teacher network are used as the targets and the objective is to minimize the difference between the targets and the class scores of the student network. Class scores are used as the targets instead of the class probabilities because the softmax layer eliminates the important information contained in the small class scores by pushing the corresponding class probabilities toward 0. Alternatively, if the softmax is configured to generate smoother class probability distribution that preserves the small class scores better, the class probabilities can be used as targets [312]. Finally, the intermediate representations of the teacher network can also be incorporated as the extra hints to train the student network [313].

Knowledge distillation can also be combined with aforementioned co-design methods in the previous chapters to improve accuracy-efficiency trade-off. For instance, Apprentice [314] combines knowledge distillation with reduced precision by using a large teacher network (e.g., ResNet-101) with full precision to train a smaller student network (e.g., ResNet-50) with reduced precision. Knowledge distillation increases the accuracy of the reduced precision network by 1.5% to 3%, which helps close the gap between reduced precision and full precision networks.

---

[5] Also commonly referred to as logits.

9.4   DESIGN CONSIDERATIONS FOR EFFICIENT DNN MODELS

There are several important factors to consider when applying techniques to design efficient network architectures for DNN models.

First and foremost is the impact on the accuracy. This is usually difficult to evaluate because the impact may vary from application to application and dataset to dataset. For example, reducing the spatial resolution of feature maps may cause less accuracy degradation for image classification compared to image segmentation, which requires high spatial resolution for accurate per-pixel labels. Moreover, the accuracy highly depends on the training procedure and hyperparameters, such as the learning rate, the usage of knowledge distillation, the degree of regularization, the data pre-processing, and the selection of optimization algorithm for training. Therefore, when comparing different efficient DNN model design approaches, the impact on accuracy should be evaluated on the same target application and dataset with the same training procedure and hyperparameters.

Second, it is important to correctly evaluate the hardware metrics, such as latency and energy consumption. While the number of weights and MAC operations provide an easy and quick approximation to the hardware metrics, a reduction in the number of weights and MAC operations may not directly translate into a proportional reduction in latency or energy consumption. In addition, different hardware platforms have different characteristics. For example, feature map movement in the memory hierarchy can consume a significant amount of energy consumption for some hardware platforms, such as *processing-in-memory* accelerators discussed in Section 10.2. If the goal is to minimize energy consumption on this type of hardware, the number of MAC operations is not a good approximation because it does not consider the memory hierarchy and data reuse. Therefore, directly measuring the target metric on the target hardware instead of using proxy metrics can enable more informed design decisions. For instance, processing-in-memory accelerators may prefer network architectures with larger filters and fewer layers to reduce feature map movement [315], which differs from conventional digital accelerators; however, solely evaluating the number of MAC operations would not provide this insight.

Third, the design time and effort for applying a given technique to achieve the desirable performance should be considered. This can include the time required for tuning the hyperparameters as explained in Section 9.2. More hyperparameters and higher degree of uncertainty in relationship between the hyperparameters and impact on performance can increase the time it takes to tune the DNN model. Therefore, the ease of use of a given technique is an important consideration, but unfortunately is often overlooked.

Finally, some techniques may require additional hardware in order to realize the potential latency and energy consumption benefits. For instance, efficient network architectures may require the use of a more diverse range of layer shapes, which means a hardware accelerator may require additional hardware to provide the flexibility to support these shapes. However, given a fixed area budget, any extra hardware overhead would result in a reduction in other hardware resources (e.g., on-chip storage or number of PEs), which can counteractively degrade performance. Therefore, one should consider whether the extra hardware cost exceeds or significantly degrades the overall benefits of a given approach.

C H A P T E R   10

Advanced Technologies

As highlighted throughout the previous chapters, data movement dominates energy consumption. The energy is consumed both in the access to the memory as well as the transfer of the data. The associated physical factors also limit the bandwidth available to deliver data between memory and compute, and thus limits the throughput of the overall system. This is commonly referred to by computer architects as the "memory wall."[1]

To address the challenges associated with data movement, there have been various efforts to bring compute and memory closer together. Chapters 5 and 6 primarily focus on how to design spatial architectures that distribute the on-chip memory closer to the computation (e.g., scratch pad memory in the PE). This chapter will describe various other architectures that use *advanced memory*, *process*, and *fabrication technologies* to bring the compute and memory together.

First, we will describe efforts to bring the off-chip high-density memory (e.g., DRAM) closer to the computation. These approaches are often referred to as *processing near memory* or *near-data processing*, and include memory technologies such as embedded DRAM and 3-D stacked DRAM.

Next, we will describe efforts to integrate the computation *into* the memory itself. These approaches are often referred to as *processing in memory* or *in-memory computing*, and include memory technologies such as Static Random Access Memories (SRAM), Dynamic Random Access Memories (DRAM), and emerging non-volatile memory (NVM). Since these approaches rely on mixed-signal circuit design to enable processing in the analog domain, we will also discuss the design challenges related to handling the increased sensitivity to circuit and device non-idealities (e.g., nonlinearity, process and temperature variations), as well as the impact on area density, which is critical for memory.

Significant data movement also occurs between the sensor that collects the data and the DNN processor. The same principles that are used to bring compute near the memory, where the weights are stored, can be used to bring the compute *near* the sensor, where the input data is collected. Therefore, we will also discuss how to integrate some of the compute *into* the sensor.

Finally, since photons travel much faster than electrons and the cost of moving a photon can be *independent* of distance, processing in the optical domain using light may provide significant improvements in energy efficiency and throughput over the electrical domain. Accordingly, we will conclude this chapter by discussing the recent work that performs DNN processing in the optical domain, referred to as *Optical Neural Networks*.

---

[1]Specifically, the memory wall refers to data moving between the off-chip memory (e.g., DRAM) and the processor.

Table 10.1: Example of recent works that explore processing near memory. For I/O, TSV refers to through-silicon vias, while TCI refers to ThruChip Interface which uses inductive coupling. For bandwidth, *ch* refers to number of parallel communication channels, which can be the number of tiles (for eDRAM) or the number of vaults (for stacked memory). The size of stacked DRAM is based on Hybrid Memory Cube (HMC) Gen2 specifications.

	Technology	Size	I/O	Bandwidth	Evaluation
DaDianNao [152]	eDRAM	32 MB	On-chip	18 ch × 310 GB/s = 5580 GB/s	Simulated
Neurocube [316]	Stacked DRAM	2 GB	TSV	16 ch × 10 GB/s = 160 GB/s	Simulated
Tetris [317]	Stacked DRAM	2 GB	TSV	16 ch × 8 GB/s = 128 GB/s	Simulated
Quest [318]	Stacked SRAM	96 MB	TCI	24 ch × 1.2 GB/s = 28.8 GB/s	Measured
N3XT [319]	Monolithic 3-D	4 GB	ILV	16 ch × 48 GB/s = 768 GB/S	Simulated

10.1   PROCESSING NEAR MEMORY

High-density memories typically require a different process technology than processors and as a result are often fabricated as separate chips; as a result, accessing high-density memories requires going off-chip. The bandwidth and energy cost of accessing high-density off-chip memories are often limited by the number of I/O pads per chip and the off-chip interconnect channel characteristics (i.e., its resistance, inductance, and capacitance). Processing near memory aims to overcome these limitations by bringing the compute near the high-density memory to reduce access energy and increase memory bandwidth. The reduction in access energy is achieved by reducing the length of the interconnect between the memory and compute, while the increase in bandwidth is primarily enabled by increasing the number of bits that can be accessed per cycle by allowing for a wider interconnect and, to a lesser extent, by increasing the clock frequency, which is made possible by the reduced interconnect length.

Various recent advanced memory technologies aim to enable processing near memory with differing integration costs. Table 10.1 summarizes some of these efforts, where high-density memories on the order of tens of megabytes to gigabytes are connected to the compute engine at bandwidths of tens to hundreds of gigabytes per second. Note that currently most academic evaluations of DNN systems using advanced memory technologies have been based on simulations rather than fabrication and measurements.

In this section, we will describe the cost and benefits of each technology and provide examples of how they have been used to process DNNs. The architectural design challenges of using processing-near-memory include how to allocate data to memory since the access patterns for high-density memories are often limited (e.g., data needs to be divided into different banks and vaults in the DRAM or stacked DRAM, respectively), how to design the network-on-chip between the memory and PEs, how to allocate the chip area between on-chip memory and compute now that off-chip communication less expensive, and how to design the memory hierarchy and dataflow now that the data movement costs are different.

10.1.1   EMBEDDED HIGH-DENSITY MEMORIES

Accessing data from off-chip memory can result in high energy cost as well as limited memory bandwidth (due to limited data bus width due to number of I/O pads, and signaling frequency due to the channel characteristics of the off-chip routing). Therefore, there has been a significant amount of effort toward embedding high-density memory on-chip. This includes technology such as *embedded DRAM (eDRAM)* [320] as well as *embedded non-volatile (eNVM)* [321], which includes embedded Flash (eFlash) [322], magnetic random-access memory (MRAM) [323], resistive random-access memory (RRAM) [324, 325], and phase change memory (PCRAM) [326].

In DNN processing, these high-density memories can be used to store tens of megabytes of weights and activations on chip to reduce off-chip access. For instance, DaDianNao [152] uses 16×2MB eDRAM tiles to store the weights and 2×2MB eDRAM tiles to store the input and output activations; furthermore, all these tiles (each with 4096-bit rows) can be accessed in parallel, which gives extremely high memory bandwidth.[2] The downside of eDRAM is that it has a lower density than off-chip DRAM and can increase the fabrication cost of the chip. In addition, it has been reported that eDRAM scaling is slower than SRAM scaling [327], and thus the density advantage of eDRAM over SRAM will reduce over time. In contrast, eNVMs have gained popularity in recent years due to its increased density as well as its non-volatility properties and reduction in standby power (e.g., leakage, refresh, etc.) compared to eDRAM [327].

10.1.2   STACKED MEMORY (3-D MEMORY)

Rather than integrating DRAM into the chip itself, the DRAM can also be stacked on top of the chip using through-silicon vias (TSVs). This technology is often referred to as *3-D memory*,[3] and has been commercialized in the form of Hybrid Memory Cube (HMC) [328] and High Bandwidth Memory (HBM) [122]. 3-D memory delivers an order of magnitude higher bandwidth and reduces access energy by up to 5× relative to existing 2-D DRAMs, as TSVs have lower capacitance than typical off-chip interconnects.

Recent works have explored the use of HMC for efficient DNN processing in a variety of ways. For instance, Neurocube [316], shown in Figure 10.1a, uses HMC to bring the memory and computation closer together. Each DRAM vault (vertically stacked DRAM banks) is connected to a PE containing a buffer and several MACs. A 2-D mesh network-on-chip (NoC) is

---

[2]DaDianNao [152] assumes that the DNN model can fit into the 32MB of eDRAM allocated to the weights. In practice, this implies that the design either limits the size of DNN model, or requires access to off-chip memory if the size of the DNN model exceeds the capacity of the eDRAM.

[3]Also referred to as "in-package" memory since both the memory and compute can be integrated into the same package.

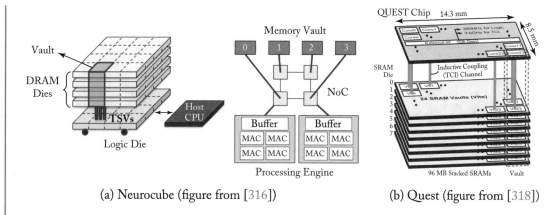

(a) Neurocube (figure from [316])   (b) Quest (figure from [318])

Figure 10.1: Stacked memory systems. (a) DRAM using through-silicon vias (TSV) and (b) SRAM using inductive coupling.

used to connect the different PEs, allowing the PEs to access data from different vaults. One major design decision involves determining how to distribute the weights and activations across the different vaults to reduce the traffic on the NoC.

Another example that uses HMC is Tetris [317], which explores the use of HMC with the Eyeriss spatial architecture and row-stationary dataflow. It proposes allocating more area to computation than on-chip memory (i.e., larger PE array and smaller global buffer) in order to exploit the low-energy and high-throughput properties of the HMC. It also adapts the dataflow to account for the HMC and smaller on-chip memory.

SRAM can also be stacked on top of the chip to provide 10× lower latency compared to DRAM [318]. For instance, Quest [318], shown in Figure 10.1b, uses eight 3-D stacked SRAM dies to store both the weights and the activations of the intermediate feature maps when processing layer by layer. The SRAM dies are connected to the chip using inductive-coupling die-to-die wireless communication technology, known as a ThruChip Interface (TCI) [330], which has lower integration cost than TSV.

The above 3-D memory designs involve using TSV or TCI to connect memory and logic dies that have been separately fabricated. Recent breakthroughs in nanotechnology have made it feasible to directly fabricate thin layers of logic and memory devices on top of each other, referred to as monolithic 3-D integration. Interlayer vias (ILVs), which have several orders of magnitude denser vertical connectivity than TSV, can then be used to connect the memory and compute. Current monolithic 3-D integration systems, such as N3XT, use on-chip non-volatile memory (e.g., resistive RAM (RRAM), spin-transfer torque RAM (STT-RAM)/magnetic RAM (MRAM), phase change RAM (PCRAM)), and carbon nanotube logic (CNFET). Based on

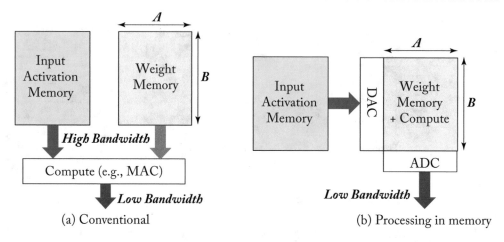

Figure 10.2: Comparison of conventional processing and processing in memory.

simulations, the energy-delay product of ILVs can be up to two orders of magnitude lower than 2-D systems on deep neural network workloads, compared to 8× for TSV [319].[4]

In order to fully understand the impact of near memory processing it is important to analyze the impact that the added storage layer has on the mappings that are now available. Specifically, the new memories are faster, but also smaller, so optimal mappings will be different.

10.2 PROCESSING IN MEMORY

While the previous section discussed methods to bring the compute near the memory, this section discusses *processing in memory*, which brings the compute *into* the memory. We will first highlight the differences between processing in memory and conventional architectures, then describe how processing in memory can be performed using different memory technologies including NVM, SRAM, and DRAM. Finally, we will highlight various design challenges associated with processing-in-memory accelerators that are commonly found across technologies.

DNN processing can be performed using matrix-vector multiplication (see Figures 4.2 and 4.3), as discussed in Chapter 4. For conventional architectures, both the input activation vector and the weight matrix are read out from their respective memories and processed by a MAC array, as shown in Figure 10.2a; the number of weights that can be read at once is limited by the memory interface (e.g., the read out logic and the number of memory ports). This limited memory bandwidth for the weights (e.g., a row of A weights per cycle in Figure 10.2b) can also limit the number of MAC operations that can be performed in parallel (i.e., operations per cycle) and thus the overall throughput (i.e., operations per second).

---

[4]The savings are highest for DNN models and configurations with low amounts of data reuse (e.g., FC layers with small batch size) resulting in more data movement across ILV.

Processing-in-memory architectures propose moving the compute into the memory that stores the weight matrix, as shown in Figure 10.2b. This can help reduce the data movement of the weights by avoiding the cost of reading the weight matrix; rather than reading the weights, only the computed results such as the partial sums or the final output activations are read out of the memory. Furthermore, processing in memory architectures can also increase the memory bandwidth, as the number of weights that can be accessed in parallel is no longer limited by the memory interface; in theory, the entire weight matrix (e.g., $A \times B$ in Figure 10.2b) could be read and processed in parallel.

Figure 10.3 shows a weight-stationary dataflow architecture that is typically used for processing in memory. The word lines (WLs) are used to deliver the input activations to the storage elements, and the bit lines (BLs) are used to read the computed output activations or partial sums. The MAC array is implemented using the storage elements (that store the weights), where a multiplication is performed at each storage element, and the accumulation across multiple storage elements on the same column is performed using the bit line. In theory, a MAC array of B rows of A elements can access all $A \times B$ weights at the same time, and perform up to A dot products in parallel, where each sums B elements (i.e., $A \times B$ MAC operations per cycle).

Similar to other weight-stationary architectures, the input activations can be reused across the different columns (up to A times for the example given in Figure 10.3), which reduces number of input activation reads. In addition, since a storage element tends to be smaller in area than the logic for a digital MAC (10 to 100× smaller in area and 3 to 10× smaller in edge length [331]), the routing capacitance to deliver the input activations can also be reduced, which further reduces the energy cost of delivering the input activations. Depending on the format of the inputs and outputs to the array, digital-to-analog converters (DACs) and analog-to-digital converters (ADCs) may also be required to convert the word line and bit line values, respectively; the cost of the DAC scales with the precision of the input activations driven on the word line, while the cost of the ADC scales with the precision of the partial sums, which depends on the precision of the weights and input activations, and the number of values accumulated on the bit line (up to B).[5]

An alternative way to view processing in memory is to use the loop nest representation introduced in Chapter 5. Design 10.20 illustrates a processing-in-memory design for an FC layer with M output channels and where the input activations are flattened along the input channel, height and width dimensions (CHW). The computation take place in one cycle computing all the results in a single cycle in line 7. For this design, some of the mapping constraints are that

---

[5]The number of bits that an ADC can correctly resolve also depends on its thermal noise (typically some multiple of kT/C, where k is the Boltzmann constant, T is the temperature, and C is the capacitance of the sampling capacitor). For instance, an N-bit ADC has 2^{N-1} decision boundaries (see Section 7.2.1). However, if the thermal noise is large, the location of the 2^{N-1} decision boundaries will move around, dynamically and randomly, and this will affect the resulting accuracy of the DNN being processed. Therefore, designing a low noise ADC is an important consideration. Note that the thermal noise of the ADC scales with the power consumption and the area of the ADC. Accordingly, it is important that the ADC's thermal noise be considered when evaluating the accuracy as demonstrated in [332–334], as the design of the ADC involves a trade-off between power, area, and accuracy.

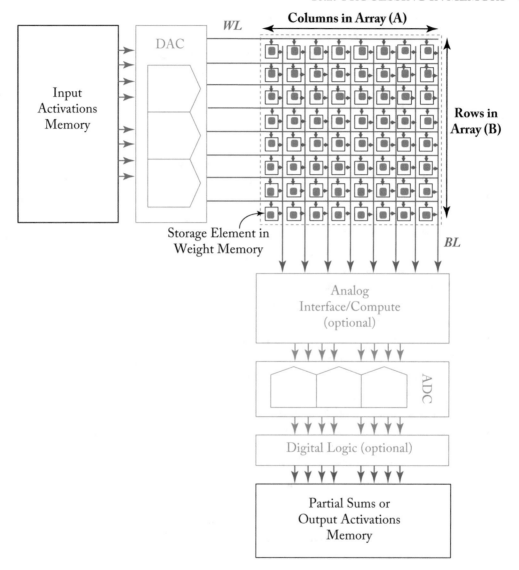

Figure 10.3: Typical dataflow for processing-in-memory accelerators.

$A \geq M$ and $B \geq C \times H \times W$.[6] Note, that when $A \neq M$ or $B \neq C \times H \times W$ under-utilization will occur, as described in Section 10.2.4.

A processing in memory design can also handle convolutions as illustrated in the loop nest in Design 10.21. Here, we show a toy design of just a 1-D convolution with multiple

---

[6]For this example, we disallow the cases where $A < M$ or $B < C \times H \times W$, since that would require multiple passes and updates of the weights, which reduces the potential benefits of processing in memory.

Design 10.20 FC layer for Processing in Memory

```
1    i = Array(CHW)              # Input   activations
2    f = Array(M, CHW)          # Filter  weights
3    o = Array(M)               # Output  partial  sums
4
5    parallel-for  m in [0, M):
6        parallel-for  chw in [0, CHW):
7                o[m] += i[chw] * f[m, chw]
```

Design 10.21 1-D Weight-Stationary Convolution Dataflow for Processing in Memory

---

```
1    i = Array(C, W)        # Input   activations
2    f = Array(M, C, S)     # Filter  weights
3    o = Array(M, Q)        # Output  partial  sums
4
5    parallel-for  m in [0, M):
6        parallel-for  s in [0, S):
7            parallel-for  c in [0, C]:
8                for q in [0, Q):
9                    w = q + s
10                   o[m, q] += i[c, w] * f[m, c, s]
```

---

input channels (C) and multiple output channels (M). The entire computation takes Q steps as the only temporal step is the **for** loop (line 8). Interpreting the activity in the body of the loop (line 10), we see that in each cycle all filter weights are used ($M \times S \times C$) each as part a distinct MAC operation, the same input activation is used multiple times ($C \times S$) and multiple output partial sums are accumulated into (M). This design reflects the Toeplitz expansion of the input activations (see Section 4.1), so the same input activations will be delivered multiple times, since the same value for the input activation index w will be generated for different qs. For the processing in memory convolution design, some of the mapping constraints are that $A \geq M$ and $B \geq C \times S$. Note, that when $A \neq M$ or $B \neq C \times S$ under-utilization will occur, as described in Section 10.2.4.

In the next few sections (Sections 10.2.1, 10.2.2, and 10.2.3), we will briefly describe how processing in memory can be performed using different memory technologies. Section 10.2.4 will then give an overview of some of the key design challenges and decisions that should be considered when designing processing-in-memory accelerators for DNNs. For instance, many of these designs are limited to reduced precision (i.e., low bit-width) due to the non-idealities of the devices and circuits used for memories.

10.2.1   NON-VOLATILE MEMORIES (NVM)

Many recent works have explored enabling processing-in-memory using *non-volatile memories (NVM)* due to their high density and thus potential for replacing off-chip memory and reducing off-chip data movement. Advanced non-volatile high-density memories use programmable resistive elements, commonly referred to as *memristors* [335], as storage elements. These NVM devices enable increased density since memory and computation can be densely packed with a similar density to DRAM [336].[7]

Non-volatile memories exploit Ohm's law by using the conductance (i.e., the inverse of the resistance) of a device to represent a filter weight and the voltage across the device to represent the input activation value. So the resulting current can be interpreted as the product (i.e., a partial sum). This is referred to as a *current-based* approach. For instance, Figure 10.4a shows how a multiplication can be performed using the conductance of the NVM device as the weight, and the voltage on the word line as the input activation, and the current output to the bit line as the product of the two. The accumulation is done by summing the currents on the bit line based on Kirchhoff's current law. Alternatively, for Flash-based NVM, the multiplication is performed using the current-voltage (IV) characteristics of the floating-gate transistor, where the threshold voltage of the floating-gate transistor is set based on the weight, as shown in Figure 10.4c. Similar to the previously described approaches, a voltage proportional to the input activation can be applied across the device, and the accumulation is performed by summing output current of the devices on the bit line.

NVM-based processing-in-memory accelerators have several unique challenges, as described in [340, 341]. First, the cost of programming the memristors (i.e., writing to non-volatile memory) can be much higher than SRAM or DRAM; thus, typical solutions in this space require that the non-volatile memory to be sufficiently large to hold *all* weights of the DNN model, rather than changing the weights in the memory for each layer or filter during processing.[8] As discussed in Chapter 3, this may reduce flexibility as it can limit the size of the DNN model that the accelerator can support.

Second, the NVM devices can also suffer from device-to-device and cycle-to-cycle variations with nonlinear conductance across the conductance range [340–342]. This affects the number of bits that can be stored per device (typically 1 to 4) and the type of signaling used for the input and output activations. For instance, rather than encoding the input activation in terms of voltage amplitude, the input can also be encoded in time using pulse width modulation

---

[7]To improve density, the resistive devices can be inserted between the cross-point of two wires and in certain cases can avoid the need for an access transistor [337]. Under this scenario, the device is commonly referred to as a cross-point element.

[8]This design choice to hold all weights of the DNN is similar to the approach taken in some of the FPGA designs such as Brainwave [209] and FINN [226], where the weights are pinned on the on-chip memory of the FPGA during synthesis.

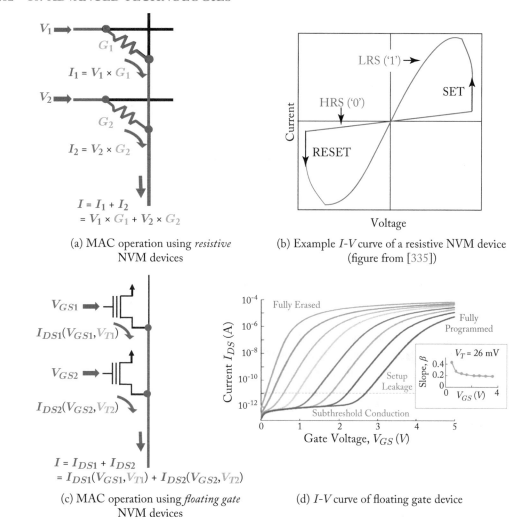

(a) MAC operation using *resistive* NVM devices

(b) Example *I-V* curve of a resistive NVM device (figure from [335])

(c) MAC operation using *floating gate* NVM devices

(d) *I-V* curve of floating gate device

Figure 10.4: Performing a multiplication and accumulation using the storage element. Input activation is encoded as a voltage amplitude (V_i). (a) For memristors, G_i is the conductance (i.e., 1/resistance) of a resistive device set according to the weight, and bit line current I is the accumulated partial sum value [329]. (b) The current-voltage (I-V) characteristics of the resistive device. The slope of the curve is inversely proportional to the resistance (recall $R = V/I$). Typically, the device can take on just two states: LRS is the low resistive state (also referred to as R_{ON}) and HRS is the high resistive state (also referred to as R_{OFF}). (c) and (d) For floating-gate transistors, the multiplication is performed using its current-voltage (I-V) characteristics, where the weight sets the threshold voltage (as illustrated by the different color lines representing different threshold voltages), and bit line current I is the accumulated partial sum value [339].

with a *fixed* voltage (i.e., a unary coding), and the resulting current can be accumulated over time on a capacitor to generate the output voltage [343].

Finally, the NVM devices cannot have negative resistance, which presents a challenge for supporting negative weights. One approach is to represent signed weights using differential signaling that requires two storage elements per weight; accordingly, the weights are often stored using two separate arrays [344]. Another approach is to avoid using signed weights. For instance, in the case of binary weights, rather than representing the weights as $[-1, 1]$ and performing binary multiplications, the weights can be represented as $[0, 1]$ and perform XNOR logic operations, as discussed in Chapter 7, or NAND logic operations, as discussed in [345].

There are several popular candidates for NVM devices including phase change RAM (PCRAM), resistive RAM (RRAM or ReRAM), conductive bridge RAM (CBRAM), and spin transfer torque magnetic RAM (STT-MRAM) [346]. These devices have different trade-offs in terms of endurance (i.e., how many times it can be written), retention time (i.e., how often it needs to be refreshed and thus how frequently it needs to be written), write current (i.e., how much power is required to perform a write), area density (i.e., cell size), variations, and speed. An in-depth discussion of how these device properties affect the performance of DNN processing can be found in [341]; Gokmen et al. [343] flips the problem and describes how these devices should be designed such that they can be better suited for DNN processing.[9]

Recent works on NVM-based processing-in-memory accelerators have reported results from both simulation [329, 338, 347, 348] as well as fabricated test chips [344, 349]. While works based on simulation demonstrate functionality on large DNN models such as variants of VGGNet [73] for image classification on ImageNet, works based on fabricated test chips still demonstrate functionality on simple DNN models for digit classification on MNIST [344, 349]. Simulations often project capabilities beyond the current state-of-the-art. For instance, while works based on simulation often assume that all 128 or 256 rows can be activated at the same time, works based on fabricated test chips only activate up to 32 rows at once to account for process variations and limitations in the read out circuits (e.g., ADC); these limitations will be discussed more in Section 10.2.4. It should also be noted that fabricated test chips typically only use one bit per memristor [344, 349, 350].

10.2.2   STATIC RANDOM ACCESS MEMORIES (SRAM)

Many recent works have explored the use of the SRAM bit cell to perform computation. They can be loosely classified into current-based and charge-based designs.

Current-based designs use the current-voltage (IV) characteristics of the bit cell to perform a multiplication, which is similar to the NVM current-based approach described in Section 10.2.1. For instance, Figure 10.5a shows how the input activation can be encoded as a voltage amplitude on the word line that controls the current through the pull-down network of

---

[9][341, 343] also describe how these devices might be used for training DNNs if the weights can be updated in parallel and in place within the memristor array.

a bit cell (I_{BC}) resulting in a voltage drop (V_{BL}) proportional to the word line voltage [351]. The current from multiple bit cells (across different rows on the same column) add together on the bit line to perform the accumulation [351]. The resulting voltage drop on the bit line is then proportional to the dot product of the weights and activations of the column.

The above current-based approach is susceptible to the variability and nonlinearity of the word line voltage-to-current relationship of the pull-down network in the bit cell; this create challenges in representing the weights precisely. Charge-based approaches avoid this by using *charge sharing* for the multiplication, where the computation based on the capacitance ratio between capacitors, which tends to be more linear and less sensitive to variations.

Figure 10.5b shows how a binary multiplication (i.e., XNOR) via charge sharing can be performed by conditionally charging up a local capacitor within a bit cell, based on the XNOR between the weight value stored in the bit cell and the input activation value that determines the word line voltage [352]. Accumulation can then be performed using charge sharing across the local capacitors of the bit cells on a bit line [352]. Other variants of this approach include performing the multiplication directly with the bit line [353], and charge sharing across different bit lines to perform the accumulation [353–355].

One particular challenge that exists for SRAM-based processing-in-memory accelerators is maintaining bit cell stability. Specifically, the voltage swing on the bit line typically needs to be kept low in order to avoid a read disturb (i.e., accidentally flipping the value stored in the bit cell when reading). This limits the voltage swing on the bit line, which affects the number of bits that can be accumulated on the bit line for the partial sum; conventional SRAMs only resolve one bit on the bit line. One way to address this is by adding extra transistors to isolate the storage node in the bit cell from the large swing of the bit line [353]; however, this would increase the bit cell area and consequently reduce the overall area density.

Recent works on SRAM-based processing-in-memory accelerators have reported results from fabricated test chips [351–355]. In these works, they demonstrate functionality on simple DNN models for digit classification on MNIST, often using layer-by-layer processing, where the weights are updated in the SRAM for each layer. Note that in these works, the layer shapes of the DNN model are often custom designed to fit the array size of the SRAM to increase utilization; this may pose a challenge in terms of flexibility, as discussed in Chapter 3.[10]

10.2.3   DYNAMIC RANDOM ACCESS MEMORIES (DRAM)

Recent works have explored how processing in memory may be feasible using DRAM by performing bit-wise logic operations when reading multiple bit cells. For instance, Figure 10.6 shows how AND and OR operations can be performed by accessing three rows in parallel [356]. When three rows are accessed at the same time, the output bit line voltage will depend on the

---

[10]It should be noted that since SRAM is less dense than typical off-chip memory (e.g., DRAM), they are not designed to replace off-chip memory or specifically addressing the "memory wall," which pertains to off-chip memory bandwidth; instead, most SRAM-based processing-in-memory accelerators focus on reducing the on-chip data movement.

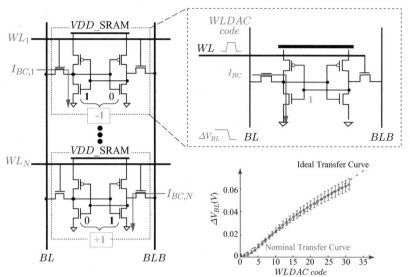

(a) Multiplication using a 6T SRAM bit-cell and accumulation by current summing on bit lines (figure adapted from [348])

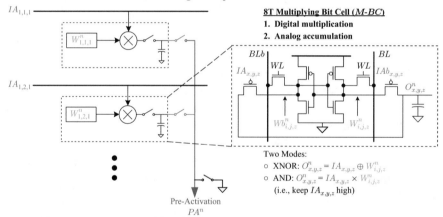

(b) Multiplication using a 8T SRAM bit-cell and a local capacitor and accumulation using charge sharing across local capacitors (figure adapted from [349])

Figure 10.5: Performing a multiplication and accumulation using the storage element. (a) Multiplication can be performed using a SRAM bit-cell by encoding the input activation as a voltage amplitude on the word line that controls the current through the pull-down network of the bit cell (I_{BC}) resulting in a voltage drop (V_{BL}) proportional to the word line voltage. If a zero (weight value of -1) is stored in the bit cell, the voltage drop occurs on BL, while if a one (weight value of $+1$) is stored the voltage drop occurs on BLB. The current from multiple bit-cells within a column add together. (b) Binary multiplication (XNOR) is performed by connection transistors and local capacitor. Accumulation is performed by charge sharing across local capacitors in bit-cells from the same column.

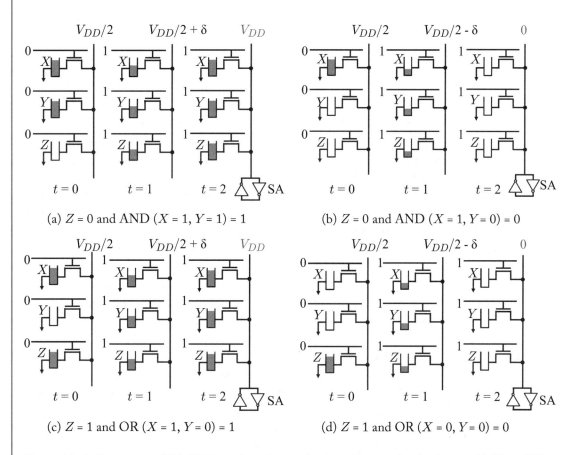

(a) $Z = 0$ and AND $(X = 1, Y = 1) = 1$

(b) $Z = 0$ and AND $(X = 1, Y = 0) = 0$

(c) $Z = 1$ and OR $(X = 1, Y = 0) = 1$

(d) $Z = 1$ and OR $(X = 0, Y = 0) = 0$

Figure 10.6: Compute in DRAM based on charge sharing. Z controls whether an AND or OR is performed on input X and Y. At time $t = 0$, the local capacitor of the bit cells for X, Y, and Z are charged to V_{DD} for one and 0 for zero, and the bit line is pre-charged to $V_{DD}/2$. At time $t = 1$, the accessed transistors to the bit cells are enabled, and the capacitors are shorted together with the bit line. Charge sharing distributes the charge between the capacitors to ensure that the voltage across each capacitor is the same; therefore the resulting voltage on the bit line is proportional to the average charge across the three capacitors. If the majority of the capacitors stored at one (i.e., V_{DD}), then the voltage on the bit line would be above $V_{DD}/2$ (i.e., $+\delta$); otherwise, the voltage on the bit line drops below $V_{DD}/2$ (i.e., $-\delta$). At time $t = 2$, the sense amplifiers (SA) on the bit line amplify the voltage to full swing (i.e., $V_{DD}/2 + \delta$ becomes V_{DD} or $V_{DD}/2 - \delta$ becomes 0), such that the output of the logic function $XY + YZ + ZX$ can be resolved on the bit line. Note that this form of computing is destructive, so we need to copy data beforehand.

average of the charge stored in the capacitors of the bit cells in three rows (note that the charge stored in capacitor of a bit cell depends on if the bit cell is storing a one or zero). Therefore, if the majority of the values of the bit cells are one (at least two out of three), then the output is a one; otherwise, the output is a zero. More precisely, if X, Y, and Z represent the logical values of the three cells, then the final state of the bit line is $XY + YZ + ZX$. If $Z = 1$, then this is effectively an OR operation between X and Y; if $Z = 0$, then this is effectively an AND operation between X and Y. The bit-wise logic operations can be built up into MAC operations across multiple cycles [357], similar to bit-serial processing described in Chapter 7.

It is important to note that the architecture of processing in memory with DRAM differs from the processing in memory with NVM and SRAM (described in Sections 10.2.1 and 10.2.2, respectively) in that: (1) for DRAM, a bit-wise operation requires three storage elements from different rows, whereas for NVM and SRAM, a MAC operation can be performed with a single storage element; and (2) for DRAM, only one bit-wise operation is performed per bit line and the accumulation occurs over time, whereas for NVM and SRAM, the accumulation of multiple MAC operations is performed on the bit line.[11] As a result, for DRAM the parallel processing can only be enabled across bit lines (A in Figure 10.3), since only one operation can be performed per bit line, whereas for NVM and SRAM, the parallel processing can be enabled across both the bit lines and the word lines (A and B in Figure 10.3), since multiple operations can be performed per bit line. In addition, for DRAM, multiple cycles are required to build up a MAC operation from a bit-wise logic operation, which reduces throughput. Thus, a challenge for DRAM-based processing-in-memory accelerators is to ensure that there is sufficient parallelism across bit lines (A) to achieve the desired improvements in throughput.

Other challenges for DRAM-based processing-in-memory accelerators include variations in the capacitance in the different bit cells, changing charge in capacitor of bit cell over time due to leakage, and detecting small changes in the bit line voltage. In addition, additional hardware may be required in the memory controller to access multiple rows at once and/or to convert the bit-wise logic operations to MAC operation, all of which can contribute to energy and area overhead.

While many of the recent works on DRAM-based processing-in-memory accelerators have been based on simulation [356, 357], it should be noted that performing AND and OR operations have been demonstrated on off-the-shelf, unmodified, commercial DRAM [359]. This was achieved by violating the nominal timing specification and activating multiple rows in rapid succession, which leaves multiple rows open simultaneously and enables charge sharing on the bit line.

10.2.4 DESIGN CHALLENGES

Processing-in-memory accelerators offer many potential benefits including reduced data movement of weights, higher memory bandwidth by reading multiple weights in parallel, higher

---

[11]This bit-wise (bit-serial) approach has also been explored for SRAM [358].

Design 10.22 Toy matrix multiply loop nest

```
1    i = Array(CHW)              # Input   activations
2    f = Array(CHW, M)           # Filter  weights
3    o = Array(M)                # Output  partial  sums
4
5    parallel-for  m in [0, M):
6        parallel-for  chw in [0, CHW):
7            o[m] += i[chw] * f[chw, m]
```

---

throughput by performing multiple computations in parallel, and lower input activation delivery cost due to increased density of compute. However, there are several key design challenges and decisions that need to be considered in practice. Analog processing is typically required to bring the computation into the array of storage elements or into its peripheral circuits; therefore the major challenges for processing in memory are its sensitivity to circuit and device non-idealities (i.e., nonlinearity and process, voltage and temperature variations).[12] Solutions to these challenges often require trade offs between energy efficiency, throughput, area density, and accuracy,[13] which reduce the achievable gains over conventional architectures. Architecture-level energy and area estimation tools such as Accelergy can be used to help evaluate some of these trade offs [360].

In this section, when applicable we will use a toy example of a matrix vector multiplication based on a FC layer shown in Figure 10.7. A loop-nest representation of the design is shown in Design 10.22, where $CHW = M = 4$. In theory, the entire computation should only require one cycle as all the 16 weights can be accessed in parallel and all the 16 MAC operations can be performed in parallel.

Number of Storage Elements per Weight
Ideally, it would be desirable to be able to use one storage element (i.e., one device or bit cell) per weight to maximize density. In practice, multiple storage elements are required per weight due to the limited precision of each device or bit cell (typically on the order of 1 to 4 bits). As a result, multiple low-precision storage elements are used to represent a higher precision weight. Figure 10.8 shows how this applies to our toy example.

---

[12]Note that per chip training (i.e., different DNN weights per chip instance) may help address nonlinearity and chip to chip variability, but is expensive in practice. In addition, while adapting the weights can help address *static* variability, *dynamic* variability, such as a change in temperature, remains a challenge.

[13]It should be noted that the loss in accuracy might not only be due to the reduced precision of the computations in the DNN model (discussed in Chapter 7), which can be replicated on a conventional processor, but also due to circuit/device non-idealities and limitations, including ADC precision and thermal noise. Unfortunately, these factors have rarely been decoupled during reporting in literature, which can make it difficult to understand the design trade offs.

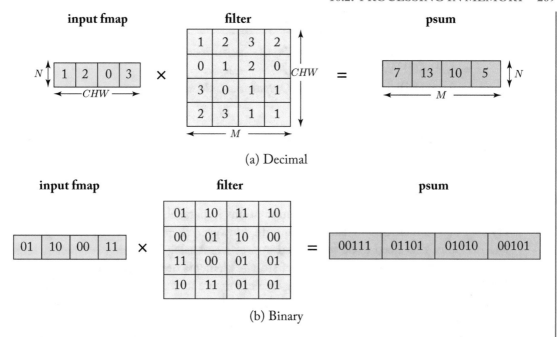

(a) Decimal

(b) Binary

Figure 10.7: Toy example of matrix vector multiplication for this section. This example uses an FC layer with $N = 1$, $CHW = 4$, and $M = 4$.

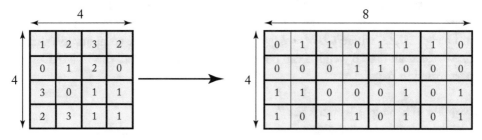

Figure 10.8: Example of multiple storage elements per weight. In our toy example we use 2 bits per weight so the storage cost goes from 4×4 to 4×8.

For non-volatile memories (e.g., RRAM), multiple storage elements can also be used per weight to reduce the effect of devices variation (e.g., average 3×3 devices per weight [342]) or to represent a signed weight (i.e., since resistance is naturally non-negative, differential coding using two arrays is often used [342]). Finally, in the case of SRAMs, often additional transistors are required in the bit cell to perform an operation, which increases the area per bit cell. All of the above factors reduce the density and/or accuracy of the system.

Array Size

Ideally, it would be desirable to have a large array size ($A \times B$) in order to allow for high weight read bandwidth and high throughput. In addition, a larger array size improves the area density by further amortizing the cost of the peripheral circuits, which can be significant (e.g., the peripheral circuits, i.e., ADC and DAC, can account for over 50% of the energy consumption of NVM-based designs [329, 349]). In practice, the size of array limited by several factors.

1. The resistance and capacitance of word line and bit line wires, which impacts robustness, speed, and energy consumption.

   For instance, the bit line capacitance impacts robustness for charge domain approaches where charge sharing is used for accumulation, as a large bit line capacitance makes it difficult to sense the charge stored on the local capacitor in the bit cell; the charge stored on the local capacitor can be an input value for DRAM-based designs or a product of weight and input activation for SRAM-based designs. An example of using charge sharing to sense the voltage across a local capacitor is shown in Figure 10.9. Specifically, the change in bit line voltage (ΔV_{BL}) is

$$\Delta V_{BL} = (V_{DD} - V_{local}) \frac{C_{local}}{C_{local} + C_{BL}}, \tag{10.1}$$

   where C_{local} and C_{BL} are the capacitance of the local capacitor and bit line, respectively, and V_{local} is the voltage across the local capacitor (due to the charge stored on the local capacitor), and V_{DD} is the supply voltage. If the local capacitor is only storing binary values, then V_{local} can either be V_{DD} or 0. ΔV_{BL} must be sufficiently large such that we can measure any change in V_{local}; the more bits we want to measure on the bit line (i.e., bits of the partial sum or output activation), the larger the required ΔV_{BL}. However, the size of C_{local} is limited by the area density of the storage element; for instance, in [352], C_{local} is limited to 1.2fF. As a result, the minimum value of ΔV_{BL} limits the size of C_{BL}, which limits the length of the bit line.

   Similarly, the bit line resistance impacts robustness for current domain approaches where current summing is used for accumulation, as a large bit line resistance makes it difficult to sense the change in the resistance in the NVM device, as shown in Figure 10.10. Specifically, the change in bit line voltage due to change on the resistance is

$$\Delta V_{BL} = V_{HIGH} - V_{LOW} = V_{in} R_{BL} \frac{R_{OFF} - R_{ON}}{(R_{ON} + R_{BL})(R_{OFF} + R_{BL})}, \tag{10.2}$$

   where R_{ON} and R_{OFF} are the minimum and maximum resistance of the NVM device (proportional to the weight), respectively, R_{BL} is the resistance of the bit line, and V_{in} is the input voltage (proportional to the input activation). The $R_{OFF} - R_{ON}$ is limited by the NVM device [342]. As a result, the minimum value of ΔV_{BL} limits the size of R_{BL}, which again limits the length of the bit line.

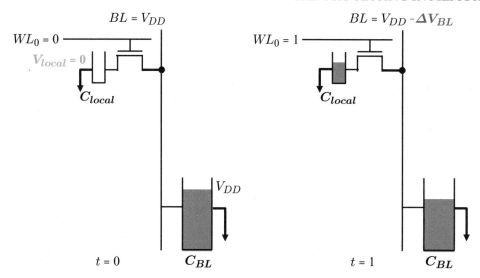

Figure 10.9: Change in bit line voltage ΔV_{BL} is proportional to $\frac{C_{local}}{C_{local}+C_{BL}}$. The bit line is precharged to V_{DD} at $t = 0$, and we read the value on the local capacitor at $t = 1$.

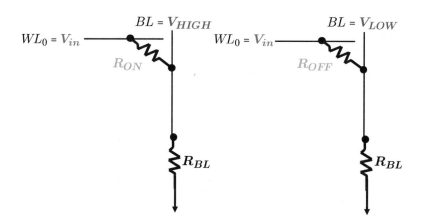

Figure 10.10: Change in bit line voltage $\Delta V_{BL} = V_{HIGH} - V_{LOW}$ is proportional to $R_{BL}\frac{R_{OFF}-R_{ON}}{(R_{ON}+R_{BL})(R_{OFF}+R_{BL})}$. R_{ON} (also referred to as LRS) and R_{OFF} (also referred to as HRS) are the minimum and maximum resistance of the NVM device, respectively.

2. The utilization of the array will drop if the workload cannot fill entire column or entire row, as shown in Figure 10.11a. If the DNN model has few weights per filter and does not require large dot products, e.g., $C \times H \times W \leq B$, where C, H, and W, are the dimensions of the filter (FC layer), and B is the number of rows in the array, then there will be $B -$

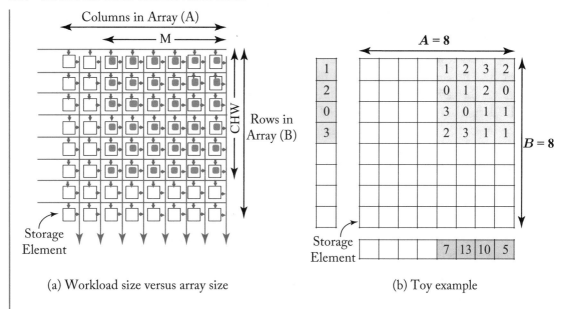

(a) Workload size versus array size   (b) Toy example

Figure 10.11: Array utilization. (a) Impact of array size on utilization. (b) Example of utilization if size of weight memory was 8×8. Even though in theory we should be able to perform 64 MAC operations in parallel, only 16 of the storage elements are used (utilization of 25%); as a result, only 16 MAC operations are performed in parallel, specifically, 4 dot products of 4 elements.

$C \times H \times W$ idle rows in the array. If the DNN model has few output channels and does not have many dot products, e.g., $M \leq A$, where M is the number of output channels and A is the number of columns in the array, then there will be $A - M$ idle columns in the array.[14] This becomes more of an issue when processing efficient DNN models as described in Chapter 9, where the trend is to reduce the number of weights per filter. In digital designs, flexible mapping can be used to increase utilization across different filter shapes, as discussed in Chapter 6; however, this is much more challenging to implement in the analog domain. One option is to redesign the DNN model specifically for processing in memory with larger filters and fewer layers [315], which increases utilization of the array and reduces input activation data movement; however, the accuracy implications of

---

[14]Note that if $C \times H \times W > B$ or $M > A$, temporal tiling will need to be applied, as discussed in Chapter 4, and multiple passes (including updating weights in the array) will be required to complete the MAC operations. Furthermore, recall that if the completed sum (final psum) can be computed within a single pass (i.e., $C \times H \times W \leq B$), then precision of the ADC can be reduced to the precision of the output activation. However, when multiple passes are needed, the ADC needs greater precision because the results of each pass need to be added together to form the completed sum; otherwise, there may be an accuracy loss.

such DNN models requires further study. Figure 10.11b shows how this applies to our toy example.

As a result, typical fabricated array sizes range from 16b × 64 [353] to 512b × 512 [352] for SRAM and from 128 × 128 to 256 × 256 [342] for NVM. This limitation in array size affects throughput, area density and energy efficiency. Multiple arrays can be used to scale up the design in order to fit the entire DNN Model and increase throughput [329, 347]. However, the impact on amortizing the peripheral cost is minimal. Furthermore, an additional NoC must be introduced between the arrays. Accordingly, the limitations on energy efficiency and area density remain.

Number of Rows Activated in Parallel

Ideally, it would be desirable to use all rows (B) at once to maximize parallelism for high bandwidth and high throughput. In practice, the number of rows that can be used at once is limited by several factors.

1. The number of bits in the ADC, since more rows means more bits are required to resolve the accumulation (i.e., the partial sums will have more bits). Some works propose using fewer bits for ADC than the maximum required [361, 362], however, this can reduce the accuracy.[15]

2. The cumulative effect of the device variations can decrease the accuracy.

3. The maximum voltage drop or accumulated current that can be tolerated by the bit line.[16] This can be particularly challenging for advanced process technologies (e.g., 7 nm and below) due to the increase in bit line resistance and increased susceptibility to electromigration issues, which limits the maximum current on the bit line.

As a result, the typical number of rows activated in parallel is 64 [342] or below [344]. A digital accumulator can be used after each ADC to accumulate across all B rows in $B/64$ cycles [342]; however, this reduces throughput and increases energy due to multiple ADC conversions. To reduce the additional ADC conversion, recent work has explored performing the accumulation in the analog domain [348]. Figure 10.12 shows how this applies to our toy example. Design 10.23 shows the corresponding loop nest, and illustrates the multiple cycles it takes to perform all the MACs.

---

[15]The number of bits required by the ADC depends on the number of values being accumulated on the bit line (i.e., number of rows activated in parallel), whether the values are sparse [361] (i.e., zero values will not contribute to the accumulated sum), and whether the accumulated sum is a partial sum or a fully accumulated sum (i.e., it only needs to go through a nonlinear function to become an output activation). Using less than the maximum required ADC bits for the fully accumulated sum has less impact on accuracy than on the partial sum, since the fully accumulated sum is typically quantized to the bit-width of the input activation for the next layer, as discussed in Chapter 7. However, the ability to fully accumulate the sum on a bit line depends on the whether the number of rows in the array is large enough to hold all the weights for a given filter (i.e., $B \geq C \times H \times W$).

[16]For instance, for a 6T SRAM bit cell, a large voltage drop on the bit line can cause the bit cell to flip (i.e, an unwanted write operation on the bit cell); using 8T bit cell can prevent this at the cost of increased area.

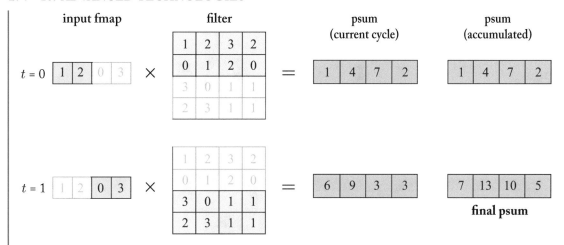

Figure 10.12: Example of limited number of rows activated in parallel. If the ADC is only 3-bits, only two rows can be used at a time. It would take two cycles (time steps) to complete the computation. There are two columns for psum in the figure: (1) psum (current cycle) corresponds to psum resulting from the dot product computed at the current cycle; (2) psum (accumulated) corresponds to the accumulated value of the psums across cycles. At $t = 1$, the psum of [6, 9, 3, 3] is computed and added (e.g., with a digital adder) to the psum at $t = 0$ of [1, 4, 7, 2] to achieve the final psum [7, 13, 10, 5], as shown in the figure.

Design 10.23 Toy matrix multiply loop nest with limited number of parallel active rows

---

```
1   i = Array(CHW)              # Input   activations
2   f = Array(CHW, M)           # Filter  weights
3   o = Array(M)                # Output partial sums
4
5   parallel-for  m in [0, M):
6       parallel-for  chw1 in [0, CHW/2):
7           for  chw0 in [0, 2):
8               chw = chw1*2 + chw0
9               o[m] += i[chw] * f[chw, m]
```

Number of Columns Activated in Parallel

Ideally, it would be desirable to use all columns (A) at once to maximize parallelism for high bandwidth and high throughput. In practice, the number of columns that can be used are limited by whether the area of ADC can pitch-match the width of the column, which is required for a compact area design; this can be challenging when using high-density storage elements

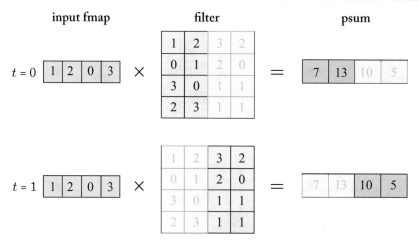

Figure 10.13: Example of limited number of columns activated in parallel. If the width of an ADC is equal to two columns, then the columns need to be time multiplexed. It would take two cycles to complete the computation. If we combined this with the previously described parallel row limitations, it would take four cycles to complete the computation.

Design 10.24 Toy matrix multiply loop nest with limited number of parallel active columns

```
1    i = Array(CHW)            # Input   activations
2    f = Array(CHW, M)         # Filter   weights
3    o = Array(M)              # Output   partial   sums
4
5    parallel-for  m1 in [0,  M/2):
6        parallel-for  chw in [0,  CHW):
7            for  m0 in [0,  2):
8                m = m1*2 + m0
9                o[m] += i[chw] * f[chw, m]
```

---

such as NVM devices. A common solution is to time multiplex the ADC across a set of eight columns, which means that only $A/8$ columns are used in parallel [342]; however, this reduces throughput. Figure 10.13 shows how this applies to our toy example, and Design 10.24 shows the corresponding loop nest.

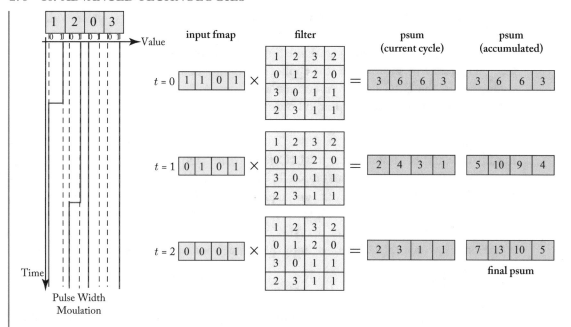

Figure 10.14: Example of performing pulse–width modulation of the input activations with a 1-bit DAC. It would take three cycles to complete the computation if all weights can be used at once. Specifically, the input activations would be signaled across time as $[1, 1, 0, 1] + [0, 1, 0, 1] + [0, 0, 0, 1] = [1, 2, 0, 3]$, where the width of the pulse in time corresponds to the value of the input. There are two columns for psum in the figure: (1) psum (current cycle) corresponds to psum resulting from the dot product computed at the current cycle; (2) psum (accumulated) corresponds to the accumulated value of the psums across cycles. Note that if we combined the limitation illustrated in this figure with the previously described parallel row and columns limitations, it would take 12 cycles to complete the computation.

Time to Deliver Input

Ideally, it would be desirable for all bits in the input activations to be encoded onto the word line in the minimum amount of time to maximize throughput; a typical approach is to use voltage amplitude modulation [351]. In practice, this can be challenging due to

1. the nonlinearity of devices makes encoding input value using voltage amplitude modulation difficult, and

2. the complexity of the DAC that drives the word line scales with the number of bits

As a result, the input activations are often encoded in time (e.g., pulse-width modulation [354, 355] or number of pulses [362][17]), with a fixed voltage (DAC is only 1-bit) where the partial sum is determined by accumulating charge over time; however, this reduces throughput.[18] Figure 10.14 shows how this applies to our toy example. One approach to reduce the complexity of the DAC or current accumulation time is to reduced the precision of the input activations, as discussed in Chapter 7; however, this will also reduce accuracy.

Time to Compute a MAC

Ideally, it would be desirable for a MAC to be computed in a single cycle. In practice, the storage element (bit cell or device) typically can only perform one-bit operations (e.g., XNOR and AND), and thus multiple cycles are required to build up to a multi-bit operation (e.g., full adder and multiplication) [331]. Figure 10.15 shows how this applies to our toy example. This also requires additional logic after the ADC to combine the one-bit operations into a multi-bit operation. However, this will reduce both the throughput, energy and density.

10.3   PROCESSING IN SENSOR

In certain applications, such as image processing, the data movement from the sensor itself can account for a significant portion of the overall energy consumption. Accordingly, there has been work on bringing the processing near or into the sensor, which is similar to the work on bringing the processing near or into memory discussed in the previous sections. In both cases, the goal is to reduce the amount of data read out of the memory/sensor and thus the number of ADC conversions, which can be expensive. Both cases also require moving the computation into the analog domain and consequently suffer from increased sensitivity to circuit non-idealities. While processing near memory and processing in memory focus on reducing data movement of the weights of the DNN model, processing near sensor and processing in sensor focus on reducing the data movement of the inputs to the DNN model.

Processing near sensor has been demonstrated for image processing applications, where computation can be performed in the analog domain before the ADC in the peripheral of the image sensor. For instance, Zhang et al. [363] and Lee et al. [364] use switched capacitors to perform 4-bit multiplications and 3-bit by 6-bit MAC operations, respectively. RedEye [365] proposes performing the entire convolution layer (including convolution, max pooling and quantization) in the analog domain before the ADC. It should be noted that the results in [365] are based on simulations, while [363, 364] report measurements from fabricated test chips.

It is also feasible to embed the computation not just before the ADC, but directly into the sensor itself (i.e., processing in sensor). For instance, in [366] an Angle Sensitive Pixels (ASP) sensor (shown in Figure 10.16) is used to compute the gradient of the input, which along

---

[17]Using pulses increases robustness to nonlinearity at the cost of increased switching activity.

[18]Alternatively, a single pulse can be used for the input activations if the weights are replicated across multiple rows (e.g., 2^{N-1} rows for an N-bit activation) [333]. This is a trade-off between time and area.

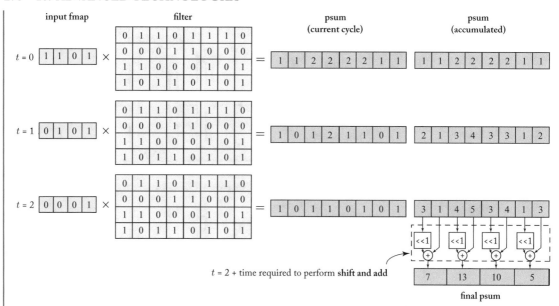

Figure 10.15: Example of time to compute MAC operation if each storage element can only perform one-bit operations. It takes three cycles to deliver the input (similar to Figure 10.14). There are two columns for psum in the figure: (1) psum (current cycle) corresponds to psum resulting from the dot product computed at the current cycle; (2) psum (accumulated) corresponds to the accumulated value of the psums across cycles. In addition, extra cycles are required at the end to combine accumulated bits from each bit line to form the final output sum. The number of cycles required to perform the shift and add would depend on the number of bit lines divide by the number of sets of shift-and-add logic.

with compression, reduces the data movement from the sensor by 10×. In addition, since the first layer of the DNN often outputs a gradient-like feature map, it may be possible to skip the computations in the first layer, which further reduces energy consumption, as discussed in [367, 368].

10.4   PROCESSING IN THE OPTICAL DOMAIN

Processing in the optical domain is an area of research that is currently being explored as an alternative to all-electronic accelerators [369]. It is motivated, in part, by the fact that photons travel much faster than electrons, and the cost of moving a photon can be *independent* of distance. Furthermore, multiplication can be performed passively (for example with optical interference [370, 371], with reconfigurable filters [373], or static phase masks [374]) and detection

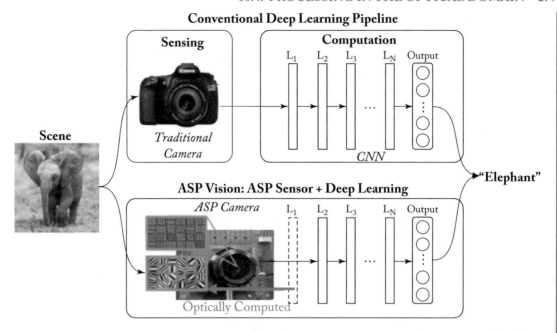

Figure 10.16: Use ASP in front end to perform processing of first layer. (Figure from [367].)

can occur at over 100 GHz. Thus, processing in the optical domain may provide significant improvements in energy efficiency and throughput over the electrical domain.

Much of the recent work in the optical computing has focused on performing matrix multiplication, which can be used for DNN processing; these works are often referred to as photonic accelerators or *optical neural networks*. For instance, Shen et al. [370] present a programmable nanophotonic processor where the input activations are encoded in the amplitudes of optical pulses (light) that travel through an array of on-chip interferometers (composed of beamsplitters) that represent the weight matrix, where the weights determine the amount of light that is passed to the output. This is effectively a weight-stationary dataflow. The accumulation is performed based on the accumulated light from various waveguides at the photodetector.

Alternatively, Hamerly et al. [371], shown in Figure 10.17b, demonstrate matrix multiplication based on coherent detection, where both the weights and activations are encoded on-the-fly into light pulses, and are interfered in free-space on a beamsplitter to perform multiplication. Since, in this scheme, there is no need for on-chip interferometers (which have a large footprint), this approach may be more scalable, at the cost of added complexity in alignment. This is effectively an output-stationary dataflow, where the output is accumulated on the photodetector as an analog electronic signal.

There is negligible power loss in the computation when processing in the optical domain. Most of the power dissipation occurs when converting between electrical and optical domains,

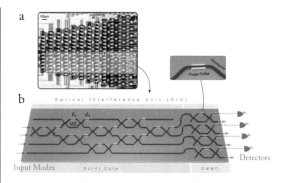

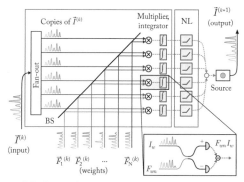

(a) Optical neural network using optical interference units (figure from [367])

(b) Optical neural network using coherent detection. BS: beamsplitter, NL: nonlinearity. (figure from [368])

Figure 10.17: Optical neural networks.

specifically, in the converter to generate the light and the detector to collect the photons. Therefore, similar to the processing in memory work, the larger the array (or in this case the matrix), the more these conversion costs can be amortized.

Note, however, that while computing in the optical domain may be energy efficient, the non-idealities in the optical devices (e.g., crosstalk between detectors, errors in phase encoding, photodetection noise) can lead to a reduction in accuracy. To address this accuracy loss, Bernstein et al. [372] propose a hybrid electronic-optics approach where the data transfer is done in the optical domain to exploit the distance-independent cost of photons, while the computation itself (i.e., MAC operation) is performed digitally in the electrical domain to avoid the non-idealities of the optical devices.

Recent works on optical neural networks have reported results based on simulations [371] or simulations based on data that has been extrapolated from experimental results [370]. These works demonstrate functionality on simple DNN models for digit classification and vowel recognition.

CHAPTER 11

Conclusion

The use of deep neural networks (DNNs) has recently seen explosive growth. They are currently widely used for many artificial intelligence (AI) applications including computer vision, speech recognition, and robotics and are often delivering better than human accuracy. However, while DNNs can deliver this outstanding accuracy, it comes at the cost of high computational complexity. With the stagnation of improvements in general-purpose computation [11], there is a movement toward more domain-specific hardware, and in particular for DNN processing. Consequently, techniques that enable efficient processing of DNNs to improve *energy-efficiency* and *throughput* without sacrificing *accuracy* with cost-effective hardware are critical to expanding the deployment of DNNs in both existing and new domains.

Creating a system for efficient DNN processing should begin with understanding the current and future applications and the specific computations required for both now and the potential evolution of those computations. Therefore, this book surveyed a number of the current applications, focusing on computer vision applications, the associated algorithms, and the data being used to drive the algorithms. These applications, algorithms, and input data are experiencing rapid change. So extrapolating these trends to determine the degree of flexibility desired to handle next generation computations becomes an important ingredient of any design project.

During the design-space exploration process, it is critical to understand and balance the important system metrics. For DNN computation these include the accuracy, energy, throughput and hardware cost. Evaluating these metrics is, of course, key, so this book surveyed the important components of a DNN workload. In specific, a DNN workload has two major components. First, the workload consists of the "network architecture" of the DNN model including the "shape" of each layer and the interconnections between layers. These can vary both within and between applications. Second, the workload consists of the specific data input to the DNN. This data will vary with the input set used for training or the data input during operation for inference.

This book also surveyed a number of avenues that prior work have taken to optimize DNN processing. Since data movement dominates energy consumption, a primary focus of some recent research has been to reduce data movement while maintaining accuracy, throughput, and cost. This means selecting architectures with favorable memory hierarchies like a spatial array, and developing dataflows that increase data reuse at the low-cost levels of the memory hierarchy. We have included a taxonomy of dataflows and an analysis of their characteristics. Understanding the throughput and energy efficiency of a DNN accelerator depends upon how each DNN workload maps to the hardware. Therefore, we discussed the process of optimally mapping workloads to the accelerator and the associated throughput and energy models.

The DNN domain also affords an excellent opportunity for hardware/algorithm co-design. Many works have aimed to save storage space and energy by changing the representation of data values in the DNN. We distill and present the key concepts from these approaches. Still other work saves energy and sometimes increases throughput by increasing and then exploiting sparsity of weights and/or activations. We presented a new abstract data representation that enables a systematic presentation of designs focused on exploiting sparsity. Co-design needs to be aware of the impact on accuracy. Therefore, to avoid losing accuracy it is often useful to modify the network or fine-tune the network's weights to accommodate these changes. Thus, this book both reviewed a variety of these techniques and discussed the frameworks that are available for describing, running and training networks.

Finally, DNNs afford the opportunity to use mixed-signal circuit design and advanced technologies to improve efficiency. These include using memristors for analog computation and 3-D stacked memory. Advanced technologies can also facilitate moving computation closer to the source by embedding computation near or within the sensor and the memories. Of course, all of these techniques should also be considered in combination, while being careful to understand their interactions and looking for opportunities for joint hardware/algorithm co-optimization.

In conclusion, although much work has been done, DNNs remain an important area of research with many promising applications and opportunities for innovation at various levels of hardware design. We hope this book provides a structured way of navigating the complex space of DNN accelerators designs that will inspire and lead to new advances in the field.

Bibliography

[1] J. Dean, Machine learning for systems and systems for machine learning, in *Conference on Neural Information Processing Systems (NeurIPS)*, 2017. xviii

[2] I. Goodfellow, Y. Bengio, A. Courville, and Y. Bengio, *Deep Learning*, vol. 1, MIT Press Cambridge, 2016. xx, 9

[3] A. Zhang, Z. C. Lipton, M. Li, and A. J. Smola, Dive into Deep Learning. http://d2l.ai/ xx

[4] F.-F. Li, A. Karpathy, and J. Johnson, Stanford CS class CS231n: Convolutional neural networks for visual recognition. http://cs231n.stanford.edu/ xx, 5, 6, 18, 35

[5] Y. LeCun, Y. Bengio, and G. Hinton, Deep learning, *Nature*, 521(7553):436–444, May 2015. DOI: 10.1038/nature14539. 3

[6] L. Deng, J. Li, J.-T. Huang, K. Yao, D. Yu, F. Seide, M. Seltzer, G. Zweig, X. He, J. Williams et al., Recent advances in deep learning for speech research at Microsoft, in *International Conference on Acoustics, Speech, and Signal Processing (ICASSP)*, 2013. DOI: 10.1109/icassp.2013.6639345. 3, 11, 13

[7] A. Krizhevsky, I. Sutskever, and G. E. Hinton, ImageNet classification with deep convolutional neural networks, in *Conference on Neural Information Processing Systems (NeurIPS)*, 2012. DOI: 10.1145/3065386. 3, 11, 12, 26, 27, 28, 29, 37

[8] C. Chen, A. Seff, A. Kornhauser, and J. Xiao, Deepdriving: Learning affordance for direct perception in autonomous driving, in *International Conference on Computer Vision (ICCV)*, 2015. DOI: 10.1109/iccv.2015.312. 3, 14

[9] A. Esteva, B. Kuprel, R. A. Novoa, J. Ko, S. M. Swetter, H. M. Blau, and S. Thrun, Dermatologist-level classification of skin cancer with deep neural networks, *Nature*, 542(7639):115–118, 2017. DOI: 10.1038/nature21056. 3, 14

[10] D. Silver, A. Huang, C. J. Maddison, A. Guez, L. Sifre, G. van den Driessche, J. Schrittwieser, I. Antonoglou, V. Panneershelvam, M. Lanctot, S. Dieleman, D. Grewe, J. Nham, N. Kalchbrenner, I. Sutskever, T. Lillicrap, M. Leach, K. Kavukcuoglu, T. Graepel, and D. Hassabis, Mastering the game of Go with deep neural networks and tree search, *Nature*, 529(7587):484–489, January 2016. DOI: 10.1038/nature16961. 3, 14, 27

[11] J. L. Hennessy and D. A. Patterson, A new golden age for computer architecture: Domain-specific hardware/software co-design, enhanced security, open instruction sets, and agile chip development, in *International Symposium on Computer Architecture (ISCA)*, 2018. DOI: 10.1109/isca.2018.00011. 3, 73, 281

[12] Arthur L. Samuel, Some studies in machine learning using the game of checkers, *IBM J. Res. Dev.*, 3(3):210–229, July 1959. 4

[13] P. A. Merolla, J. V. Arthur, R. Alvarez-Icaza, A. S. Cassidy, J. Sawada, F. Akopyan, B. L. Jackson, N. Imam, C. Guo, Y. Nakamura et al., A million spiking-neuron integrated circuit with a scalable communication network and interface, *Science*, 345(6197):668–673, 2014. DOI: 10.1126/science.1254642. 5

[14] S. K. Esser, P. A. Merolla, J. V. Arthur, A. S. Cassidy, R. Appuswamy, A. Andreopoulos, D. J. Berg, J. L. McKinstry, T. Melano, D. R. Barch et al., Convolutional networks for fast, energy-efficient neuromorphic computing, *Proc. of the National Academy of Sciences (PNAS)*, 2016. DOI: 10.1073/pnas.1604850113. 5, 165

[15] H. Lee, R. Grosse, R. Ranganath, and A. Y. Ng, Unsupervised learning of hierarchical representations with convolutional deep belief networks, *Communications of the ACM*, 54(10):95–103, 2011. DOI: 10.1145/2001269.2001295. 7

[16] C. Sun, A. Shrivastava, S. Singh, and A. Gupta, Revisiting unreasonable effectiveness of data in deep learning era, in *International Conference on Computer Vision (ICCV)*, pages 843–852, 2017. DOI: 10.1109/iccv.2017.97. 9, 39

[17] N. Srivastava, G. Hinton, A. Krizhevsky, I. Sutskever, and R. Salakhutdinov, Dropout: A simple way to prevent neural networks from overfitting, *The Journal of Machine Learning Research (JMLR)*, 15(1):1929–1958, 2014. 9

[18] D. P. Kingma and J. Ba, Adam: A method for stochastic optimization, in *International Conference on Learning Representations (ICLR)*, 2015. 9

[19] M. Mathieu, M. Henaff, and Y. LeCun, Fast training of convolutional networks through FFTs, in *International Conference on Learning Representations (ICLR)*, 2014. 10, 70

[20] Y. LeCun, L. D. Jackel, B. Boser, J. S. Denker, H. P. Graf, I. Guyon, D. Henderson, R. E. Howard, and W. Hubbard, Handwritten digit recognition: applications of neural network chips and automatic learning, *IEEE Communications Magazine*, 27(11):41–46, November 1989. DOI: 10.1109/35.41400. 11, 28

[21] B. Widrow and M. E. Hoff, Adaptive switching circuits, in *IRE WESCON Convention Record*, 1960. DOI: 10.21236/ad0241531. 11

[22] B. Widrow, Thinking about thinking: the discovery of the LMS algorithm, *IEEE Signal Processing Magazine*, 2005. DOI: 10.1109/msp.2005.1407720. 11

[23] O. Russakovsky, J. Deng, H. Su, J. Krause, S. Satheesh, S. Ma, Z. Huang, A. Karpathy, A. Khosla, M. Bernstein, A. C. Berg, and L. Fei-Fei, ImageNet large scale visual recognition challenge, *International Journal of Computer Vision (IJCV)*, 115(3):211–252, 2015. DOI: 10.1007/s11263-015-0816-y. 12, 13, 27, 28, 37, 38, 246

[24] K. He, X. Zhang, S. Ren, and J. Sun, Deep residual learning for image recognition, in *Conference on Computer Vision and Pattern Recognition (CVPR)*, 2016. DOI: 10.1109/cvpr.2016.90. 13, 20, 27, 28, 31, 33, 37, 231

[25] Complete Visual Networking Index (VNI) Forecast, Cisco, June 2016. 13

[26] J. Woodhouse, Big, big, big data: Higher and higher resolution video surveillance, technology.ihs.com, January 2016. 13

[27] R. Girshick, J. Donahue, T. Darrell, and J. Malik, Rich feature hierarchies for accurate object detection and semantic segmentation, in *Conference on Computer Vision and Pattern Recognition (CVPR)*, 2014. DOI: 10.1109/cvpr.2014.81. 13

[28] J. Long, E. Shelhamer, and T. Darrell, Fully convolutional networks for semantic segmentation, in *Conference on Computer Vision and Pattern Recognition (CVPR)*, 2015. DOI: 10.1109/cvpr.2015.7298965. 13

[29] K. Simonyan and A. Zisserman, Two-stream convolutional networks for action recognition in videos, in *Conference on Neural Information Processing Systems (NeurIPS)*, 2014. 13

[30] G. Hinton, L. Deng, D. Yu, G. E. Dahl, A.-r. Mohamed, N. Jaitly, A. Senior, V. Vanhoucke, P. Nguyen, T. N. Sainath et al., Deep neural networks for acoustic modeling in speech recognition: The shared views of four research groups, *IEEE Signal Processing Magazine*, 29(6):82–97, 2012. DOI: 10.1109/msp.2012.2205597. 13

[31] R. Collobert, J. Weston, L. Bottou, M. Karlen, K. Kavukcuoglu, and P. Kuksa, Natural language processing (almost) from scratch, *Journal of Machine Learning Research (JMLR)*, 12:2493–2537, 2011. 13

[32] A. van den Oord, S. Dieleman, H. Zen, K. Simonyan, O. Vinyals, A. Graves, N. Kalchbrenner, A. Senior, and K. Kavukcuoglu, Wavenet: A generative model for raw audio, *ISCA Speech Synthesis Workshop*, 2016. 13

[33] H. Y. Xiong, B. Alipanahi, L. J. Lee, H. Bretschneider, D. Merico, R. K. Yuen, Y. Hua, S. Gueroussov, H. S. Najafabadi, T. R. Hughes et al., The human splicing code reveals

new insights into the genetic determinants of disease, *Science*, 347(6218), p. 1254806, 2015. DOI: 10.1126/science.1254806. 13

[34] J. Zhou and O. G. Troyanskaya, Predicting effects of noncoding variants with deep learning-based sequence model, *Nature Methods*, 12(10):931–934, 2015. DOI: 10.1038/nmeth.3547.

[35] B. Alipanahi, A. Delong, M. T. Weirauch, and B. J. Frey, Predicting the sequence specificities of DNA-and RNA-binding proteins by deep learning, *Nature Biotechnology*, 33(8):831–838, 2015. DOI: 10.1038/nbt.3300.

[36] H. Zeng, M. D. Edwards, G. Liu, and D. K. Gifford, Convolutional neural network architectures for predicting DNA—protein binding, *Bioinformatics*, 32(12):i121–i127, 2016. DOI: 10.1093/bioinformatics/btw255. 14

[37] M. Jermyn, J. Desroches, J. Mercier, M.-A. Tremblay, K. St-Arnaud, M.-C. Guiot, K. Petrecca, and F. Leblond, Neural networks improve brain cancer detection with raman spectroscopy in the presence of operating room light artifacts, *Journal of Biomedical Optics*, 21(9):094002, 2016. DOI: 10.1117/1.jbo.21.9.094002. 14

[38] L. Shen, L. R. Margolies, J. H. Rothstein, et al., Deep learning to improve breast cancer detection on screening mammography, *Sci Rep 9*, 12495, 2019. https://doi.org/10.1038/s41598-019-48995-4 14

[39] L. P. Kaelbling, M. L. Littman, and A. W. Moore, Reinforcement learning: A survey, *Journal of Artificial Intelligence Research (JAIR)*, 4:237–285, 1996. DOI: 10.1613/jair.301. 14

[40] V. Mnih, K. Kavukcuoglu, D. Silver, A. Graves, I. Antonoglou, D. Wierstra, and M. Riedmiller, Playing Atari with deep reinforcement learning, in *NeurIPS Deep Learning Workshop*, 2013. 14

[41] O. Vinyals, I. Babuschkin, W. M. Czarnecki, M. Mathieu, A. Dudzik, J. Chung, D. H. Choi, R. Powell, T. Ewalds, P. Georgiev et al., Grandmaster level in StarCraft II using multi-agent reinforcement learning, *Nature*, 575(7782):350–354, 2019. DOI: 10.1038/s41586-019-1724-z. 14

[42] S. Levine, C. Finn, T. Darrell, and P. Abbeel, End-to-end training of deep visuomotor policies, *Journal of Machine Learning Research (JMLR)*, 17(39):1–40, 2016. 14, 27

[43] M. Pfeiffer, M. Schaeuble, J. Nieto, R. Siegwart, and C. Cadena, From perception to decision: A data-driven approach to end-to-end motion planning for autonomous ground robots, in *International Conference on Robotics and Automation (ICRA)*, 2017. DOI: 10.1109/icra.2017.7989182. 14

[44] S. Gupta, J. Davidson, S. Levine, R. Sukthankar, and J. Malik, Cognitive mapping and planning for visual navigation, in *Conference on Computer Vision and Pattern Recognition (CVPR)*, 2017. DOI: 10.1109/cvpr.2017.769. 14

[45] T. Zhang, G. Kahn, S. Levine, and P. Abbeel, Learning deep control policies for autonomous aerial vehicles with MPC-guided policy search, in *International Conference on Robotics and Automation (ICRA)*, 2016. DOI: 10.1109/icra.2016.7487175. 14

[46] S. Shalev-Shwartz, S. Shammah, and A. Shashua, Safe, multi-agent, reinforcement learning for autonomous driving, in *NeurIPS Workshop on Learning, Inference and Control of Multi-Agent Systems*, 2016. 14

[47] N. Hemsoth, *The Next Wave of Deep Learning Applications*, Next Platform, September 2016. 14

[48] A. Suleiman, Y.-H. Chen, J. Emer, and V. Sze, Towards closing the energy gap between HOG and CNN features for embedded vision, in *International Symposium on Circuits and Systems (ISCAS)*, 2017. DOI: 10.1109/iscas.2017.8050341. 14, 164

[49] Y. Kang, J. Hauswald, C. Gao, A. Rovinski, T. Mudge, J. Mars, and L. Tang, Neurosurgeon: Collaborative intelligence between the cloud and mobile edge, *ACM SIGARCH Computer Architecture News*, 45(1):615–629, 2017. DOI: 10.1145/3093337.3037698. 15

[50] F. V. Veen, The neural network zoo, The Asimov Institute Blog, 2016. https://www.asimovinstitute.org/neural-network-zoo/ 17

[51] F. Kjolstad, S. Kamil, S. Chou, D. Lugato, and S. Amarasinghe, The tensor algebra compiler, *Conference on Object-Oriented Programming Systems, Languages, and Applications (OOPSLA)*, 2017. DOI: 10.1145/3133901. 22

[52] V. Nair and G. E. Hinton, Rectified linear units improve restricted Boltzmann machines, in *International Conference on Machine Learning (ICML)*, 2010. 24

[53] A. L. Maas, A. Y. Hannun, and A. Y. Ng, Rectifier nonlinearities improve neural network acoustic models, in *International Conference on Machine Learning (ICML)*, 2013. 24

[54] K. He, X. Zhang, S. Ren, and J. Sun, Delving deep into rectifiers: Surpassing human-level performance on ImageNet classification, in *International Conference on Computer Vision (ICCV)*, 2015. DOI: 10.1109/iccv.2015.123. 24

[55] D.-A. Clevert, T. Unterthiner, and S. Hochreiter, Fast and accurate deep network learning by exponential linear units (ELUs), *International Conference on Learning Representations (ICLR)*, 2016. 24

[56] P. Ramachandran, B. Zoph, and Q. V. Le, Searching for activation functions, *ICLR Workshop*, 2018. 24

[57] X. Zhang, J. Trmal, D. Povey, and S. Khudanpur, Improving deep neural network acoustic models using generalized maxout networks, in *International Conference on Acoustics, Speech, and Signal Processing (ICASSP)*, 2014. DOI: 10.1109/icassp.2014.6853589. 24

[58] Y. Zhang, M. Pezeshki, P. Brakel, S. Zhang, C. Laurent, Y. Bengio, and A. Courville, Towards end-to-end speech recognition with deep convolutional neural networks, in *Interspeech*, 2016. DOI: 10.21437/interspeech.2016-1446. 24

[59] Y. Jia, E. Shelhamer, J. Donahue, S. Karayev, J. Long, R. Girshick, S. Guadarrama, and D. Trevor, Caffe: Convolutional architecture for fast feature embedding, ACM International Conference on Multimedia, 2014. 36

[60] A. Dosovitskiy, J. T. Springenberg, M. Tatarchenko, and T. Brox, Learning to generate chairs, tables and cars with convolutional networks, *IEEE Transactions on Pattern Analysis and Machine Intelligence (TPAMI)*, 39(4):692–705, 2016. DOI: 10.1109/tpami.2016.2567384. 25, 26, 173

[61] M. D. Zeiler, G. W. Taylor, and R. Fergus, Adaptive deconvolutional networks for mid and high level feature learning, in *International Conference on Computer Vision (ICCV)*, pages 2018–2025, 2011. DOI: 10.1109/iccv.2011.6126474. 25

[62] E. Shelhamer, J. Donahue, and J. Lon, *Deep Learning for Vision Using CNNs and Caffe: A Hands-On Tutorial*, Embedded Vision Alliance Tutorial, 2016. 24

[63] A. Odena, V. Dumoulin, and C. Olah, Deconvolution and checkerboard artifacts, *Distill*, 2016. http://distill.pub/2016/deconv-checkerboard DOI: 10.23915/distill.00003. 25

[64] D. Wofk, F. Ma, T. Yang, S. Karaman, and V. Sze, FastDepth: Fast monocular depth estimation on embedded systems, in *International Conference on Robotics and Automation (ICRA)*, pages 6101–6108, May 2019. DOI: 10.1109/icra.2019.8794182. 25, 34, 173, 176

[65] C. Dong, C. C. Loy, K. He, and X. Tang, Learning a deep convolutional network for image super-resolution, in *European Conference on Computer Vision (ECCV)*, 2014. DOI: 10.1007/978-3-319-10593-2_13. 25

[66] S. Ioffe and C. Szegedy, Batch normalization: Accelerating deep network training by reducing internal covariate shift, in *International Conference on Machine Learning (ICML)*, 2015. 26, 31

[67] S. Santurkar, D. Tsipras, A. Ilyas, and A. Madry, How does batch normalization help optimization? (No, it is not about internal covariate shift), in *Conference on Neural Information Processing Systems (NeurIPS)*, 2018. 26

[68] A. Vaswani, N. Shazeer, N. Parmar, J. Uszkoreit, L. Jones, A. N. Gomez, Ł. Kaiser, and I. Polosukhin, Attention is all you need, in *Conference on Neural Information Processing Systems (NeurIPS)*, 2017. 26, 34

[69] W. Shi, J. Caballero, L. Theis, F. Huszar, A. Aitken, C. Ledig, and Z. Wang, Is the deconvolution layer the same as a convolutional layer? *ArXiv Preprint ArXiv:1609.07009*, 2016. 26, 173, 176

[70] T. N. Sainath, A.-r. Mohamed, B. Kingsbury, and B. Ramabhadran, Deep convolutional neural networks for LVCSR, in *International Conference on Acoustics, Speech, and Signal Processing (ICASSP)*, 2013. 27

[71] Y. LeCun, L. Bottou, Y. Bengio, and P. Haffner, Gradient-based learning applied to document recognition, *Proc. of the IEEE*, 86(11):2278–2324, November 1998. DOI: 10.1109/5.726791. 27, 28, 37

[72] P. Sermanet, D. Eigen, X. Zhang, M. Mathieu, R. Fergus, and Y. LeCun, OverFeat: Integrated recognition, localization and detection using convolutional networks, in *International Conference on Learning Representations (ICLR)*, 2014. 29

[73] K. Simonyan and A. Zisserman, Very deep convolutional networks for large-scale image recognition, in *International Conference on Learning Representations (ICLR)*, 2015. 28, 29, 33, 37, 186, 231, 263

[74] C. Szegedy, W. Liu, Y. Jia, P. Sermanet, S. Reed, D. Anguelov, D. Erhan, V. Vanhoucke, and A. Rabinovich, Going deeper with convolutions, in *Conference on Computer Vision and Pattern Recognition (CVPR)*, 2015. DOI: 10.1109/cvpr.2015.7298594. 28, 29, 31, 33, 37, 231

[75] M. Lin, Q. Chen, and S. Yan, Network in network, in *International Conference on Learning Representations (ICLR)*, 2014. 31, 231

[76] C. Szegedy, V. Vanhoucke, S. Ioffe, J. Shlens, and Z. Wojna, Rethinking the inception architecture for computer vision, in *Conference on Computer Vision and Pattern Recognition (CVPR)*, 2016. DOI: 10.1109/cvpr.2016.308. 31, 231

[77] C. Szegedy, S. Ioffe, V. Vanhoucke, and A. Alemi, Inception-v4, Inception-ResNet and the impact of residual connections on learning, in *AAAI Conference on Artificial Intelligence*, 2017. 31

[78] Y. Bengio, P. Simard, and P. Frasconi, Learning long-term dependencies with gradient descent is difficult, *IEEE Transactions on Neural Networks*, 5(2):157–166, 1994. DOI: 10.1109/72.279181. 32

[79] G. Urban, K. J. Geras, S. E. Kahou, O. Aslan, S. Wang, R. Caruana, A. Mohamed, M. Philipose, and M. Richardson, Do deep convolutional nets really need to be deep and convolutional?, in *International Conference on Learning Representations (ICLR)*, 2017. 33

[80] Caffe LeNet MNIST. http://caffe.berkeleyvision.org/gathered/examples/mnist.html 28, 37

[81] Caffe Model Zoo. http://caffe.berkeleyvision.org/model_zoo.html 28

[82] Matconvnet Pretrained Models. http://www.vlfeat.org/matconvnet/pretrained/ 28

[83] TensorFlow-Slim image classification library. https://github.com/tensorflow/models/tree/master/slim 28, 37

[84] G. Huang, Z. Liu, L. Van Der Maaten, and K. Q. Weinberger, Densely connected convolutional networks, in *Conference on Computer Vision and Pattern Recognition (CVPR)*, 2017. DOI: 10.1109/cvpr.2017.243. 33, 237

[85] S. Zagoruyko and N. Komodakis, Wide residual networks, in *British Machine Vision Conference (BMVC)*, 2017. DOI: 10.5244/c.30.87. 34

[86] S. Xie, R. Girshick, P. Dollár, Z. Tu, and K. He, Aggregated residual transformations for deep neural networks, in *Conference on Computer Vision and Pattern Recognition (CVPR)*, 2017. DOI: 10.1109/cvpr.2017.634. 34

[87] M. Tan and Q. V. Le, EfficientNet: Rethinking model scaling for convolutional neural networks, in *International Conference on Machine Learning (ICML)*, 2019. 34, 38

[88] D. E. Rumelhart, G. E. Hinton, and R. J. Williams, Learning representations by back-propagating errors, *Nature*, 323(6088):533–536, 1986. DOI: 10.1038/323533a0. 34

[89] J. L. Elman, Finding structure in time, *Cognitive Science*, 14(2):179–211, 1990. DOI: 10.1207/s15516709cog1402_1. 34

[90] G. E. Hinton and R. R. Salakhutdinov, Reducing the dimensionality of data with neural networks, *Science*, 313(5786):504–507, 2006. DOI: 10.1126/science.1127647. 34

[91] I. Goodfellow, J. Pouget-Abadie, M. Mirza, B. Xu, D. Warde-Farley, S. Ozair, A. Courville, and Y. Bengio, Generative adversarial nets, in *Conference on Neural Information Processing Systems (NeurIPS)*, 2014. 34

[92] H. Noh, S. Hong, and B. Han, Learning deconvolution network for semantic segmentation, in *International Conference on Computer Vision (ICCV)*, 2015. DOI: 10.1109/iccv.2015.178. 34, 173

[93] A. Shrivastava, T. Pfister, O. Tuzel, J. Susskind, W. Wang, and R. Webb, Learning from simulated and unsupervised images through adversarial training, in *Conference on Computer Vision and Pattern Recognition (CVPR)*, 2017. DOI: 10.1109/cvpr.2017.241. 34

[94] J.-Y. Zhu, T. Park, P. Isola, and A. A. Efros, Unpaired image-to-image translation using cycle-consistent adversarial networks, in *International Conference on Computer Vision (ICCV)*, 2017. DOI: 10.1109/iccv.2017.244. 34

[95] S. Hochreiter and J. Schmidhuber, Long short-term memory, *Neural Computation*, 9(8):1735–1780, 1997. DOI: 10.1162/neco.1997.9.8.1735. 35

[96] J. Lee, C. Kim, S. Kang, D. Shin, S. Kim, and H.-J. Yoo, UNPU: A 50.6 TOPS/W unified deep neural network accelerator with 1b-to-16b fully-variable weight bit-precision, in *International Solid-State Circuits Conference (ISSCC)*, 2018. DOI: 10.1109/isscc.2018.8310262. 35, 163, 165

[97] J. Giraldo and M. Verhelst, Laika: A 5 uW programmable LSTM accelerator for always-on keyword spotting in 65 nm CMOS, in *European Solid-State Circuits Conference (ESSCIRC)*, pages 166–169, 2018. DOI: 10.1109/esscirc.2018.8494342. 35

[98] M. Abadi, A. Agarwal, P. Barham, E. Brevdo, Z. Chen, C. Citro, G. S. Corrado, A. Davis, J. Dean, M. Devin et al., Tensorflow: Large-scale machine learning on heterogeneous distributed systems, *ArXiv Preprint ArXiv:1603.04467*, 2016. 36, 142, 158, 159

[99] A. Paszke, S. Gross, F. Massa, A. Lerer, J. Bradbury, G. Chanan, T. Killeen, Z. Lin, N. Gimelshein, L. Antiga et al., PyTorch: An imperative style, high-performance deep learning library, in *Conference on Neural Information Processing Systems (NeurIPS)*, 2019. 36, 142

[100] Deep Learning Frameworks. https://developer.nvidia.com/deep-learning-frameworks 36

[101] Y.-H. Chen, T. Krishna, J. Emer, and V. Sze, Eyeriss: An energy-efficient reconfigurable accelerator for deep convolutional neural networks, *IEEE Journal of Solid-State Circuits (JSSC)*, 51(1), 2017. DOI: 10.1109/isscc.2016.7418007. 36, 110, 115, 195

[102] Open Neural Network Exchange (ONNX). https://onnx.ai/ 37

[103] C. J. B. Yann LeCun and C. Cortes, The MNIST database of handwritten digits. http://yann.lecun.com/exdb/mnist/ 37

[104] L. Wan, M. Zeiler, S. Zhang, Y. LeCun, and R. Fergus, Regularization of neural networks using dropconnect, in *International Conference on Machine Learning (ICML)*, 2013. 38

[105] A. Krizhevsky, V. Nair, and G. Hinton, The CIFAR-10 dataset. https://www.cs.toronto.edu/kriz/cifar.html 38, 246

[106] A. Torralba, R. Fergus, and W. T. Freeman, 80 million tiny images: A large data set for nonparametric object and scene recognition, *IEEE Transactions on Pattern Analysis and Machine Intelligence (TPAMI)*, 30(11):1958–1970, 2008. DOI: 10.1109/tpami.2008.128. 38

[107] A. Krizhevsky and G. Hinton, Convolutional deep belief networks on CIFAR-10, 40, 2010. (unpublished manuscript) 38

[108] B. Graham, Fractional max-pooling, *ArXiv Preprint ArXiv:1412.6071*, 2014. 38

[109] Pascal VOC data sets. http://host.robots.ox.ac.uk/pascal/VOC/ 39

[110] Microsoft Common Objects in Context (COCO) dataset. http://mscoco.org/ 39

[111] Google Open Images. https://github.com/openimages/dataset 39

[112] YouTube-8M. https://research.google.com/youtube8m/ 39

[113] AudioSet. https://research.google.com/audioset/index.html 39

[114] Standard Performance Evaluation Corporation (SPEC). https://www.spec.org/ 44

[115] MLPref. https://mlperf.org/ 44

[116] DeepBench. https://github.com/baidu-research/DeepBench 44

[117] R. Adolf, S. Rama, B. Reagen, G.-Y. Wei, and D. Brooks, Fathom: Reference workloads for modern deep learning methods, in *International Symposium on Workload Characterization (IISWC)*, pages 1–10, 2016. DOI: 10.1109/iiswc.2016.7581275. 44

[118] J. D. Little, A proof for the queuing formula: $L = \lambda w$, *Operations Research*, 9(3):383–387, 1961. DOI: 10.1287/opre.9.3.383. 45, 109

[119] S. Williams, A. Waterman, and D. Patterson, Roofline: An insightful visual performance model for multicore architectures, *Communications of the ACM*, 52(4):65–76, April 2009. DOI: 10.2172/1407078. 47, 135, 138

[120] B. Chen and J. M. Gilbert, Introducing the CVPR 2018 on-device visual intelligence challenge, Google AI Blog, 2018. https://ai.googleblog.com/2018/04/introducing-cvpr-2018-on-device-visual.html 50

[121] M. Horowitz, Computing's energy problem (and what we can do about it), in *International Solid-State Circuits Conference (ISSCC)*, 2014. DOI: 10.1109/isscc.2014.6757323. 52, 74, 131, 148

[122] J. Standard, High bandwidth memory (HBM) DRAM, *JESD235*, 2013. 55, 255

[123] Y. S. Shao, J. Clemons, R. Venkatesan, B. Zimmer, M. Fojtik, N. Jiang, B. Keller, A. Klinefelter, N. Pinckney, P. Raina, and et al., Simba: Scaling deep-learning inference with multi-chip-module-based architecture, in *International Symposium on Microarchitecture (MICRO)*, 2019. https://doi-org.libproxy.mit.edu/10.1145/3352460.3358302 DOI: 10.1145/3352460.3358302. 56, 106

[124] S. Lie, Wafer scale deep learning, in *Hot Chips: A Symposium on High Performance Chips*, 2019. DOI: 10.1109/hotchips.2019.8875628. 56

[125] S. Condon, Facebook unveils Big Basin, new server geared for deep learning, *ZDNet*, March 2017. 59

[126] A. Jaleel, K. B. Theobald, S. C. Steely Jr., and J. Emer, High performance cache replacement using re-reference interval prediction (RRIP), in *Annual Symposium on Computer Architecture*, 2010. DOI: 10.1145/1816038.1815971. 66

[127] M. D. Lam, E. E. Rothberg, and M. E. Wolf, The cache performance and optimizations of blocked algorithms, *SIGPLAN Not.*, 26(4):63–74, April 1991. http://doi.acm.org/10.1145/106973.106981 DOI: 10.1145/106973.106981. 66

[128] P. Cousot and N. Halbwachs, Automatic discovery of linear restraints among variables of a program, in *Symposium on Principles of Programming Languages (POPL)*, 1978. DOI: 10.1145/512760.512770. 66, 142

[129] J. Ragan-Kelley, C. Barnes, A. Adams, S. Paris, F. Durand, and S. Amarasinghe, Halide: A language and compiler for optimizing parallelism, locality, and recomputation in image processing pipelines, in *Conference on Programming Language Design and Implementation (PLDI)*, 2013. http://doi.acm.org/10.1145/2491956.2462176 DOI: 10.1145/2491956.2462176. 66

[130] T. Chen, T. Moreau, Z. Jiang, L. Zheng, E. Yan, H. Shen, M. Cowan, L. Wang, Y. Hu, L. Ceze et al., TVM: An automated end-to-end optimizing compiler for deep learning, in *Symposium on Operating Systems Design and Implementation (OSDI)*, 2018. 66, 140

[131] J. Cong and B. Xiao, Minimizing computation in convolutional neural networks, in *ICANN*, 2014. DOI: 10.1007/978-3-319-11179-7_36. 68

[132] D. H. Bailey, K. Lee, and H. D. Simon, Using Strassen's algorithm to accelerate the solution of linear systems, *The Journal of Supercomputing*, 4(4):357–371, 1991. DOI: 10.1007/bf00129836. 68

[133] S. Winograd, *Arithmetic Complexity of Computations*, SIAM, 33, 1980. DOI: 10.1137/1.9781611970364. 68

[134] A. Lavin and S. Gray, Fast algorithms for convolutional neural networks, in *Conference on Computer Vision and Pattern Recognition (CVPR)*, 2016. DOI: 10.1109/cvpr.2016.435. 68

[135] Nvidia, NVDLA Open Source Project, 2017. http://nvdla.org/ 69, 76, 92, 94, 96, 97, 113, 114

[136] C. Dubout and F. Fleuret, Exact acceleration of linear object detectors, in *European Conference on Computer Vision (ECCV)*, 2012. DOI: 10.1007/978-3-642-33712-3_22. 70

[137] J. S. Lim, *Two-Dimensional Signal and Image Processing*, p. 710, Prentice Hall, Englewood Cliffs, NJ, 1990. 70, 150, 231

[138] Intel Math Kernel Library. https://software.intel.com/en-us/mkl DOI: 10.1007/978-3-319-06486-4_7. 71

[139] S. Chetlur, C. Woolley, P. Vandermersch, J. Cohen, J. Tran, B. Catanzaro, and E. Shelhamer, cuDNN: Efficient primitives for deep learning, *ArXiv Preprint ArXiv:1410.0759*, 2014. 71

[140] G. E. Moore, Cramming more components onto integrated circuits, *Electronics*, 38(8), 1965. DOI: 10.1109/jproc.1998.658762. 73

[141] C. E. Leiserson, N. C. Thompson, J. S. Emer, B. C. Kuszmaul, B. W. Lampson, D. Sanchez, T. B. Schardl, There's plenty of room at the Top: What will drive computer performance after Moore's law?, *Science*, 2020. 73

[142] Y.-H. Chen, J. Emer, and V. Sze, Eyeriss: A spatial architecture for energy-efficient dataflow for convolutional neural networks, in *International Symposium on Computer Architecture (ISCA)*, 2016. DOI: 10.1109/isca.2016.40. 76, 92, 98, 99, 100, 101, 102, 103, 105, 110, 136

[143] Y. Harata, Y. Nakamura, H. Nagase, M. Takigawa, and N. Takagi, A high-speed multiplier using a redundant binary adder tree, *IEEE Journal of Solid-State Circuits (JSSC)*, 22(1):28–34, 1987. DOI: 10.1109/jssc.1987.1052667. 76

[144] C.-E. Lee, Y. S. Shao, A. Parashar, J. Emer, S. W. Keckler, and Z. Zhang, Stitch-X: An accelerator architecture for exploiting unstructured sparsity in deep neural networks, in *Conference on Machine Learning and Systems (MLSys)*, 2018. 76

[145] N. P. Jouppi, C. Young, N. Patil, D. Patterson, G. Agrawal, R. Bajwa, S. Bates, S. Bhatia, N. Boden, A. Borchers et al., In-datacenter performance analysis of a tensor processing unit, in *International Symposium on Computer Architecture (ISCA)*, 2017. 92, 94, 113, 114

[146] V. Gokhale, J. Jin, A. Dundar, B. Martini, and E. Culurciello, A 240 G-ops/s mobile coprocessor for deep neural networks, in *CVPR Workshop*, 2014. DOI: 10.1109/cvprw.2014.106. 92, 93, 95

[147] M. Sankaradas, V. Jakkula, S. Cadambi, S. Chakradhar, I. Durdanovic, E. Cosatto, and H. P. Graf, A massively parallel coprocessor for convolutional neural networks, in *International Conference on Application-specific Systems, Architectures and Processors (ASAP)*, 2009. DOI: 10.1109/asap.2009.25. 92, 94

[148] S. Park, K. Bong, D. Shin, J. Lee, S. Choi, and H.-J. Yoo, A 1.93TOPS/W scalable deep learning/inference processor with tetra-parallel MIMD architecture for big-data applications, in *International Solid-State Circuits Conference (ISSCC)*, 2015. DOI: 10.1109/isscc.2015.7062935. 92

[149] S. Chakradhar, M. Sankaradas, V. Jakkula, and S. Cadambi, A dynamically configurable coprocessor for convolutional neural networks, in *International Symposium on Computer Architecture (ISCA)*, 2010. DOI: 10.1145/1815961.1815993. 92

[150] V. Sriram, D. Cox, K. H. Tsoi, and W. Luk, Towards an embedded biologically-inspired machine vision processor, in *International Conference on Field-Programmable Technology (FPT)*, 2010. DOI: 10.1109/fpt.2010.5681487. 92

[151] L. Cavigelli, D. Gschwend, C. Mayer, S. Willi, B. Muheim, and L. Benini, Origami: A convolutional network accelerator, in *Great Lakes Symposium on VLSI (GLSVLSI)*, 2015. DOI: 10.1145/2742060.2743766. 92, 94

[152] Y. Chen, T. Luo, S. Liu, S. Zhang, L. He, J. Wang, L. Li, T. Chen, Z. Xu, N. Sun, and O. Temam, DaDianNao: A machine-learning supercomputer, in *International Symposium on Microarchitecture (MICRO)*, 2014. DOI: 10.1109/micro.2014.58. 92, 110, 255

[153] T. Chen, Z. Du, N. Sun, J. Wang, C. Wu, Y. Chen, and O. Temam, DianNao: A small-footprint high-throughput accelerator for ubiquitous machine-learning, in *Architectural Support for Programming Languages and Operating Systems (ASPLOS)*, 2014. DOI: 10.1145/2541940.2541967. 92

[154] C. Zhang, P. Li, G. Sun, Y. Guan, B. Xiao, and J. Cong, Optimizing FPGA-based accelerator design for deep convolutional neural networks, in *International Symposium on Field Programmable Gate Arrays (FPGA)*, 2015. DOI: 10.1145/2684746.2689060. 92

[155] B. Moons and M. Verhelst, A 0.3–2.6 TOPS/W precision-scalable processor for real-time large-scale ConvNets, in *Symposium on VLSI Circuits*, 2016. DOI: 10.1109/vlsic.2016.7573525. 92, 95, 96, 98, 160

[156] Z. Du, R. Fasthuber, T. Chen, P. Ienne, L. Li, T. Luo, X. Feng, Y. Chen, and O. Temam, ShiDianNao: Shifting vision processing closer to the sensor, in *International Symposium on Computer Architecture (ISCA)*, 2015. DOI: 10.1145/2749469.2750389. 92, 95, 96, 97, 113, 114

[157] S. Gupta, A. Agrawal, K. Gopalakrishnan, and P. Narayanan, Deep learning with limited numerical precision, in *International Conference on Machine Learning (ICML)*, 2015. 92, 95, 113, 114, 159

[158] M. Peemen, A. A. A. Setio, B. Mesman, and H. Corporaal, Memory-centric accelerator design for convolutional neural networks, in *International Conference on Computer Design (ICCD)*, 2013. DOI: 10.1109/iccd.2013.6657019. 92, 95, 96

[159] A. Parashar, M. Rhu, A. Mukkara, A. Puglielli, R. Venkatesan, B. Khailany, J. Emer, S. W. Keckler, and W. J. Dally, SCNN: An accelerator for compressed-sparse convolutional neural networks, in *International Symposium on Computer Architecture (ISCA)*, 2017. DOI: 10.1145/3140659.3080254. 92, 97, 99, 110, 136, 201, 218

[160] Y.-H. Chen, T. Krishna, J. S. Emer, and V. Sze, Eyeriss: An energy-efficient reconfigurable accelerator for deep convolutional neural networks, in *International Solid-State Circuits Conference (ISSCC)*, 2016. DOI: 10.1109/isscc.2016.7418007. 92, 103, 104, 200

[161] Y.-H. Chen, T.-J. Yang, J. Emer, and V. Sze, Eyeriss v2: A flexible accelerator for emerging deep neural networks on mobile devices, *IEEE Journal on Emerging and Selected Topics in Circuits and Systems (JETCAS)*, 2019. DOI: 10.1109/jetcas.2019.2910232. 92, 116, 221

[162] W. Lu, G. Yan, J. Li, S. Gong, Y. Han, and X. Li, Flexflow: A flexible dataflow accelerator architecture for convolutional neural networks, in *International Symposium on High-Performance Computer Architecture (HPCA)*, pages 553–564, 2017. DOI: 10.1109/hpca.2017.29. 105

[163] F. Tu, S. Yin, P. Ouyang, S. Tang, L. Liu, and S. Wei, Deep convolutional neural network architecture with reconfigurable computation patterns, *IEEE Transactions*

on Very Large Scale Integration (VLSI) Systems, 25(8):2220–2233, 2017. DOI: 10.1109/tvlsi.2017.2688340. 105

[164] H. Kwon, A. Samajdar, and T. Krishna, MAERI: Enabling flexible dataflow mapping over DNN accelerators via reconfigurable interconnects, in *Architectural Support for Programming Languages and Operating Systems (ASPLOS)*, pages 461–475, 2018. DOI: 10.1145/3173162.3173176. 105

[165] M. Alwani, H. Chen, M. Ferdman, and P. Milder, Fused-layer CNN accelerators, in *International Symposium on Microarchitecture (MICRO)*, 2016. DOI: 10.1109/micro.2016.7783725. 106

[166] A. Azizimazreah and L. Chen, Shortcut mining: Exploiting cross-layer shortcut reuse in DCNN accelerators, in *International Symposium on High-Performance Computer Architecture (HPCA)*, 2019. DOI: 10.1109/hpca.2019.00030. 106

[167] K. Ando, K. Ueyoshi, K. Orimo, H. Yonekawa, S. Sato, H. Nakahara, M. Ikebe, T. Asai, S. Takamaeda-Yamazaki, M. Kuroda, and T. Motomura, BRein memory: A 13-layer 4.2 K Neuron/0.8 M synapse binary/ternary reconfigurable in-memory deep neural network accelerator in 65 nm CMOS, in *Symposium on VLSI Circuits*, 2017. DOI: 10.23919/vlsic.2017.8008533. 106, 164

[168] J. Fowers, K. Ovtcharov, M. Papamichael, T. Massengill, M. Liu, D. Lo, S. Alkalay, M. Haselman, L. Adams, M. Ghandi, S. Heil, P. Patel, A. Sapek, G. Weisz, L. Woods, S. Lanka, S. K. Reinhardt, A. M. Caulfield, E. S. Chung, and D. Burger, A configurable cloud-scale DNN processor for real-time AI, in *International Symposium on Computer Architecture (ISCA)*, 2018. DOI: 10.1109/isca.2018.00012. 106, 157

[169] M. Gao, X. Yang, J. Pu, M. Horowitz, and C. Kozyrakis, Tangram: Optimized coarse-grained dataflow for scalable NN accelerators, in *Architectural Support for Programming Languages and Operating Systems (ASPLOS)*, 2019. https://doi-org.libproxy.mit.edu/10.1145/3297858.3304014 DOI: 10.1145/3297858.3304014. 106

[170] M. Pellauer, Y. S. Shao, J. Clemons, N. Crago, K. Hegde, R. Venkatesan, S. W. Keckler, C. W. Fletcher, and J. Emer, Buffets: An efficient and composable storage idiom for explicit decoupled data orchestration, in *Architectural Support for Programming Languages and Operating Systems (ASPLOS)*, 2019. DOI: 10.1145/3297858.3304025. 107, 108, 110

[171] J. Nickolls and W. J. Dally, The GPU computing era, *IEEE Micro*, 30(2):56–69, March 2010. DOI: 10.1109/mm.2010.41. 108

[172] E. G. Cota, P. Mantovani, G. D. Guglielmo, and L. P. Carloni, An analysis of accelerator coupling in heterogeneous architectures, in *Design Automation Conference (DAC)*, pages 1–6, June 2015. 109

[173] J. Cong, M. A. Ghodrat, M. Gill, B. Grigorian, K. Gururaj, and G. Reinman, Accelerator-rich architectures: Opportunities and progresses, in *Design Automation Conference (DAC)*, 2014. DOI: 10.1145/2593069.2596667. 109

[174] J. E. Smith, Decoupled access/execute computer architectures, in *International Symposium on Computer Architecture (ISCA)*, pages 112–119, April 1982. DOI: 10.1145/285930.285982. 109

[175] *FIFO Generator v13.1:LogiCORE IP Product Guide, Vivado Design Suite*, PG057, Xilinx, April 5, 2017. 110

[176] *FIFO: Intel FPGA IP User Guide*, Updated for Intel Quartus Prime Design Suite: 18.0, Intel, 2018. 110

[177] A. Yazdanbakhsh, H. Falahati, P. J. Wolfe, K. Samadi, N. S. Kim, and H. Esmaeilzadeh, Ganax: A unified MIMD-SIMD acceleration for generative adversarial networks, in *International Symposium on Computer Architecture (ISCA)*, 2018. DOI: 10.1109/isca.2018.00060. 110

[178] J. Fowers, K. Ovtcharov, M. Papamichael, T. Massengill, M. Liu, D. Lo, S. Alkalay, M. Haselman, L. Adams, M. Ghandi, S. Heil, P. Patel, A. Sapek, G. Weisz, L. Woods, S. Lanka, S. K. Reinhardt, A. M. Caulfield, E. S. Chung, and D. Burger, A configurable cloud-scale DNN processor for real-time AI, in *International Symposium on Computer Architecture (ISCA)*, 2018. DOI: 10.1109/isca.2018.00012. 110

[179] T. J. Ham, J. L. Aragón, and M. Martonosi, Desc: Decoupled supply-compute communication management for heterogeneous architectures, in *International Symposium on Microarchitecture (MICRO)*, pages 191–203, December 2015. DOI: 10.1145/2830772.2830800. 110

[180] B. Moons, R. Uytterhoeven, W. Dehaene, and M. Verhelst, Envision: A 0.26-to-10TOPS/W subword-parallel dynamic-voltage-accuracy-frequency-scalable convolutional neural network processor in 28 nm FDSOI, in *International Solid-State Circuits Conference (ISSCC)*, pages 246–247, 2017. DOI: 10.1109/isscc.2017.7870353. 113, 114, 161

[181] S. Yin, P. Ouyang, S. Tang, F. Tu, X. Li, L. Liu, and S. Wei, A 1.06-to-5.09 TOPS/W reconfigurable hybrid-neural-network processor for deep learning applications, in *Symposium on VLSI Circuits*, pages C26–C27, IEEE, 2017. DOI: 10.23919/vlsic.2017.8008534.

[182] J. Lee, C. Kim, S. Kang, D. Shin, S. Kim, and H. Yoo, UNPU: A 50.6TOPS/W unified deep neural network accelerator with 1b-to-16b fully-variable weight bit-precision, in *International Solid-State Circuits Conference (ISSCC)*, pages 218–220, 2018. DOI: 10.1109/isscc.2018.8310262. 113, 114

[183] A. G. Howard, M. Zhu, B. Chen, D. Kalenichenko, W. Wang, T. Weyand, M. Andreetto, and H. Adam, Mobilenets: Efficient convolutional neural networks for mobile vision applications, *ArXiv Preprint ArXiv:1704.04861*, 2017. 115, 233

[184] Y.-H. Chen, J. Emer, and V. Sze, Using dataflow to optimize energy efficiency of deep neural network accelerators, *IEEE Micro's Top Picks from the Computer Architecture Conferences*, 37(3), May–June 2017. DOI: 10.1109/mm.2017.54. 123, 124, 136, 140

[185] R. M. Karp, R. E. Miller, and S. Winograd, The organization of computations for uniform recurrence equations, *Journal of the ACM*, 14(3):563–590, 1967. DOI: 10.1145/321406.321418. 126

[186] L. Lamport, The parallel execution of DO loops, *Communications of the ACM*, 17(2):83–93, February 1974. DOI: 10.1145/360827.360844. 126

[187] Y. N. Wu, J. S. Emer, and V. Sze, Accelergy: An architecture-level energy estimation methodology for accelerator designs, in *International Conference on Computer Aided Design (ICCAD)*, 2019. DOI: 10.1109/iccad45719.2019.8942149. 130, 142, 143, 181

[188] B. Dally, Power, programmability, and granularity: The challenges of ExaScale computing, in *International Parallel and Distributed Processing Symposium (IPDPS)*, 2011. DOI: 10.1109/ipdps.2011.420. 131

[189] K. T. Malladi, B. C. Lee, F. A. Nothaft, C. Kozyrakis, K. Periyathambi, and M. Horowitz, Towards energy-proportional datacenter memory with mobile DRAM, in *International Symposium on Computer Architecture (ISCA)*, 2012. DOI: 10.1109/isca.2012.6237004. 131

[190] H. Kwon, P. Chatarasi, M. Pellauer, A. Parashar, V. Sarkar, and T. Krishna, Understanding reuse, performance, and hardware cost of DNN dataflows: A data-centric approach, *International Symposium on Microarchitecture (MICRO)*, 2019. 140

[191] B. Pradelle, B. Meister, M. Baskaran, J. Springer, and R. Lethin, Polyhedral optimization of TensorFlow computation graphs, in *Workshop on Extreme-Scale Programming Tools (ESPT)*, November 2017. DOI: 10.1007/978-3-030-17872-7_5. 140

[192] A. Parashar, P. Raina, Y. S. Shao, Y.-H. Chen, V. A. Ying, A. Mukkara, R. Venkatesan, B. Khailany, S. W. Keckler, and J. Emer, Timeloop: A systematic approach to DNN accelerator evaluation, in *International Symposium on Performance Analysis of Systems and Software (ISPASS)*, 2019. DOI: 10.1109/ispass.2019.00042. 140, 143

[193] J. Ragan-Kelley, C. Barnes, A. Adams, S. Paris, F. Durand, and S. Amarasinghe, Halide: A language and compiler for optimizing parallelism, locality, and recomputation in image processing pipelines, *Conference on Programming Language Design and Implementation (PLDI)*, 2013. https://doi.org/10.1145/2499370.2462176 DOI: 10.1145/2499370.2462176. 140

[194] R. Baghdadi, J. Ray, M. B. Romdhane, E. D. Sozzo, A. Akkas, Y. Zhang, P. Suriana, S. Kamil, and S. P. Amarasinghe, Tiramisu: A polyhedral compiler for expressing fast and portable code, in *International Symposium on Code Generation and Optimization (CGO)*, 2019. https://doi.org/10.1109/CGO.2019.8661197 DOI: 10.1109/cgo.2019.8661197. 140

[195] J. M. Rabaey, A. P. Chandrakasan, and B. Nikolić, *Digital Integrated Circuits: A Design Perspective*, vol. 7, Pearson Education Upper Saddle River, NJ, 2003. 148

[196] B. Ramkumar and H. M. Kittur, Low-power and area-efficient carry select adder, *IEEE Transactions on Very Large Scale Integration (VLSI) Systems*, 20(2):371–375, February 2012. DOI: 10.1109/tvlsi.2010.2101621. 148

[197] S. Han, H. Mao, and W. J. Dally, Deep compression: Compressing deep neural networks with pruning, trained quantization and Huffman coding, in *International Conference on Learning Representations (ICLR)*, 2016. 151, 152

[198] D. Miyashita, E. H. Lee, and B. Murmann, Convolutional neural networks using logarithmic data representation, *ArXiv Preprint ArXiv:1603.01025*, 2016. 151, 152

[199] E. H. Lee, D. Miyashita, E. Chai, B. Murmann, and S. S. Wong, Lognet: Energy-efficient neural networks using logarithmic computations, in *International Conference on Acoustics, Speech, and Signal Processing (ICASSP)*, 2017. DOI: 10.1109/icassp.2017.7953288. 151, 152

[200] P. Gysel, M. Motamedi, and S. Ghiasi, Hardware-oriented approximation of convolutional neural networks, in *International Conference on Learning Representations (ICLR)*, 2016. 152

[201] W. Chen, J. T. Wilson, S. Tyree, K. Q. Weinberger, and Y. Chen, Compressing neural networks with the hashing trick, in *International Conference on Machine Learning (ICML)*, 2015. 152

[202] W. Dally, High-Performance Hardware for Machine Learning, Tutorial at NeurIPS 2015. https://media.nips.cc/Conferences/2015/tutorialslides/Dally-NIPS-Tutorial-2015.pdf 156

[203] J. H. Wilkinson, *Rounding Errors in Algebraic Processes*, Prentice Hall, 1963. 157

[204] D. Williamson, Dynamically scaled fixed point arithmetic, in *Pacific Rim Conference on Communications, Computers and Signal Processing*, 1991. DOI: 10.1109/pacrim.1991.160742. 157

[205] U. Köster, T. Webb, X. Wang, M. Nassar, A. K. Bansal, W. Constable, O. Elibol, S. Gray, S. Hall, L. Hornof, A. Khosrowshahi, C. Kloss, R. Pai, and N. Rao, Flexpoint: An adaptive numerical format for efficient training of deep neural networks, in *Conference on Neural Information Processing Systems (NeurIPS)*, 2017. 157, 159

[206] T. P. Morgan, Nvidia Pushes Deep Learning Inference With New Pascal GPUs, Next Platform, September 2016. 157

[207] S. Higginbotham, Google Takes Unconventional Route with Homegrown Machine Learning Chips, Next Platform, May 2016. 157

[208] D. Kalamkar, D. Mudigere, N. Mellempudi, D. Das, K. Banerjee, S. Avancha, D. T. Vooturi, N. Jammalamadaka, J. Huang, H. Yuen et al., A study of bfloat16 for deep learning training, *ArXiv Preprint ArXiv:1905.12322*, 2019. 159

[209] E. Chung, J. Fowers, K. Ovtcharov, M. Papamichael, A. Caulfield, T. Massengill, M. Liu, D. Lo, S. Alkalay, M. Haselman et al., Serving DNNs in real time at datacenter scale with project brainwave, *IEEE Micro*, 38(2):8–20, 2018. DOI: 10.1109/mm.2018.022071131. 159, 261

[210] I. Hubara, M. Courbariaux, D. Soudry, R. El-Yaniv, and Y. Bengio, Quantized neural networks: Training neural networks with low precision weights and activations, *The Journal of Machine Learning Research (JMLR)*, 18(1):6869–6898, 2017. 159, 164

[211] S. Zhou, Y. Wu, Z. Ni, X. Zhou, H. Wen, and Y. Zou, DoReFa-Net: Training low bitwidth convolutional neural networks with low bitwidth gradients, *ArXiv Preprint ArXiv:1606.06160*, 2016. 159, 164

[212] V. Camus, L. Mei, C. Enz, and M. Verhelst, Review and benchmarking of precision-scalable multiply-accumulate unit architectures for embedded neural-network processing, *IEEE Journal on Emerging and Selected Topics in Circuits and Systems (JETCAS)*, 2019. DOI: 10.1109/jetcas.2019.2950386. 160, 161, 162, 163

[213] D. Shin, J. Lee, J. Lee, J. Lee, and H.-J. Yoo, DNPU: An energy-efficient deep-learning processor with heterogeneous multi-core architecture, *IEEE Micro*, 38(5):85–93, 2018. DOI: 10.1109/mm.2018.053631145. 161

[214] H. Sharma, J. Park, N. Suda, L. Lai, B. Chau, V. Chandra, and H. Esmaeilzadeh, Bit fusion: Bit-level dynamically composable architecture for accelerating deep neural network, in *International Symposium on Computer Architecture (ISCA)*, 2018. DOI: 10.1109/isca.2018.00069. 161

[215] L. Mei, M. Dandekar, D. Rodopoulos, J. Constantin, P. Debacker, R. Lauwereins, and M. Verhelst, Sub-word parallel precision-scalable MAC engines for efficient embedded DNN inference, in *International Conference on Artificial Intelligence Circuits and Systems (AICAS)*, pages 6–10, 2019. DOI: 10.1109/aicas.2019.8771481. 161

[216] P. Judd, J. Albericio, T. Hetherington, T. M. Aamodt, and A. Moshovos, Stripes: Bit-serial deep neural network computing, in *International Symposium on Microarchitecture (MICRO)*, 2016. DOI: 10.1109/lca.2016.2597140. 163

[217] S. Sharify, A. D. Lascorz, K. Siu, P. Judd, and A. Moshovos, Loom: Exploiting weight and activation precisions to accelerate convolutional neural networks, in *Design Automation Conference (DAC)*, pages 1–6, 2018. DOI: 10.1109/dac.2018.8465915. 163

[218] J. Albericio, A. Delmás, P. Judd, S. Sharify, G. O'Leary, R. Genov, and A. Moshovos, Bit-pragmatic deep neural network computing, in *International Symposium on Microarchitecture (MICRO)*, 2017. DOI: 10.1145/3123939.3123982. 163

[219] S. Ryu, H. Kim, W. Yi, and J.-J. Kim, BitBlade: Area and energy-efficient precision-scalable neural network accelerator with bitwise summation, in *Design Automation Conference (DAC)*, 2019. DOI: 10.1145/3316781.3317784. 163

[220] M. Courbariaux and Y. Bengio, Binarynet: Training deep neural networks with weights and activations constrained to +1 or −1, *ArXiv Preprint ArXiv:1602.02830*, 2016. 164

[221] M. Rastegari, V. Ordonez, J. Redmon, and A. Farhadi, XNOR-Net: ImageNet classification using binary convolutional neural networks, in *European Conference on Computer Vision (ECCV)*, 2016. DOI: 10.1007/978-3-319-46493-0_32. 164

[222] F. Li and B. Liu, Ternary weight networks, in *NeurIPS Workshop on Efficient Methods for Deep Neural Networks*, 2016. 164

[223] Z. Cai, X. He, J. Sun, and N. Vasconcelos, Deep learning with low precision by half-wave Gaussian quantization, in *Conference on Computer Vision and Pattern Recognition (CVPR)*, 2017. DOI: 10.1109/cvpr.2017.574. 164

[224] S. Yin, P. Ouyang, J. Yang, T. Lu, X. Li, L. Liu, and S. Wei, An ultra-high energy-efficient reconfigurable processor for deep neural networks with binary/ternary weights in 28 nm CMOS, in *Symposium on VLSI Circuits*, 2018. DOI: 10.1109/vlsic.2018.8502388. 164, 165

[225] R. Andri, L. Cavigelli, D. Rossi, and L. Benini, YodaNN: An ultra-low power convolutional neural network accelerator based on binary weights, in *IEEE Computer Society Annual Symposium on VLSI (ISVLSI)*, 2016. DOI: 10.1109/isvlsi.2016.111. 164, 165

[226] Y. Umuroglu, N. J. Fraser, G. Gambardella, M. Blott, P. Leong, M. Jahre, and K. Vissers, Finn: A framework for fast, scalable binarized neural network inference, in *International Symposium on Field-Programmable Gate Arrays (FPGA)*, 2017. DOI: 10.1145/3020078.3021744. 165, 261

[227] A. Mishra, E. Nurvitadhi, J. J. Cook, and D. Marr, WRPN: Wide reduced-precision networks, in *International Conference on Learning Representations (ICLR)*, 2018. 165

[228] J. Albericio, P. Judd, T. Hetherington, T. Aamodt, N. E. Jerger, and A. Moshovos, Cnvlutin: Ineffectual-neuron-free deep neural network computing, in *International Symposium on Computer Architecture (ISCA)*, 2016. DOI: 10.1109/isca.2016.11. 168, 169, 215

[229] Y. Lin, C. Sakr, Y. Kim, and N. Shanbhag, PredictiveNet: An energy-efficient convolutional neural network via zero prediction, in *International Symposium on Circuits and Systems (ISCAS)*, 2017. DOI: 10.1109/iscas.2017.8050797. 169

[230] M. Song, J. Zhao, Y. Hu, J. Zhang, and T. Li, Prediction based execution on deep neural networks, in *International Symposium on Computer Architecture (ISCA)*, 2018. DOI: 10.1109/isca.2018.00068. 169

[231] V. Akhlaghi, A. Yazdanbakhsh, K. Samadi, R. K. Gupta, and H. Esmaeilzadeh, SnaPEA: Predictive early activation for reducing computation in deep convolutional neural networks, in *International Symposium on Computer Architecture (ISCA)*, 2018. DOI: 10.1109/isca.2018.00061. 169, 170

[232] W. B. Pennebaker and J. L. Mitchell, *JPEG: Still Image Data Compression Standard*, Springer Science & Business Media, 1992. 170

[233] I. T. U. (ITU), Recommendation ITU-T H.264: Advanced video coding for generic audiovisual services, *ITU-T, Tech. Rep.*, 2003. 170

[234] I. T. U. (ITU), Recommendation ITU-T H.265: High efficiency video coding, *ITU-T, Tech. Rep.*, 2013. 170

[235] M. Mahmoud, K. Siu, and A. Moshovos, Diffy: A deja vu-free differential deep neural network accelerator , in *International Symposium on Microarchitecture (MICRO)*, 2018. DOI: 10.1109/micro.2018.00020. 171, 173

[236] M. Riera, J. Maria Arnau, and A. Gonzalez, Computation reuse in DNNs by exploiting input similarity, in *International Symposium on Computer Architecture (ISCA)*, 2018. DOI: 10.1109/isca.2018.00016. 171, 174

[237] M. Buckler, P. Bedoukian, S. Jayasuriya, and A. Sampson, Eva²: Exploiting temporal redundancy in live computer vision, in *International Symposium on Computer Architecture (ISCA)*, 2018. DOI: 10.1109/isca.2018.00051. 171, 173, 175

[238] Y. Zhu, A. Samajdar, M. Mattina, and P. Whatmough, Euphrates: Algorithm-SoC co-design for low-power mobile continuous vision, in *International Symposium on Computer Architecture (ISCA)*, 2018. DOI: 10.1109/isca.2018.00052. 172, 175

[239] Z. Zhang and V. Sze, FAST: A framework to accelerate super-resolution processing on compressed videos, in *CVPR Workshop on New Trends in Image Restoration and Enhancement*, 2017. DOI: 10.1109/cvprw.2017.138. 173, 175

[240] L.-C. Chen, G. Papandreou, I. Kokkinos, K. Murphy, and A. L. Yuille, Deeplab: Semantic image segmentation with deep convolutional nets, atrous convolution, and fully connected CRFS, *IEEE Transactions on Pattern Analysis and Machine Intelligence (TPAMI)*, 40(4):834–848, 2017. DOI: 10.1109/tpami.2017.2699184. 173, 187

[241] T.-J. Yang, M. D. Collins, Y. Zhu, J.-J. Hwang, T. Liu, X. Zhang, V. Sze, G. Papandreou, and L.-C. Chen, Deeperlab: Single-shot image parser, *ArXiv Preprint ArXiv:1902.05093*, 2019. 173, 237

[242] I. Laina, C. Rupprecht, V. Belagiannis, F. Tombari, and N. Navab, Deeper depth prediction with fully convolutional residual networks, in *International Conference on 3D Vision (3DV)*, pages 239–248, 2016. DOI: 10.1109/3dv.2016.32. 173, 176, 177

[243] C. Dong, C. C. Loy, K. He, and X. Tang, Image super-resolution using deep convolutional networks, *IEEE Transactions on Pattern Analysis and Machine Intelligence (TPAMI)*, 38(2):295–307, 2015. DOI: 10.1109/tpami.2015.2439281. 173

[244] C. Dong, C. C. Loy, and X. Tang, Accelerating the super-resolution convolutional neural network, in *European Conference on Computer Vision (ECCV)*, 2016. DOI: 10.1007/978-3-319-46475-6_25.

[245] W. Shi, J. Caballero, F. Huszár, J. Totz, A. P. Aitken, R. Bishop, D. Rueckert, and Z. Wang, Real-time single image and video super-resolution using an efficient sub-pixel convolutional neural network, in *Conference on Computer Vision and Pattern Recognition (CVPR)*, 2016. DOI: 10.1109/cvpr.2016.207. 173

[246] J. Johnson, A. Alahi, and L. Fei-Fei, Perceptual losses for real-time style transfer and super-resolution, in *European Conference on Computer Vision (ECCV)*, 2016. DOI: 10.1007/978-3-319-46475-6_43. 173

[247] K. Hegde, J. Yu, R. Agrawal, M. Yan, M. Pellauer, and C. Fletcher, UCNN: Exploiting computational reuse in deep neural networks via weight repetition, in *International Symposium on Computer Architecture (ISCA)*, 2018. DOI: 10.1109/isca.2018.00062. 177

[248] D. Blalock, J. J. Gonzalez Ortiz, J. Frankle, and J. Guttag, What is the state of neural network pruning? in *Conference on Machine Learning and Systems (MLSys)*, 2020. 178, 185, 186

[249] M. C. Mozer and P. Smolensky, Using relevance to reduce network size automatically, *Connection Science*, 1(1):3–16, 1989. DOI: 10.1080/09540098908915626. 178

[250] S. A. Janowsky, Pruning versus clipping in neural networks, *Physical Review A*, 39(12):6600, 1989. DOI: 10.1103/physreva.39.6600. 179, 183

[251] E. D. Karnin, A simple procedure for pruning back-propagation trained neural networks, *IEEE Transactions on Neural Networks*, 1(2):239–242, 1990. DOI: 10.1109/72.80236.

[252] R. Reed, Pruning algorithms—a survey, *IEEE Transactions on Neural Networks*, 4(5):740–747, 1993. DOI: 10.1109/72.248452. 178

[253] T.-J. Yang, Y.-H. Chen, and V. Sze, Designing energy-efficient convolutional neural networks using energy-aware pruning, in *Conference on Computer Vision and Pattern Recognition (CVPR)*, 2017. DOI: 10.1109/cvpr.2017.643. 178, 180, 181, 183, 184, 185, 246

[254] T. Gale, E. Elsen, and S. Hooker, The state of sparsity in deep neural networks, *ArXiv Preprint ArXiv:1902.09574*, 2019. 184

[255] J. Frankle and M. Carbin, The lottery ticket hypothesis: Finding sparse, trainable neural networks, in *International Conference on Learning Representations (ICLR)*, 2019. 178

[256] Y. LeCun, J. S. Denker, and S. A. Solla, Optimal brain damage, in *Conference on Neural Information Processing Systems (NeurIPS)*, 1990. 179, 183

[257] S. Han, J. Pool, J. Tran, and W. J. Dally, Learning both weights and connections for efficient neural networks, in *Conference on Neural Information Processing Systems (NeurIPS)*, 2015. 179, 183, 184

[258] T.-J. Yang, A. Howard, B. Chen, X. Zhang, A. Go, M. Sandler, V. Sze, and H. Adam, NetAdapt: Platform-aware neural network adaptation for mobile applications, in *European Conference on Computer Vision (ECCV)*, 2018. DOI: 10.1007/978-3-030-01249-6_18. 181, 182, 184, 185, 243, 245, 246, 247, 248

[259] T.-J. Yang, Y.-H. Chen, J. Emer, and V. Sze, A method to estimate the energy consumption of deep neural networks, in *Asilomar Conference on Signals, Systems, and Computers*, 2017. DOI: 10.1109/acssc.2017.8335698. 181

[260] E. Cai, D.-C. Juan, D. Stamoulis, and D. Marculescu, NeuralPower: Predict and deploy energy-efficient convolutional neural networks, in *Asian Conference on Machine Learning*, 2017. 181

[261] W. Wen, C. Wu, Y. Wang, Y. Chen, and H. Li, Learning structured sparsity in deep neural networks, in *Conference on Neural Information Processing Systems (NeurIPS)*, 2016. 182

[262] H. Li, A. Kadav, I. Durdanovic, H. Samet, and H. P. Graf, Pruning filters for efficient convnets, *International Conference on Learning Representations (ICLR)*, 2017.

[263] Y. He, X. Zhang, and J. Sun, Channel pruning for accelerating very deep neural networks, in *International Conference on Computer Vision (ICCV)*, 2017. DOI: 10.1109/iccv.2017.155.

[264] J.-H. Luo, J. Wu, and W. Lin, Thinet: A filter level pruning method for deep neural network compression, in *Conference on Computer Vision and Pattern Recognition (CVPR)*, 2017. DOI: 10.1109/iccv.2017.541. 182

[265] J. Yu, A. Lukefahr, D. Palframan, G. Dasika, R. Das, and S. Mahlke, Scalpel: Customizing DNN pruning to the underlying hardware parallelism, in *International Symposium on Computer Architecture (ISCA)*, 2017. DOI: 10.1145/3140659.3080215. 183

[266] X. Wang, J. Yu, C. Augustine, R. Iyer, and R. Das, Bit prudent in-cache acceleration of deep convolutional neural networks, in *International Symposium on High-Performance Computer Architecture (HPCA)*, 2019. DOI: 10.1109/hpca.2019.00029. 183

[267] H. Kung, B. McDanel, and S. Q. Zhang, Packing sparse convolutional neural networks for efficient systolic array implementations: Column combining under joint optimization, in *Architectural Support for Programming Languages and Operating Systems (ASPLOS)*, 2019. DOI: 10.1145/3297858.3304028. 183

[268] A. Renda, J. Frankle, and M. Carbin, Comparing rewinding and fine-tuning in neural network pruning, in *International Conference on Learning Representations (ICLR)*, 2020. 183

[269] V. Tresp, R. Neuneier, and H.-G. Zimmermann, Early brain damage, in *Conference on Neural Information Processing Systems (NeurIPS)*, 1997. 184

[270] X. Jin, X. Yuan, J. Feng, and S. Yan, Training skinny deep neural networks with iterative hard thresholding methods, *ArXiv Preprint ArXiv:1607.05423*, 2016.

[271] Y. Guo, A. Yao, and Y. Chen, Dynamic network surgery for efficient DNNs, in *Conference on Neural Information Processing Systems (NeurIPS)*, 2016. 184

[272] Z. Liu, M. Sun, T. Zhou, G. Huang, and T. Darrell, Rethinking the value of network pruning, in *International Conference on Learning Representations (ICLR)*, 2019. 184

[273] F. Yu and V. Koltun, Multi-scale context aggregation by dilated convolutions, in *International Conference on Learning Representations (ICLR)*, 2016. 187

[274] S. Chou, F. Kjolstad, and S. Amarasinghe, Format abstraction for sparse tensor algebra compilers, *Conference on Object-Oriented Programming Systems, Languages, and Applications (OOPSLA)*, 2018. DOI: 10.1145/3276493. 187, 199

[275] S. Smith and G. Karypis, Tensor-matrix products with a compressed sparse tensor, in *Workshop on Irregular Applications: Architectures and Algorithms*, pages 1–7, ACM, 2015. DOI: 10.1145/2833179.2833183. 189, 194

[276] K. Hegde, H. Asghari-Moghaddam, M. Pellauer, N. Crago, A. Jaleel, E. Solomonik, J. Emer, and C. W. Fletcher, ExTensor: An accelerator for sparse tensor algebra, in *International Symposium on Microarchitecture (MICRO)*, 2019. DOI: 10.1145/3352460.3358275. 189, 199, 220, 226

[277] N. Sato and W. F. Tinney, Techniques for exploiting the sparsity of the network admittance matrix, *IEEE Transactions on Power Apparatus and Systems*, 82(69):944–950, 1963. DOI: 10.1109/tpas.1963.291477. 194, 198

[278] A. Buluç and J. R. Gilbert, On the representation and multiplication of hypersparse matrices, in *International Symposium on Parallel and Distributed Processing, (IPDPS)*, pages 1–11, April 2008. DOI: 10.1109/ipdps.2008.4536313. 194

[279] C. E. Shannon, A mathematical theory of communication, *Bell System Technical Journal*, 27(3):379–423, July 1948. DOI: 10.1002/j.1538-7305.1948.tb00917.x. 195

[280] A. Gondimalla, N. Chesnut, M. Thottethodi, and T. Vijaykumar, Sparten: A sparse tensor accelerator for convolutional neural networks, in *International Symposium on Microarchitecture (MICRO)*, 2019. DOI: 10.1145/3352460.3358291. 221

[281] S. Zhang, Z. Du, L. Zhang, H. Lan, S. Liu, L. Li, Q. Guo, T. Chen, and Y. Chen, Cambricon-x: An accelerator for sparse neural networks, in *International Symposium on Computer Architecture (ISCA)*, 2016. DOI: 10.1109/micro.2016.7783723. 209

[282] S. Han, X. Liu, H. Mao, J. Pu, A. Pedram, M. A. Horowitz, and W. J. Dally, EIE: Efficient inference engine on compressed deep neural network, in *International Symposium on Computer Architecture (ISCA)*, 2016. DOI: 10.1109/isca.2016.30. 223

[283] A. Delmas Lascorz, P. Judd, D. M. Stuart, Z. Poulos, M. Mahmoud, S. Sharify, M. Nikolic, K. Siu, and A. Moshovos, Bit-tactical: A software/hardware approach to exploiting value and bit sparsity in neural networks, in *Architectural Support for Programming Languages and Operating Systems (ASPLOS)*, 2019. DOI: 10.1145/3297858.3304041. 216

[284] F. N. Iandola, M. W. Moskewicz, K. Ashraf, S. Han, W. J. Dally, and K. Keutzer, SqueezeNet: AlexNet-level accuracy with 50x fewer parameters and <1 MB model size, in *International Conference on Learning Representations (ICLR)*, 2017. 232

[285] X. Zhang, X. Zhou, M. Lin, and J. Sun, Shufflenet: An extremely efficient convolutional neural network for mobile devices, in *Conference on Computer Vision and Pattern Recognition (CVPR)*, 2018. DOI: 10.1109/cvpr.2018.00716. 233

[286] J. Hu, L. Shen, and G. Sun, Squeeze-and-excitation networks, in *The IEEE Conference on Computer Vision and Pattern Recognition (CVPR)*, 2018. DOI: 10.1109/cvpr.2018.00745. 233

[287] F. Yu, D. Wang, E. Shelhamer, and T. Darrell, Deep layer aggregation, in *Conference on Computer Vision and Pattern Recognition (CVPR)*, 2018. DOI: 10.1109/cvpr.2018.00255. 237

[288] E. Denton, W. Zaremba, J. Bruna, Y. LeCun, and R. Fergus, Exploiting linear structure within convolutional networks for efficient evaluation, in *Conference on Neural Information Processing Systems (NeurIPS)*, 2014. 240

[289] V. Lebedev, Y. Ganin, M. Rakhuba1, I. Oseledets, and V. Lempitsky, Speeding-up convolutional neural networks using fine-tuned CP-decomposition, in *International Conference on Learning Representations (ICLR)*, 2015. 240

[290] Y.-D. Kim, E. Park, S. Yoo, T. Choi, L. Yang, and D. Shin, Compression of deep convolutional neural networks for fast and low power mobile applications, in *International Conference on Learning Representations (ICLR)*, 2016. 240

[291] B. Zoph and Q. V. Le, Neural architecture search with reinforcement learning, in *International Conference on Learning Representations (ICLR)*, 2017. 243, 246

[292] E. Real, S. Moore, A. Selle, S. Saxena, Y. L. Suematsu, J. Tan, Q. V. Le, and A. Kurakin, Large-scale evolution of image classifiers, in *International Conference on Machine Learning (ICML)*, 2017. 243, 245, 246

[293] H. Liu, K. Simonyan, and Y. Yang, DARTS: Differentiable architecture search, in *International Conference on Learning Representations (ICLR)*, 2019. 243, 245

[294] G. Bender, P.-J. Kindermans, B. Zoph, V. Vasudevan, and Q. Le, Understanding and simplifying one-shot architecture search, in *International Conference on Machine Learning (ICML)*, 2018. 246

[295] M. Tan, B. Chen, R. Pang, V. Vasudevan, and Q. V. Le, Mnasnet: Platform-aware neural architecture search for mobile, in *Conference on Computer Vision and Pattern Recognition (CVPR)*, 2019. DOI: 10.1109/cvpr.2019.00293. 246

[296] L.-C. Chen, M. D. Collins, Y. Zhu, G. Papandreou, B. Zoph, F. Schroff, H. Adam, and J. Shlens, Searching for efficient multi-scale architectures for dense image prediction, in *Conference on Neural Information Processing Systems (NeurIPS)*, 2018. 244, 246

[297] B. Zoph, V. Vasudevan, J. Shlens, and Q. V. Le, Learning transferable architectures for scalable image recognition, in *Conference on Computer Vision and Pattern Recognition (CVPR)*, 2018. DOI: 10.1109/cvpr.2018.00907. 246

[298] Z. Zhong, J. Yan, W. Wu, J. Shao, and C.-L. Liu, Practical block-wise neural network architecture generation, in *Conference on Computer Vision and Pattern Recognition (CVPR)*, 2018. DOI: 10.1109/cvpr.2018.00257. 246

[299] H. Cai, T. Chen, W. Zhang, Y. Yu, and J. Wang, Efficient architecture search by network transformation, in *AAAI Conference on Artificial Intelligence*, 2018. 246

[300] H. Liu, K. Simonyan, O. Vinyals, C. Fernando, and K. Kavukcuoglu, Hierarchical representations for efficient architecture search, in *International Conference on Learning Representations (ICLR)*, 2018. 245, 246

[301] A. Zela, A. Klein, S. Falkner, and F. Hutter, Towards automated deep learning: Efficient joint neural architecture and hyperparameter search, in *ICML AutoML Workshop*, July 2018. 246

[302] E. Real, A. Aggarwal, Y. Huang, and Q. V. Le, Regularized evolution for image classifier architecture search, in *AAAI Conference on Artificial Intelligence*, 2019. DOI: 10.1609/aaai.v33i01.33014780. 244, 245, 246

[303] C. Liu, L.-C. Chen, F. Schroff, H. Adam, W. Hua, A. Yuille, and F.-F. Li, Auto-deeplab: Hierarchical neural architecture search for semantic image segmentation, in *Conference on Computer Vision and Pattern Recognition (CVPR)*, 2019. DOI: 10.1109/cvpr.2019.00017. 243, 245, 246

[304] H. Mendoza, A. Klein, M. Feurer, J. T. Springenberg, and F. Hutter, Towards automatically-tuned neural networks, in *ICML 2016 AutoML Workshop*, 2016. 243, 246

[305] A. Gordon, E. Eban, O. Nachum, B. Chen, H. Wu, T.-J. Yang, and E. Choi, Morphnet: Fast and simple resource-constrained structure learning of deep networks, in *Conference on Computer Vision and Pattern Recognition (CVPR)*, 2018. DOI: 10.1109/cvpr.2018.00171. 245, 246, 247

[306] X. Dai, P. Zhang, B. Wu, H. Yin, F. Sun, Y. Wang, M. Dukhan, Y. Hu, Y. Wu, Y. Jia, P. Vajda, M. Uyttendaele, and N. K. Jha, Chamnet: Towards efficient network design through platform-aware model adaptation, in *Conference on Computer Vision and Pattern Recognition (CVPR)*, 2019. DOI: 10.1109/cvpr.2019.01166. 245, 246, 248

[307] B. Wu, X. Dai, P. Zhang, Y. Wang, F. Sun, Y. Wu, Y. Tian, P. Vajda, Y. Jia, and K. Keutzer, Fbnet: Hardware-aware efficient convnet design via differentiable neural architecture search, Conference on Computer Vision and Pattern Recognition (CVPR), 2019. DOI: 10.1109/cvpr.2019.01099. 245, 246, 248

[308] H. Cai, L. Zhu, and S. Han, ProxylessNAS: Direct neural architecture search on target task and hardware, in *International Conference on Learning Representations (ICLR)*, 2019. 243, 245, 246

[309] A. Howard, M. Sandler, G. Chu, L.-C. Chen, B. Chen, M. Tan, W. Wang, Y. Zhu, R. Pang, V. Vasudevan, Q. V. Le, and H. Adam, Searching for mobilenetv3, in *International Conference on Computer Vision (ICCV)*, 2019. DOI: 10.1109/iccv.2019.00140. 248, 249

[310] C. Buciluă, R. Caruana, and A. Niculescu-Mizil, Model compression, in *Special Interest Group on Knowledge Discovery and Data Mining (SIGKDD)*, 2006. DOI: 10.1145/1150402.1150464. 249

[311] L. Ba and R. Caurana, Do deep nets really need to be deep? *Conference on Neural Information Processing Systems (NeurIPS)*, 2014. 249

[312] G. Hinton, O. Vinyals, and J. Dean, Distilling the knowledge in a neural network, in *NeurIPS Deep Learning Workshop*, 2014. 249, 250

[313] A. Romero, N. Ballas, S. E. Kahou, A. Chassang, C. Gatta, and Y. Bengio, Fitnets: Hints for thin deep nets, in *International Conference on Learning Representations (ICLR)*, 2015. 250

[314] A. Mishra and D. Marr, Apprentice: Using knowledge distillation techniques to improve low-precision network accuracy, in *International Conference on Learning Representations (ICLR)*, 2018. 250

[315] T.-J. Yang and V. Sze, Design considerations for efficient deep neural networks on processing-in-memory accelerators, in *International Electron Devices Meeting (IEDM)*, 2019. DOI: 10.1109/iedm19573.2019.8993662. 251, 272

[316] D. Kim, J. Kung, S. Chai, S. Yalamanchili, and S. Mukhopadhyay, Neurocube: A programmable digital neuromorphic architecture with high-density 3D memory, in *International Symposium on Computer Architecture (ISCA)*, 2016. DOI: 10.1109/isca.2016.41. 255

[317] M. Gao, J. Pu, X. Yang, M. Horowitz, and C. Kozyrakis, TETRIS: Scalable and efficient neural network acceleration with 3D memory, in *Architectural Support for Programming Languages and Operating Systems (ASPLOS)*, 2017. DOI: 10.1145/3037697.3037702. 256

[318] K. Ueyoshi, K. Ando, K. Hirose, S. Takamaeda-Yamazaki, M. Hamada, T. Kuroda, and M. Motomura, QUEST: Multi-purpose log-quantized DNN inference engine stacked on 96-MB 3-D SRAM using inductive coupling technology in 40 nm CMOS, *IEEE Journal of Solid-State Circuits (JSSC)*, pages 1–11, 2018. DOI: 10.1109/jssc.2018.2871623. 256

[319] M. M. S. Aly, T. F. Wu, A. Bartolo, Y. H. Malviya, W. Hwang, G. Hills, I. Markov, M. Wootters, M. M. Shulaker, H.-S. P. Wong et al., The N3XT approach to energy-efficient abundant-data computing, *Proc. of the IEEE*, 107(1):19–48, 2019. DOI: 10.1109/jproc.2018.2882603. 257

[320] D. Keitel-Schulz and N. Wehn, Embedded DRAM development: Technology, physical design, and application issues, *IEEE Design and Test of Computers*, 18(3):7–15, 2001. DOI: 10.1109/54.922799. 255

[321] A. Chen, A review of emerging non-volatile memory (NVM) technologies and applications, *Solid-State Electronics*, 125:25–38, 2016. DOI: 10.1016/j.sse.2016.07.006. 255

[322] P. Cappelletti, C. Golla, P. Olivo, and E. Zanoni, *Flash Memories*, Springer Science & Business Media, 2013. DOI: 10.1007/978-1-4615-5015-0. 255

[323] O. Golonzka, J.-G. Alzate, U. Arslan, M. Bohr, P. Bai, J. Brockman, B. Buford, C. Connor, N. Das, B. Doyle et al., MRAM as embedded non-volatile memory solution for 22FFL FinFET technology, in *International Electron Devices Meeting (IEDM)*, 2018. DOI: 10.1109/iedm.2018.8614620. 255

[324] S.-S. Sheu, M.-F. Chang, K.-F. Lin, C.-W. Wu, Y.-S. Chen, P.-F. Chiu, C.-C. Kuo, Y.-S. Yang, P.-C. Chiang, W.-P. Lin et al., A 4 Mb embedded SLC resistive-RAM macro with 7.2 ns read-write random-access time and 160 ns MLC-access capability, in *International Solid-State Circuits Conference (ISSCC)*, 2011. DOI: 10.1109/isscc.2011.5746281. 255

[325] O. Golonzka, U. Arslan, P. Bai, M. Bohr, O. Baykan, Y. Chang, A. Chaudhari, A. Chen, N. Das, C. English et al., Non-volatile RRAM embedded into 22FFL FinFET technology, in *Symposium on VLSI Technology*, 2019. DOI: 10.23919/vlsit.2019.8776570. 255

[326] G. De Sandre, L. Bettini, A. Pirola, L. Marmonier, M. Pasotti, M. Borghi, P. Mattavelli, P. Zuliani, L. Scotti, G. Mastracchio et al., A 90 nm 4 Mb embedded phase-change memory with 1.2 V 12 ns read access time and 1 MB/s write throughput, in *International Solid-State Circuits Conference (ISSCC)*, 2010. DOI: 10.1109/isscc.2010.5433911. 255

[327] J. T. Pawlowski, Vision of processor-memory systems, Keynote at MICRO-48, 2015. https://www.microarch.org/micro48/files/slides/Keynote-III.pdf 255

[328] J. Jeddeloh and B. Keeth, Hybrid memory cube new DRAM architecture increases density and performance, in *Symposium on VLSI Technology*, 2012. DOI: 10.1109/vlsit.2012.6242474. 255

[329] A. Shafiee, A. Nag, N. Muralimanohar, R. Balasubramonian, J. P. Strachan, M. Hu, R. S. Williams, and V. Srikumar, ISAAC: A convolutional neural network accelerator with in-situ analog arithmetic in crossbars, in *International Symposium on Computer Architecture (ISCA)*, 2016. DOI: 10.1109/isca.2016.12. 262, 263, 270, 273

[330] D. Ditzel, T. Kuroda, and S. Lee, Low-cost 3D chip stacking with thruchip wireless connections, in *Hot Chips: A Symposium on High Performance Chips*, 2014. DOI: 10.1109/HOTCHIPS.2014.7478813. 256

[331] N. Verma, H. Jia, H. Valavi, Y. Tang, M. Ozatay, L.-Y. Chen, B. Zhang, and P. Deaville, In-memory computing: Advances and prospects, *IEEE Solid-State Circuits Magazine*, 11(3):43–55, 2019. DOI: 10.1109/mssc.2019.2922889. 258, 277

[332] D. Bankman, L. Yang, B. Moons, M. Verhelst, and B. Murmann, An always-on 3.8 μJ/86% CIFAR-10 mixed-signal binary CNN processor with all memory on chip in 28 nm CMOS, *IEEE Journal of Solid-State Circuits (JSSC)*, 54(1):158–172, 2018. DOI: 10.1109/JSSC.2018.2869150. 258

[333] D. Bankman, J. Messner, A. Gural, and B. Murmann, RRAM-based in-memory computing for embedded deep neural networks, in *Asilomar Conference on Signals, Systems, and Computers*, 2019. DOI: 10.1109/ieeeconf44664.2019.9048704. 277

[334] S. Ma, D. Brooks, and G.-Y. Wei, A binary-activation, multi-level weight RNN and training algorithm for processing-in-memory inference with eNVM, *ArXiv Preprint ArXiv:1912.00106*, 2019. 258

[335] L. Chua, Memristor-the missing circuit element, *IEEE Transactions on Circuit Theory*, 18(5):507–519, 1971. DOI: 10.1109/tct.1971.1083337. 261

[336] L. Wilson, International technology roadmap for semiconductors (ITRS), *Semiconductor Industry Association*, 2013. 261

[337] J. Liang and H.-S. P. Wong, Cross-point memory array without cell selectors—device characteristics and data storage pattern dependencies, *IEEE Transactions on Electron Devices*, 57(10):2531–2538, 2010. DOI: 10.1109/ted.2010.2062187. 261

[338] P. Chi, S. Li, C. Xu, T. Zhang, J. Zhao, Y. Liu, Y. Wang, and Y. Xie, PRIME: A novel processing-in-memory architecture for neural network computation in ReRAM-based main memory, in *International Symposium on Computer Architecture (ISCA)*, 2016. DOI: 10.1109/isca.2016.13. 263

[339] X. Guo, F. M. Bayat, M. Bavandpour, M. Klachko, M. Mahmoodi, M. Prezioso, K. Likharev, and D. Strukov, Fast, energy-efficient, robust, and reproducible mixed-signal neuromorphic classifier based on embedded nor flash memory technology, in *International Electron Devices Meeting (IEDM)*, 2017. DOI: 10.1109/iedm.2017.8268341. 262

[340] S. B. Eryilmaz, S. Joshi, E. Neftci, W. Wan, G. Cauwenberghs, and H.-S. P. Wong, Neuromorphic architectures with electronic synapses, in *International Symposium on Quality Electronic Design (ISQED)*, 2016. DOI: 10.1109/isqed.2016.7479186. 261

[341] W. Haensch, T. Gokmen, and R. Puri, The next generation of deep learning hardware: Analog computing, *Proc. of the IEEE*, pages 1–15, 2018. DOI: 10.1109/jproc.2018.2871057. 261, 263

[342] S. Yu, Neuro-inspired computing with emerging nonvolatile memorys, *Proc. of the IEEE*, 106(2):260–285, 2018. DOI: 10.1109/jproc.2018.2790840. 261, 269, 270, 273, 275

[343] T. Gokmen and Y. Vlasov, Acceleration of deep neural network training with resistive cross-point devices: Design considerations, *Frontiers in Neuroscience*, 10:333, 2016. DOI: 10.3389/fnins.2016.00333. 263

[344] W. Chen, K. Li, W. Lin, K. Hsu, P. Li, C. Yang, C. Xue, E. Yang, Y. Chen, Y. Chang, T. Hsu, Y. King, C. Lin, R. Liu, C. Hsieh, K. Tang, and M. Chang, A 65 nm 1 Mb nonvolatile computing-in-memory ReRAM macro with sub-16 ns multiply-and-accumulate for binary DNN AI edge processors, in *International Solid-State Circuits Conference (ISSCC)*, 2018. DOI: 10.1109/isscc.2018.8310400. 263, 273

[345] H. Kim, J. Sim, Y. Choi, and L.-S. Kim, NAND-Net: Minimizing computational complexity of in-memory processing for binary neural networks, in *International Symposium on High-Performance Computer Architecture (HPCA)*, 2019. DOI: 10.1109/hpca.2019.00017. 263

[346] L. Darsen, Tutorial on emerging memory devices, *International Symposium on Microarchitecture (MICRO)*, 2016. 263

[347] L. Song, X. Qian, H. Li, and Y. Chen, Pipelayer: A pipelined ReRAM-based accelerator for deep learning, in *International Symposium on High-Performance Computer Architecture (HPCA)*, 2017. DOI: 10.1109/hpca.2017.55. 263, 273

[348] T. Chou, W. Tang, J. Botimer, and Z. Zhang, CASCADE: Connecting RRAMs to extend analog dataflow in an end-to-end in-memory processing paradigm, in *International Symposium on Microarchitecture (MICRO)*, 2019. DOI: 10.1145/3352460.3358328. 263, 273

[349] F. Su, W.-H. Chen, L. Xia, C.-P. Lo, T. Tang, Z. Wang, K.-H. Hsu, M. Cheng, J.-Y. Li, Y. Xie et al., A 462GOPs/J RRAM-based nonvolatile intelligent processor for energy harvesting IoE system featuring nonvolatile logics and processing-in-memory, in *Symposium on VLSI Technology*, 2017. DOI: 10.23919/vlsit.2017.7998149. 263, 270

[350] M. Prezioso, F. Merrikh-Bayat, B. Hoskins, G. Adam, K. K. Likharev, and D. B. Strukov, Training and operation of an integrated neuromorphic network based on metal-oxide memristors, *Nature*, 521(7550):61–64, 2015. DOI: 10.1038/nature14441. 263

[351] J. Zhang, Z. Wang, and N. Verma, A machine-learning classifier implemented in a standard 6T SRAM array, in *Symposium on VLSI Circuits*, 2016. DOI: 10.1109/vlsic.2016.7573556. 264, 276

[352] H. Valavi, P. J. Ramadge, E. Nestler, and N. Verma, A mixed-signal binarized convolutional-neural-network accelerator integrating dense weight storage and multiplication for reduced data movement, in *Symposium on VLSI Circuits*, 2018. DOI: 10.1109/vlsic.2018.8502421. 264, 270, 273

[353] A. Biswas and A. P. Chandrakasan, Conv-RAM: An energy-efficient SRAM with embedded convolution computation for low-power CNN-based machine learning applications, in *International Solid-State Circuits Conference (ISSCC)*, pages 488–490, 2018. DOI: 10.1109/isscc.2018.8310397. 264, 273

[354] M. Kang, S. K. Gonugondla, A. Patil, and N. R. Shanbhag, A multi-functional in-memory inference processor using a standard 6T SRAM array, *IEEE Journal of Solid-State Circuits*, 53(2):642–655, 2018. DOI: 10.1109/jssc.2017.2782087. 277

[355] M. Kang, S. Lim, S. Gonugondla, and N. R. Shanbhag, An in-memory VLSI architecture for convolutional neural networks, *IEEE Journal on Emerging and Selected Topics in Circuits and Systems (JETCAS)*, 2018. DOI: 10.1109/jetcas.2018.2829522. 264, 277

[356] V. Seshadri, D. Lee, T. Mullins, H. Hassan, A. Boroumand, J. Kim, M. A. Kozuch, O. Mutlu, P. B. Gibbons, and T. C. Mowry, Ambit: In-memory accelerator for bulk bitwise operations using commodity dram technology, in *International Symposium on Microarchitecture (MICRO)*, 2017. DOI: 10.1145/3123939.3124544. 264, 267

[357] S. Li, D. Niu, K. T. Malladi, H. Zheng, B. Brennan, and Y. Xie, DRISA: A DRAM-based reconfigurable in-situ accelerator, in *International Symposium on Microarchitecture (MICRO)*, 2017. DOI: 10.1145/3123939.3123977. 267

[358] C. Eckert, X. Wang, J. Wang, A. Subramaniyan, R. Iyer, D. Sylvester, D. Blaaauw, and R. Das, Neural cache: Bit-serial in-cache acceleration of deep neural networks, in *International Symposium on Computer Architecture (ISCA)*, 2018. DOI: 10.1109/isca.2018.00040. 267

[359] F. Gao, G. Tziantzioulis, and D. Wentzlaff, ComputeDRAM: In-memory compute using off-the-shelf DRAMs, in *International Symposium on Microarchitecture (MICRO)*, 2019. DOI: 10.1145/3352460.3358260. 267

[360] Y. N. Wu, V. Sze, and J. S. Emer, An architecture-level energy and area estimator for processing-in-memory accelerator designs, in *International Symposium on Performance Analysis of Systems and Software (ISPASS)*, 2020. 268

[361] H. Jia, Y. Tang, H. Valavi, J. Zhang, and N. Verma, A microprocessor implemented in 65 nm CMOS with configurable and bit-scalable accelerator for programmable in-memory computing, *ArXiv Preprint ArXiv:1811.04047*, 2018. 273

[362] Q. Dong, M. E. Sinangil, B. Erbagci, D. Sun, W.-S. Khwa, H.-J. Liao, Y. Wang, and J. Chang, A 351TOPS/W and 372.4GOPS compute-in-memory SRAM macro in 7 nm FinFET CMOS for machine-learning applications, in *International Solid–State Circuits Conference (ISSCC)*, 2020. DOI: 10.1109/isscc19947.2020.9062985. 273, 277

[363] J. Zhang, Z. Wang, and N. Verma, A matrix-multiplying ADC implementing a machine-learning classifier directly with data conversion, in *International Solid-State Circuits Conference (ISSCC)*, 2015. DOI: 10.1109/isscc.2015.7063061. 277

[364] E. H. Lee and S. S. Wong, A 2.5 GHz 7.7 TOPS/W switched-capacitor matrix multiplier with co-designed local memory in 40 nm, in *International Solid–State Circuits Conference (ISSCC)*, 2016. DOI: 10.1109/isscc.2016.7418085. 277

[365] R. LiKamWa, Y. Hou, J. Gao, M. Polansky, and L. Zhong, RedEye: Analog ConvNet image sensor architecture for continuous mobile vision, in *International Symposium on Computer Architecture (ISCA)*, 2016. DOI: 10.1109/isca.2016.31. 277

[366] A. Wang, S. Sivaramakrishnan, and A. Molnar, A 180 nm CMOS image sensor with on-chip optoelectronic image compression, in *Custom Integrated Circuits Conference (CICC)*, 2012. DOI: 10.1109/cicc.2012.6330604. 277

[367] H. Chen, S. Jayasuriya, J. Yang, J. Stephen, S. Sivaramakrishnan, A. Veeraraghavan, and A. Molnar, ASP vision: Optically computing the first layer of convolutional neural

networks using angle sensitive pixels, in *Conference on Computer Vision and Pattern Recognition (CVPR)*, 2016. DOI: 10.1109/cvpr.2016.104. 278, 279

[368] A. Suleiman and V. Sze, Energy-efficient HOG-based object detection at 1080HD 60 fps with multi-scale support, in *International Workshop on Signal Processing Systems (SiPS)*, 2014. DOI: 10.1109/sips.2014.6986096. 278

[369] Q. Cheng, J. Kwon, M. Glick, M. Bahadori, L. P. Carloni, and K. Bergman, Silicon photonics codesign for deep learning, *Proc. of the IEEE*, pages 1–22, 2020. DOI: 10.1109/jproc.2020.2968184. 278

[370] Y. Shen, N. C. Harris, S. Skirlo, M. Prabhu, T. Baehr-Jones, M. Hochberg, X. Sun, S. Zhao, H. Larochelle, D. Englund, and Marin Soljačić, Deep learning with coherent nanophotonic circuits, *Nature Photonics*, 11(7):441, 2017. DOI: 10.1109/phosst.2017.8012714. 278, 279, 280

[371] R. Hamerly, L. Bernstein, A. Sludds, M. Soljačić, and D. Englund, Large-scale optical neural networks based on photoelectric multiplication, *Physical Review X*, 9(2):021–032, 2019. DOI: 10.1103/physrevx.9.021032. 278, 279, 280

[372] L. Bernstein, A. Sludds, R. Hamerly, V. Sze, J. Emer, and D. Englund, Digital optical neural networks for large-scale machine learning, *Conference on Lasers and Electro-Optics (CLEO)*, 2020. 280

[373] A. N. Tait, M. A. Nahmias, B. J. Shastri, and P. R. Prucnal, Broadcast and weight: An integrated network for scalable photonic spike processing, *Journal of Lightwave Technology*, 32(21):4029–4041, 2014. DOI: 10.1109/jlt.2014.2345652. 278

[374] X. Lin, Y. Rivenson, N. T. Yardimci, M. Veli, Y. Luo, M. Jarrahi, A. Ozcan, All-optical machine learning using diffractive deep neural networks, *Science*, 361(6406):1004–1008, American Association for the Advancement of Science, 2018. 278

Authors' Biographies

VIVIENNE SZE

Vivienne Sze received a B.A.Sc. (Hons) degree in Electrical Engineering from the University of Toronto, Toronto, ON, Canada, in 2004, and S.M. and Ph.D. degrees in Electrical Engineering from the Massachusetts Institute of Technology (MIT), Cambridge, MA, in 2006 and 2010, respectively. In 2011, she received the Jin-Au Kong Outstanding Doctoral Thesis Prize in Electrical Engineering at MIT.

She is an Associate Professor at MIT in the Electrical Engineering and Computer Science Department. Her research interests include energy-aware signal processing algorithms, and low-power circuit and system design for portable multimedia applications including computer vision, deep learning, autonomous navigation, image processing, and video compression. Prior to joining MIT, she was a Member of Technical Staff in the Systems and Applications R&D Center at Texas Instruments (TI), Dallas, TX, where she designed low-power algorithms and architectures for video coding. She also represented TI in the JCT-VC committee of ITU-T and ISO/IEC standards body during the development of High Efficiency Video Coding (HEVC), which received a Primetime Engineering Emmy Award. Within the committee, she was the primary coordinator of the core experiment on coefficient scanning and coding, and she chaired/vice-chaired several ad hoc groups on entropy coding. She is a co-editor of *High Efficiency Video Coding (HEVC): Algorithms and Architectures* (Springer, 2014).

Prof. Sze is a recipient of the inaugural ACM-W Rising Star Award, the 2019 Edgerton Faculty Achievement Award at MIT, the 2018 Facebook Faculty Award, the 2018 & 2017 Qualcomm Faculty Award, the 2018 & 2016 Google Faculty Research Award, the 2016 AFOSR Young Investigator Research Program (YIP) Award, the 2016 3M Non-Tenured Faculty Award, the 2014 DARPA Young Faculty Award, the 2007 DAC/ISSCC Student Design Contest Award and a co-recipient of the 2018 VLSI Best Student Paper Award, the 2017 CICC Outstanding Invited Paper Award, the 2016 IEEE Micro Top Picks Award, and the 2008 A-SSCC Outstanding Design Award. She currently serves on the technical program committee for the International Solid-State Circuits Conference (ISSCC) and the SSCS Advisory Committee (AdCom). She has served on the technical program committees for VLSI Circuits Symposium, Micro and the Conference on Machine Learning and Systems (MLSys), as a guest editor for the IEEE Transactions on Circuits and Systems for Video Technology (TCSVT), and as a Distinguished Lecturer for the IEEE Solid-State Circuits Society (SSCS). Prof. Sze was Program

Co-chair of the 2020 Conference on Machine Learning and Systems (MLSys) and teaches the MIT Professional Education course on Designing Efficient Deep Learning Systems.

For more information about Prof. Sze's research, please visit the Energy-Efficient Multimedia Systems group at MIT: http://www.rle.mit.edu/eems/.

YU-HSIN CHEN

Yu-Hsin Chen received a B.S. degree in Electrical Engineering from National Taiwan University, Taipei, Taiwan, in 2009, and M.S. and Ph.D. degrees in Electrical Engineering and Computer Science (EECS) from Massachusetts Institute of Technology (MIT), Cambridge, MA, in 2013 and 2018, respectively. He received the 2018 Jin-Au Kong Outstanding Doctoral Thesis Prize in Electrical Engineering at MIT and the 2019 ACM SIGARCH/IEEE-CS TCCA Outstanding Dissertation Award. He is currently a Research Scientist at Facebook focusing on hardware/software co-design to enable on-device AI for AR/VR systems. Previously, he was a Research Scientist in Nvidia's Architecture Research Group.

He was the recipient of the 2015 Nvidia Graduate Fellowship, 2015 ADI Outstanding Student Designer Award, and 2017 IEEE SSCS Predoctoral Achievement Award. His work on the dataflows for CNN accelerators was selected as one of the Top Picks in Computer Architecture in 2016. He also co-taught a tutorial on "Hardware Architectures for Deep Neural Networks" at MICRO-49, ISCA2017, MICRO-50, and ISCA2019.

TIEN-JU YANG

Tien-Ju Yang received a B.S. degree in Electrical Engineering from National Taiwan University (NTU), Taipei, Taiwan, in 2010, and an M.S. degree in Electronics Engineering from NTU in 2012. Between 2012 and 2015, he worked in the Intelligent Vision Processing Group, MediaTek Inc., Hsinchu, Taiwan, as an engineer. He is currently a Ph.D. candidate in Electrical Engineering and Computer Science at Massachusetts Institute of Technology, Cambridge, MA, working on efficient deep neural network design. His research interest spans the area of deep learning, computer vision, machine learning, image/video processing, and VLSI system design. He won first place in the 2011 National Taiwan University Innovation Contest. He also co-taught a tutorial on "Efficient Image Processing with Deep Neural Networks" at IEEE International Conference on Image Processing 2019.

JOEL S. EMER

Joel S. Emer received B.S. (Hons.) and M.S. degrees in Electrical Engineering from Purdue University, West Lafayette, IN, USA, in 1974 and 1975, respectively, and a Ph.D. degree in Electrical Engineering from the University of Illinois at Urbana-Champaign, Champaign, IL, USA, in 1979.

He is currently a Senior Distinguished Research Scientist with Nvidia's Architecture Research Group, Westford, MA, USA, where he is responsible for exploration of future architectures and modeling and analysis methodologies. He is also a Professor of the Practice at the Massachusetts Institute of Technology, Cambridge, MA, USA. Previously, he was with Intel, where he was an Intel Fellow and the Director of Microarchitecture Research. At Intel, he led the VSSAD Group, which he had previously been a member of at Compaq and Digital Equipment Corporation. Over his career, he has held various research and advanced development positions investigating processor micro-architecture and developing performance modeling and evaluation techniques. He has made architectural contributions to a number of VAX, Alpha, and X86 processors and is recognized as one of the developers of the widely employed quantitative approach to processor performance evaluation. He has been recognized for his contributions in the advancement of simultaneous multithreading technology, processor reliability analysis, cache organization, pipelined processor organization and spatial architectures for deep learning.

Dr. Emer is a Fellow of the ACM and IEEE and a member of the NAE. He has been a recipient of numerous public recognitions. In 2009, he received the Eckert-Mauchly Award for lifetime contributions in computer architecture. He received the Purdue University Outstanding Electrical and Computer Engineer Alumni Award and the University of Illinois Electrical and Computer Engineering Distinguished Alumni Award in 2010 and 2011, respectively. His 1996 paper on simultaneous multithreading received the ACM/SIGARCH-IEEE-CS/TCCA: Most Influential Paper Award in 2011. He was named to the ISCA and MICRO Halls of Fame in 2005 and 2015, respectively. He has had six papers selected for the IEEE Micro's Top Picks in Computer Architecture in 2003, 2004, 2007, 2013, 2015, and 2016. He was the Program Chair of the International Symposium on Computer Architecture (ISCA) in 2000 and the International Symposium on Microarchitecture (MICRO) in 2017.

Table 2.2: Summary of popular CNNs [7, 24, 71, 73, 74]. †Accuracy is measured based on Top-5 error on ImageNet [23] using multiple crops. ‡This version of LeNet-5 has 431k weights for the filters and requires 2.3M MACs per image, and uses ReLU rather than sigmoid.

Metrics	LeNet 5	AlexNet	Overfeat Fast	VGG 16	GoogLeNet V1	ResNet 50
Top-5 error†	n/a	16.4	14.2	7.4	6.7	5.3
Top-5 error (single crop)†	n/a	19.8	17.0	8.8	10.7	7.0
Input size	28×28	227×227	231×231	224×224	224×224	224×224
Number of CONV layers	2	5	5	13	57	53
Depth in number of CONV layers	2	5	5	13	21	49
Filter sizes	5	3, 5, 11	2, 5, 11	3	1, 3, 5, 7	1, 3, 7
Number of channels	1, 20	3–256	3–1,024	3–512	3–832	3–2,048
Number of filters	20, 50	96–384	96–1,024	64–512	16–384	64–2,048
Stride	1	1, 4	1, 4	1	1, 2	1, 2
Weights	2.6 k	2.3 M	16 M	14.7 M	6.0 M	23.5 M
MACs	283 k	666 M	2.67 G	15.3 G	1.43 G	3.86 G
Number of FC layers	2	3	3	3	1	1
Filter sizes	1, 4	1, 6	1, 6, 12	1, 7	1	1
Number of channels	50, 500	256–4,096	1,024–4,096	512–4,096	1,024	2,048
Number of filters	10, 500	1,000–4,096	1,000–4,096	1,000–4,096	1,000	1,000
Weights	58 k	58.6 M	130 M	124 M	1 M	2 M
MACS	58 K	58.6 M	130 M	124 M	1 M	2 M
Total weights	60 k	61 M	146 M	138 M	7 M	25.5 M
Total MACs	341 k	724 M	2.8 G	15.5 G	1.43 G	3.9 G
Pretrained model website	[77]‡	[78, 79]	n/a	[78, 79, 80]	[78, 79, 80]	[78, 79, 80]

(i.e., the CNN is run only once), which is more consistent with what would likely be deployed in real-time and/or energy-constrained applications.

   LeNet [20] was one of the first CNN approaches introduced in 1989. It was designed for the task of digit classification in grayscale images of size 28×28. The most well known version, LeNet-5, contains two CONV layers followed by two FC layers [71]. Each CONV layer uses filters of size 5×5 (1 channel per filter) with 6 filters in the first layer and 16 filters in the second layer. Average pooling of 2×2 is used after each convolution and a sigmoid is used for the non-linearity. In total, LeNet requires 60k weights and 341k multiply-and-accumulates (MACs) per

Table 3.1: Classification of factors that affect inferences per second

Factor	Hardware	DNN Model	Input Data
Operations per inference		✓	
Operations per cycle	✓		
Cycles per second	✓		
Number of PEs	✓		
Number of active PEs	✓	✓	
Utilization of active PEs	✓	✓	
Effectual operations out of (total) operations		✓	✓
Effectual operations plus unexploited ineffectual operations per cycle	✓		

However, exploiting sparsity requires additional hardware to identify when inputs to the MAC are zero to avoid performing unnecessary MAC operations. The additional hardware can increase the critical path, which decreases cycles per second, and also increase the area of the PE, which reduces the number of PEs for a given area. Both of these factors can reduce the *operations per second*, as shown in Equation (3.2). Therefore, the complexity of the additional hardware can result in a trade off between reducing the number of *unexploited ineffectual operations* and increasing critical path or reducing the number of PEs.

Finally, designing hardware and DNN models that support reduced precision (i.e., fewer bits per operand and per operations), which is discussed in Chapter 7, can also increase the number of *operations per second*. Fewer bits per operand means that the memory bandwidth required to support a given operation is reduced, which can increase the utilization of active PEs since they are less likely to be starved for data. In addition, the area of each PE can be reduced, which can increase the number of PEs for a given area. Both of these factors can increase the *operations per second*, as shown in Equation (3.2). Note, however, that if *multiple* levels of precision need to be supported, additional hardware is required, which can, once again, increase the critical path and also increase the area of the PE, both of which can reduce the *operations per second*, as shown in Equation (3.2).

In this section, we discussed multiple factors that affect the number of inferences per second. Table 3.1 classifies whether the factors are dictated by the hardware, by the DNN model or both.

In summary, the number of MAC operations in the DNN model alone is not sufficient for evaluating the throughput and latency. While the DNN model can affect the number of MAC operations per inference based on the network architecture (i.e., layer shapes) and the sparsity of the weights and activations, the overall impact that the DNN model has on throughput and

Table 5.1: Classification of recent work by dataflow

Dataflow	Recent Work
Weight Stationary (Section 5.7.1)	NVDLA [132], TPU [142], neuFlow [143], Sankaradas et al. [144], Park et al. [145], Chakradhar et al. [146], Sriram et al. [147], Origami [148]
Output Stationary (Section 5.7.2)	DaDianNao [149], DianNao [150], Zhang et al. [151], Moons et al. [152], ShiDianNao [153], Gupta et al. [154], Peeman et al. [155]
Input Stationary (Section 5.7.3)	SCNN [156]
Row Stationary (Section 5.7.4)	Eyeriss v1 [157, 139], Eyeriss v2 [158]

A dataflow only defines the following aspects of a loop nest: (1) the specific order of the loops to prioritize the data types; (2) the number of loops for each data dimension to describe the tiling; and (3) whether each of the loops is temporal (`for`) or spatial (`parallel-for`). The maximum number of loops that each data dimension can have is capped by the number of storage levels in the hierarchy that the specific data type can utilize.

The specific loop bounds in the loop nest, e.g., $S0$ and $S1$ in Figure 5.11, are not defined by the dataflow. However, the maximum value of each loop bound can be limited by a variety of factors, including: the storage capacity for the temporal loops, by the number of reachable consumers through the multicast network for the spatial loops (i.e., `parallel-for`), or by the size of the data dimension. Determination of the specific values of the loop bounds to use for a particular workload are determined by the optimization process that finds the optimal mapping as will be discussed in more depth in Chapter 6.

5.7   DATAFLOW TAXONOMY

In Sections 5.4 and 5.5, we have introduced several techniques to exploit data reuse. While there are many ways to apply these techniques, there are several commonly used design patterns that can be categorized into a taxonomy of dataflows: Weight Stationary (WS), Output Stationary (OS), Input Stationary (IS), and Row Stationary (RS). These dataflows can be seen in many recent works of DNN accelerator design, as shown in Table 5.1. In this section, we will use a generic architecture to describe how these dataflows are used in the recent works. As shown in Figure 5.14, this architecture consists of an array of PEs, with each PE having some local storage called a register file (RF), and the array of PEs shares a common storage level called the global buffer.

```
Input Fmaps:     i[W]
Filter Weights:  f[S]
Output Fmaps:    o[Q]

for (q2=0; q2<Q2; q2++) {
    for(s2=0; s2<S2; s2++) {
        parallel-for (q1=0; q1<Q1; q1++) {
            parallel-for (s1=0; s1<S1; s1++) {
                for(q0=0; q0<Q0; q0++) {
                    for(s0=0; s0<S0; s0++) {
                        o[q2*Q1*Q0+q1*Q0+q0] +=
                            i[q2*Q1*Q0+q1*Q0+q0 + s2*S1*S0+s1*S0+s0] ×
                            f[s2*S1*S0+s1*S0+s0];
}}}}}}
```

Figure 6.9: An example dataflow for a 1-D convolution.

Table 6.1: Reuse parameters for the 1-D convolution dataflow in Figure 6.9

Data Type	Reuse Parameters			
	a	b	c	d
Input activation	$S/(S0 \times S1 \times S2)$	$S2$	$S1$	$S0$
Weight	$Q/(Q0 \times Q1 \times Q2)$	$Q2$	$Q1$	$Q0$
Partial sums	$S/(S0 \times S1 \times S2)$	$S2$	$S1$	$S0$

architectural factors such as system setup and technology differences, Eyexam provides a step-by-step process that associates a certain amount of performance loss to each architectural design decision (e.g., dataflow, number of PEs, NoC, etc.) as well as the properties of the workload, which for DNNs is dictated by the layer shape and size (e.g., filter shape, feature map size, batch size, etc.).

Eyexam focuses on two main factors that affect performance: (1) the *number of active PEs* due to the mapping as constrained by the dataflow; and (2) the *utilization of active PEs*, i.e., percentage of active cycles for the PE, based on whether the NoC has sufficient bandwidth to deliver data to PEs to keep them active. The product of these two components can be used to compute the *utilization of the PE array* as follows:

$$\text{utilization of the PE array} = \text{number of active PEs} \times \text{utilization of active PEs}. \qquad (6.3)$$

Later in this section, we will see how this approach can use an adapted form of the well-known roofline model [119] for the analysis of DNN processors.

We will perform this analysis on a generic DNN processor architecture based on a spatial architecture that consists of a global buffer and an array of PEs. Each PE can have its own register

Table 6.2: Summary of steps in Eyexam

Step	Constraint	Type	New Performance Bound	Reason for Performance Loss
1	Layer size and shape	Workload	Max workload parallelism	Finite workload size
2	Dataflow loop nest	Architectural	Max dataflow parallelism	Restricted dataflow mapping space by defined by loop nest
3	Number of PEs	Architectural	Max PE parallelism	Additional restriction to mapping space due to shape fragmentation
4	Physical dimensions of PEs array	Architectural	Number of active PEs	Additional restriction to mapping space due to shape fragmentation for each dimension
5	Fixed storage capacity	Architectural	Number of active PEs	Additional restriction to mapping space due to storage of intermediate data (depends on dataflow)
6	Fixed data bandwidth	Microarchitectural	Max data bandwidth to active PEs	Insufficient average bandwidth to active PEs
7	Varying data access patterns	Microarchitectural	Actual measure performance	Insufficient instant bandwidth to active PEs

riss' row-stationary dataflow. In order to support a wider range of target designs others accept a *template* that describes a DNN accelerator architecture.[11] These templates include factors such as the dataflows supported, the number of PEs and levels of storage hierarchy and the storage sizes and network connectivity. The MAESTRO, Timeloop, and TVM mappers all accept such templates. Some tools, (e.g., TVM) extend to target general purpose processors by characterizing certain hardware functionality (e.g., vectors or Nvidia's tensor cores) as highly constrained templates.

The range of problem specifications accepted by these tools also varies. Some, like the Eyeriss mapper, assume the standard expression for CONV/FC layers (see Chapter 2). Others, accept expressions equivalent to the tensor index expressions described in Chapter 2. These in-

[11]Architecture templates will typically be expressed in a human-readable configuration language, such as YAML.

Table 7.1: Example values for uniform and scale factor-based quantization with $L = 16$ quantization levels. For uniform, the quantized values are $q_i = 44{,}000 \times i/16$. For scale factor quantization, the quantized values uniform within each scale as computed with $q_i = (4^{i/2+1} - 4^{i/2}) \times (i\%2)/8 + 4^{i/2} - 1$.

i	q_i - Uniform	q_i - Scale Factor
0	0	0
1	2750	1.5
2	5500	3
3	8250	9
4	11000	15
5	13750	39
6	16500	63
7	19250	159
8	22000	255
9	24749	639
10	27499	1023
11	30249	2559
12	32999	4095
13	35749	10239
14	38499	16383
15	41249	40959

quantization is a function of the magnitude of the original full precision value. The benefit of using such quantization is illustrated in Figure 7.7, where the quantization error for multiplication of values is much lower for the scale factor quantization than uniform quantization.

7.2.2   STANDARD COMPONENTS OF THE BIT WIDTH

Once we have determined the quantized values that we want to represent at reduced precision, we need to map them to a numerical representation in the form of bits. This can be done using the standard format for numerical representations, where the bit width is determined based on several factors:

- **The range of the values that are represented.** As discussed in Section 7.2.1, adding a scale factor can be used to better represent data with a large range. As a result, representing values with a large range (e.g., 10^{-38} to 10^{38}) often require more bits to support the scale

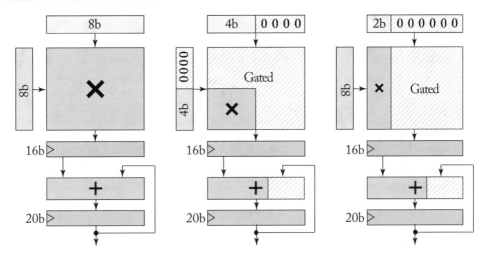

Figure 7.11: Example of a data-gated MAC. For simplicity, in this example only one input is precision scalable (weights). (Figure adapted from [212].)

of the DNN model (e.g., different layers or different weights), then additional hardware support may be required, which will be discussed in the next section.

7.4   VARYING PRECISION: CHANGE PRECISION FOR DIFFERENT PARTS OF THE DNN

Just as different data types have different distributions, their distributions can also vary across different parts of the DNN (e.g., layer, filter, channels). Therefore, to even more aggressively reduce the bit width, the quantization method can adapt to the varying distribution. Allowing the precision to vary across the different parts of the DNN model is commonly referred to as *varying precision*.

Although some systems simply build separate MAC units per precision, varying precision requires the use of a precision-scalable MAC in order to translate the reduced precision into improvement in energy-efficiency or throughput without a significant increase in area. A conventional approach is to use a data-gated MAC, where the unused logic (e.g., full adders) are gated, as shown in Figure 7.11. This reduces unnecessary switching activity, and consequently reduces energy consumption. The data-gated MAC can also be combined with voltage scaling to exploit the shortened critical path for additional energy savings [155].

While a data-gated MAC is a simple approach, it leaves many idle gates without increasing the throughput, making it inefficient in terms of throughput per area. Accordingly, there has been a lot of recent works that look at adding logic gates to increase the utilization of the full adders for higher throughput per area. One of the key challenges is to reduce the overhead of the

Table 8.1: Fiber representations

Label	Description	Coordinates	Compressed	Example
U	Uncompressed array	Implicit	No	DianNao [150]
R	Run-length-encoded (RLE) stream	Implicit	Yes	Eyeriss [98]
B	Bitmask of non-zero coordinates	Implicit	Yes	SparTen [277]
C	Coordinate/payload list	Explicit	Yes	SCNN [156]
H_f	Hash table per fiber	Explicit	Yes	
H_r	Hash table per rank	Explicit	Yes	

their own unique labels. However, when two levels of fiber with known representations are flattened, we use a notation like U^2, P^2, or (RU) to describe the flattened combination.

Given the above fiber representations, the combination of all the choices for representing a fiber's coordinates and payloads leads to a large number of implementation choices. This is multiplied by the fact that each rank of a tensor might use a different representation.

To provide a specific example, we will show how the well-known compressed sparse row (CSR) format [277] can be represented as a concrete representation of a fiber tree. Figure 8.23 illustrates this by showing the matrix of Figure 8.18 in CSR format. The figure shows CSR as a two rank tree that uses an uncompressed array as it top rank fiber (H), and a coordinate/payload list as its bottom rank (W). Thus, the rows are compressed.

In CSR, each position in the upper rank (which is also its coordinate since it is uncompressed) has a payload consisting of a open range that points at a fiber in the bottom rank.[12] And each fiber in lower rank consists of a list of explicit coordinates each of whose position is an implicit payload that is the position of the value in the value array.

So in the figure, if we want to find the value at coordinate (2,1) we start by looking at position (and coordinate) 2 of the upper rank and find its payload is the open range [2, 4). Note how the open range is cleverly encoded with information from two successive positions in the array. Then looking at the fiber in the bottom rank at positions 2 and 3 (i.e., in the open range [2,4)), we search for the coordinate 1. Finding that it is at position 3, we know that the value for coordinate (2,1) is the value "h" at position 3 in the value array.

The CSC representation is the dual of CSR and is basically the same just with the rank order reversed. These schemes statically pick a representation per rank, but an even more complex approach would be to dynamically choose the representation used for each fiber. That choice could be made at the rank level, so that the rank would have a tag indicating the representation for all its fibers or the choice could be made at the individual fiber level.

---

[12]Note, that the CSR representation depends on the fibers in the lower rank being consecutive in memory so the payloads of the upper rank can point to a position in the lower rank.

Table 8.2: Tensor representations

Specification	Name
Tensor < *U+*) > (…	Standard multi-dimensional array
Tensor <UC> (*H,W*)	Compressed sparse row (CSR) [274]
Tensor <UC> (*W,H*)	Compressed sparse column (CSC) [275]
Tensor <C +> (…)	Compressed sparse fiber (CSF) [272]
Tensor <C^n> (…)	Coordinate format (COO) [274]

be considered in conjunction with the computation sequencing, which is described in the next section.

8.3   SPARSE DATAFLOW

In Section 8.2, we discussed the opportunity that sparsity presents to compress the tensors used in DNN computation. This provides obvious benefits savings in storage space, access energy costs and data movement costs by storing and moving compressed data. However, we also recognize that sparsity means that individual values of activations or weights (or entire multi-dimensional tiles) are zero, that the multiplication of anything by zero is zero, and furthermore the addition with zero simply preserves the other input operand. Such operations therefore become *ineffectual* (i.e., doing the operation had no effect on the result). As a consequence, when performing the pervasive sum of product operations (i.e., dot products) in DNN computations there is an opportunity to exploit these ineffectual operations.

The simplest way to exploit ineffectual operations is to save accessing operands and avoid executing the multiplication when an operand is zero. The Eyeriss design saved energy by avoiding reading operand values and running the multiplier when an activation was zero [160]. That eliminated activity (in energy) includes the accesses to input operands, writes/updates to output operands, and arithmetic computation so it can enhance the energy efficiency/power consumption metrics discussed in Section 3.3. However, this only saved energy, not time, for ineffectual operations.

When the hardware can recognize the zeros in the products terms, the amount of time spent performing the dot product can be reduced by eliminating all the time spent on ineffectual activity related to product terms with any zeros as an operand. This will not only improve energy, but improve operations per cycle, as described in Equation (3.2).

Figure 8.24 gives an indication of the potential for reducing computation time by showing the density (proportion of non-zeros or 1 - sparsity) in both weights and input activations for the layers of VGGNet. From those statistics, an architecture that could optimally exploit weight or activation sparsity could provide speedups of 2 to 5×. If one assumes that weight and activation

Design 8.7 Sparse 2-D Summation

```
1    t = Tensor(H, W)
2    sum = 0
3
4    for (h, t_w) in t:
5        for (w, t_val) in t_w:
6            sum += t_val
```

Table 8.3: Sparse dataflow roadmap

Section	Dataflow Description	Examples
8.3.1	Convolution with sparse weights	Cambricon-X [278]
8.3.2	Convolution with sparse activations	Cnvlutin [225]
8.3.3	Convolution with sparse weights and activations	SCNN [156], SparTen [277], Eyeriss V2 [158]
8.3.4	Fully connected with sparse weights and activations	EIE [279], ExTensor [273]

length of the fibers in the bottom rank (W) of t. Similar performance estimates can be made for the other dataflows presented in this section.

Given this notation, which focuses on the dataflow without the complexity of dealing with a specific tensor representation, we can succinctly express a variety of dataflows that exploit sparsity. The following sections explore various dataflows that are designed to exploit sparsity in weights and/or activations for both CONV and FC layers. In those sections, we will both describe designs in terms of loop nests and block diagrams whose structure is implied by the loop nests.

Figure 8.25 shows a key for the components used in the design block diagrams. The block diagrams generally illustrate a single storage-level design that process untiled computations. Therefore, the storage elements are assumed to hold the entire data sets. In actual designs, those storage elements would be implemented using one of the hierarchical buffering schemes described in Section 5.8, such as caches, scratchpads, or explicit decoupled data orchestration (EDDO) units, like buffets.

Table 8.3 gives a roadmap of the dataflows that will be explored.

Table 9.1: This table summarizes which terms in Equation (9.2) can be improved by improving each of the three main components of NAS

	$\text{size}_{search\_space}$	$\text{num}_{samples}$		$\text{time}_{samples}$	
		efficiency_{alg}	num_{alg_tuning}	time_{eval}	time_{train}
Search Space	✓				
Optimization Algorithm		✓	✓		
Performance Evaluation				✓	✓

The selection of the optimization algorithm also influences performance and search time. A more efficient optimization algorithm (efficiency_{alg}) can better utilize the samples and hence reduce the number of required samples. However, it is also important to consider the difficulty of tuning the hyperparameters of the optimization algorithm itself (e.g., network architecture of the reinforcement learning agent). Optimization algorithms that are difficult to tune may require multiple iterations before they can enable effective search (num_{alg_tuning}). This critical factor is often overlooked.

The time required for evaluating a sample (time_{sample}) includes the time required for evaluating the network performance (time_{eval}). Once a given network is sampled, it may need to be trained to get the precise accuracy numbers, which leads to the training time (time_{train}).

Researchers improve NAS algorithms by introducing innovations in the three main components, where each improves different terms in Equation (9.2) (summarized in Table 9.1):

- **Shrinking the search space**, which reduces $\text{size}_{search\_space}$.

- **Improving the optimization algorithm**, which increases efficiency_{alg} and reduces num_{alg_tuning}.

- **Accelerating the performance evaluation**, which reduces time_{eval} and time_{train}.

It is important to note that these three components are not independent of each other, and a change in one component may involve a change in another component. For example, some optimization algorithms cannot support hardware metrics (e.g., latency and energy consumption) and thus can only use proxy metrics (e.g., number of MACs and number of weights).

9.2.1   SHRINKING THE SEARCH SPACE

Shrinking the search space increases the search speed by limiting the discoverable network architectures. The idea is to only search a subset of the network architectures in the network architecture universe out of all the possible network architectures. Although this class of methods can effectively reduce the required number of samples, it may irrecoverably limit the achievable network performance and needs to be carried out carefully. For example, the optimal network

Table 10.1: Example of recent works that explore processing near memory. For I/O, TSV refers to through-silicon vias, while TCI refers to ThruChip Interface which uses inductive coupling. For bandwidth, *ch* refers to number of parallel communication channels, which can be the number of tiles (for eDRAM) or the number of vaults (for stacked memory). The size of stacked DRAM is based on Hybrid Memory Cube (HMC) Gen2 specifications.

	Technology	Size	I/O	Bandwidth	Evaluation
DaDianNao [152]	eDRAM	32 MB	On-chip	18 ch × 310 GB/s = 5580 GB/s	Simulated
Neurocube [316]	Stacked DRAM	2 GB	TSV	16 ch × 10 GB/s = 160 GB/s	Simulated
Tetris [317]	Stacked DRAM	2 GB	TSV	16 ch × 8 GB/s = 128 GB/s	Simulated
Quest [318]	Stacked SRAM	96 MB	TCI	24 ch × 1.2 GB/s = 28.8 GB/s	Measured
N3XT [319]	Monolithic 3-D	4 GB	ILV	16 ch × 48 GB/s = 768 GB/S	Simulated

10.1   PROCESSING NEAR MEMORY

High-density memories typically require a different process technology than processors and as a result are often fabricated as separate chips; as a result, accessing high-density memories requires going off-chip. The bandwidth and energy cost of accessing high-density off-chip memories are often limited by the number of I/O pads per chip and the off-chip interconnect channel characteristics (i.e., its resistance, inductance, and capacitance). Processing near memory aims to overcome these limitations by bringing the compute near the high-density memory to reduce access energy and increase memory bandwidth. The reduction in access energy is achieved by reducing the length of the interconnect between the memory and compute, while the increase in bandwidth is primarily enabled by increasing the number of bits that can be accessed per cycle by allowing for a wider interconnect and, to a lesser extent, by increasing the clock frequency, which is made possible by the reduced interconnect length.

Various recent advanced memory technologies aim to enable processing near memory with differing integration costs. Table 10.1 summarizes some of these efforts, where high-density memories on the order of tens of megabytes to gigabytes are connected to the compute engine at bandwidths of tens to hundreds of gigabytes per second. Note that currently most academic evaluations of DNN systems using advanced memory technologies have been based on simulations rather than fabrication and measurements.

In this section, we will describe the cost and benefits of each technology and provide examples of how they have been used to process DNNs. The architectural design challenges of using processing-near-memory include how to allocate data to memory since the access patterns for high-density memories are often limited (e.g., data needs to be divided into different banks and vaults in the DRAM or stacked DRAM, respectively), how to design the network-on-chip between the memory and PEs, how to allocate the chip area between on-chip memory

Printed in the United States
by Baker & Taylor Publisher Services